IET CONTROL, ROBOTICS AND SENSORS SERIES 97

Sensory Systems for Robotic Applications

The IET International Book Series on Sensors

IET International Book Series on Sensors—Call for Authors

The use of sensors has increased dramatically in all industries. They are fundamental in a wide range of applications from communication to monitoring, remote operation, process control, precision and safety, and robotics and automation. These developments have brought new challenges such as demands for robustness and reliability in networks, security in the communications interface, and close management of energy consumption. This book series covers the research and applications of sensor technologies in the fields of ICTs, security, tracking, detection, monitoring, control and automation, robotics, machine learning, smart technologies, production and manufacturing, photonics, environment, energy, and transport.

Proposals for coherently integrated international multiauthored edited or coauthored handbooks and research monographs will be considered for this book series. Each proposal will be reviewed by the IET Book Series Editorial Board members with additional external reviews from independent reviewers. Please e-mail your book proposal to: vmoliere@theiet.org or author_support@theiet.org.

Sensory Systems for Robotic Applications

Edited by
Ravinder Dahiya, Oliver Ozioko and Gordon Cheng

The Institution of Engineering and Technology

Published by The Institution of Engineering and Technology, London, United Kingdom

The Institution of Engineering and Technology is registered as a Charity in England & Wales (no. 211014) and Scotland (no. SC038698).

First published 2022

The Institution of Engineering and Technology
Futures Place
Kings Way, Stevenage
Herts, SG1 2UA., United Kingdom

www.theiet.org

British Library Cataloguing in Publication Data
A catalogue record for this product is available from the British Library

ISBN 978-1-84919-948-3 (hardback)
ISBN 978-1-84919-949-0 (PDF)

Typeset in India by Exeter Premedia Services Private Limited
Printed in the UK by CPI Group (UK) Ltd, Croydon
Cover Image copyrights: Astrid Eckert / TUM / ICS_H-1_!22018_6_jpg

Contents

About the Editors

Ravinder Dahiya is Professor in Electrical and Computer Engineering Department at Northeastern University, Boston, USA. His group (Bendable Electronics and Sustainable Technologies (BEST)) conducts fundamental research in electronic skin, flexible printed electronics and their applications in robotics, prosthetics, wearables, augmented/virtual reality and similar interactive systems. He has authored or co-authored more than 500 publications, books and submitted/granted patents and disclosures. He has led or contributed to many international projects. Prof. Dahiya is President of IEEE Sensors Council. He is the Founding Editor-in-Chief of *IEEE Journal on Flexible Electronics* (J-FLEX). He has been recipient of EPSRC Fellowship, Marie Curie Fellowship and Japanese Monbusho Fellowship. He has received several awards, including Technical Achievement award from IEEE Sensors Council, Young Investigator Award from Elsevier, and 12 best journal/conference paper awards as author/co-author. He is Fellow of IEEE and the Royal Society of Edinburgh.

Oliver Ozioko is a lecturer in electrical and electronic engineering at the University of Derby, UK. Prior to joining the University of Derby, he worked as a postdoctoral researcher at the University of Glasgow. He holds a PhD Degree in Electrical and Electronic Engineering from the University of Glasgow. His research focuses on sensors and intelligent systems, electronic skin, haptics, assistive technologies, smart systems, as well as self-powered wearable and portable systems. He has authored or co-authored over 29 technical publications. He is the 2023 YP chair for IEEE Sensors council.

Gordon Cheng is chair professor and director of the Institute for Cognitive Systems and is the coordinator of the Center of Competence Neuro-Engineering, Technical University of Munich, Germany. He is also the founding director of the Elite Master of Science program in Neuroengineering (MSNE) of the Elite Network of Bavaria, Germany. For more than 20 years, he has made pioneering contributions in humanoid robotics, neuroengineering and artificial intelligence. He founded the department of humanoid robotics and computational neuroscience at the Institute for Advanced Telecommunications Research in Kyoto, Japan, where he was department head from 2003 to 2008. In addition, from 2007 to 2008 he was a project manager at the National Institute of Information and Communications Technology, Japan, and the Japanese Science and Technology Agency, where he was responsible for the Computational Brain project (2004-2008). He is the co-inventor of 20 patents and co-authored over 350 technical publications. He was acknowledged as an IEEE Fellow in 2017 for his "contributions in humanoid robotic systems and neurorobotics". He holds a Doctorate Degree in Systems Engineering from The Australian National University.

Chapter 1

Development of tactile sensors for intelligent robotics research

Yoshiyuki Ohmura[1], Akihiko Nagakubo[2], and Yasuo Kuniyoshi[1,3]

Our goal is to reveal and reconstruct human intelligence, and we believe that physical and informational interactions with other humans and the environment are essential. To study the physical interaction in the real world, we have developed systems such as humanoid robots and sensor suits, all with tactile sensors based on unique concepts. Since the development of a device is critically influenced by the objectives and constraints of each user, it is important to explain "what it is for" and "how it will be used." Therefore, we will discuss not only developed tactile sensors but also the research and background related to them. First, we will introduce developed tactile sensors and systems, including a sensor module with many features such as scalability and the ability to fit on three-dimensional (3D)-curved surfaces, a sensor glove with approximately 1,000 detection points per hand while adapting to wrinkles, and a highly stretchable tactile sensor based on inverse problem analysis. Next, research on robots with tactile sensors and tactile information processing is presented, including the lifting of large objects by a humanoid equipped with full-body tactile sensors, a dynamic roll-and-rise motion using contact data from the back of the humanoid, learning in-hand grasp of different shaped objects based on tactile recognition, and object's posture recognition while the state of the object continues to change through interaction. Finally, we will discuss the relationship between the whole-body movement, haptic exploration, and general creative activity, as an important milestone in our research.

1.1 Introduction

Our long-term goal is to reconstruct human-like intelligence using computers and robots via investigation and understanding of human behavior and intelligence [1]. A particular emphasis in our intelligence research is the interaction between the self

[1]Department of Mechano-informatics, The University of Tokyo, Tokyo, Japan
[2]Department of Information Technology and Human Factors, National Institute of Advanced Industrial Science and Technology (AIST), Tokyo, Japan
[3]Next Generation AI Research Center, The University of Tokyo, Tokyo, Japan

and the external world (environment, other humans, etc.). Therefore, tactile sensors that can sense physical interactions have been used in various experiments for humanoid robots and measurements of human behaviors.

In this chapter, we present unique tactile sensors that we developed for such experiments. Though we have developed several tactile sensors, current technology does not allow us to create a sensor that is equivalent to the human skin. Therefore, what should be compromised and what should be emphasized depends on the objectives and constraints of each user. Thus, we would like to explain the interaction that we consider important and that forms the background of our research using tactile sensors.

Now let us imagine that you are moving through a rubble field where your footing is unstable. You will address this difficult situation in a variety of ways, such as choosing the rubble that seems safe to step on, adjusting your force to avoid losing your footing, using sticks to assist your balance, re-stacking rubbles, planning detours, and sometimes jumping or crawling on all fours. In this way, humans have the ability to achieve their goals by adjusting motion, refining strategy and sometimes unique solutions.

Although adults can perform a certain level of sophisticated strategies and behaviors, this is the result of various experiences and practices; climbing trees and playing in the sand during childhood, and crawling and toddling during infancy. Of course, not all of these adaptabilities were acquired after birth. Evolutionarily acquired 'seeds' of intelligence are innately embedded in the embodiment; the structure of brain circuits and the body. And it is thought that these 'seeds' not only serve as primitives for behavior but also to direct and constrain exploration and learning.

When encountering a new situation or object, we interpret it based on the knowledge and rules already acquired and try to respond toward the desired result. However, our knowledge and rules can always contain various levels of insufficiency and inconsistency. Even if the situation is seemingly the same, since the state of affairs is always slightly different in the real world, some level of ongoing adjustment is necessary. Furthermore, when failures or unexpected incidents occur in the process, it is necessary to make adjustments or changes in strategy. In this way, processes consisting of exploring information, reinterpreting the situation, refining strategies, adjusting actions, and learning a model of the situation, are thought to be performed repeatedly for immediate adaptation. All of these take place in continuous physical interaction with the environment. We believe that a mechanism that enables the emergence of such adaptive activity is exactly the essential one for human intelligence.

In physical interaction, visual and tactile information are very important. While visual information is obtained relatively passively, in order to get tactile information, it is highly necessary to generate active and intentional interaction. The information acquired in such interactions can be a mixture of both the information generated by one's actions and the information generated by reactions and motions of the target. Therefore, the ability to discriminate between them is very important for interaction-based intelligence. This ability of discrimination could potentially be the basis for the ability to separate self and others in social interactions. Furthermore, we think it could lead to the ability to recognize oneself objectively.

It is interesting and challenging to realize those themes such as the emergence of adaptive activities and objectification of the self, though there is still a long way to achieve them. In this chapter, we introduce our research on tactile information processing and the tactile sensors wherefore as parts of our research in the above-discussed context.

1.2 Developed tactile sensors and implementations

Assuming not only physical contact with the environment and humans but also intellectual interaction, such as interpretation and imitation of human behavior, we began developing an adult-sized humanoid robot [2]. We envisioned physical contact with various parts of the body, such as sitting and lying on the sofa, holding a large package with both hands, and placing hands and knees on the wall or floor when losing balance. Therefore, to avoid harming humans or the environment, as well as to avoid damaging themselves, robots were designed to have smooth outlines similar to a mannequin, and they were covered with soft materials such as sponge or gel. At that time, bipedal and upper body humanoid robots existed, but none of them were designed for this kind of full-body contact. Thus, it was challenging research. At about the same time, since HONDA presented a highly sophisticated bipedal walking robot, the concept of full-body contact was reinforced.

Our initially developed humanoid also focused on face-to-face interaction with humans. However, as we have come to recognize the fundamental importance of the sense of touch from an evolutionary and developmental perspective, we became interested in research that involved a more proactive physical interaction through full-body movements and contact. This led to the development of an infant humanoid robot, and research on proactive physical interaction and communication with parents and the environment during the infant stage. In addition, a wide range of related research has been conducted, such as simulations on the self-organization of movements during the infant period [3] and the fetal period [4].

1.2.1 Conformable and scalable tactile sensor for 3D-curved surfaces

Initially, we were considering using a custom-made tactile sensor from a commercial product, but we had to develop our own because of the technical problems and the extremely high cost of a custom-made sensor. Developed tactile sensor modules [5] are shown in Figure 1.1.

1.2.1.1 Ability to fit on 3D-curved surfaces due to branch-shaped sheet

Typical film-type tactile sensors can fit on simple curved surfaces such as cylinders and cones, but not spherical and three-dimensional (3D)-curved surfaces such as the human shoulder. Therefore, we adopted an approach to cut the film into a

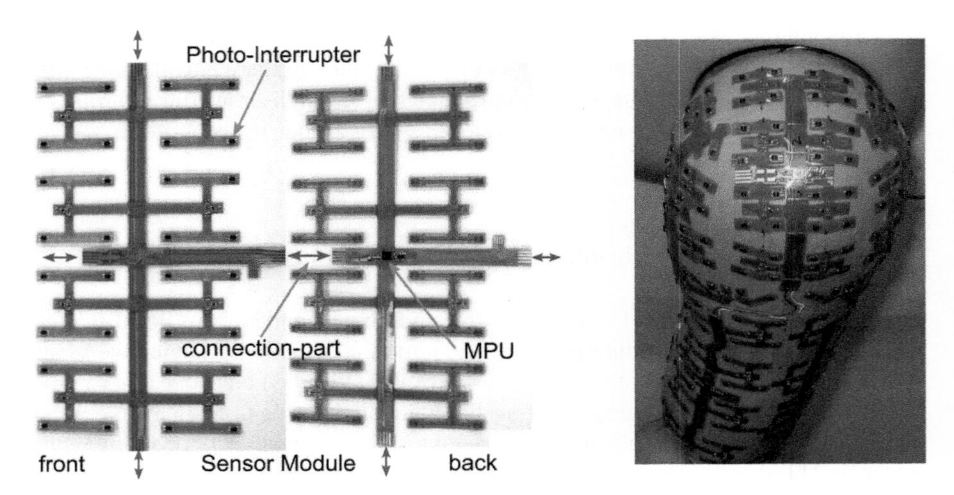

Figure 1.1 *The left shows the branch-shaped tactile sensor modules. The right shows an example of implementation on a 3D-curved surface of the upper arm.*

branch-shaped outline. Pressure sensing elements are placed at the tip or middle of the branch.

1.2.1.2 Customizability by free cutting

When the branch-shaped substrate is fitted to a curved surface, branches occasionally overlap each other. Therefore, we designed the electrical circuit so that cutting off interfering branches would not affect the acquisition of other sensor signals. In addition, we also found a way of bending the branches to adjust the position and resolution of the elements.

In order to cover the body with sensor sheets, it is necessary to combine various shaped sheets, such as sewing patterns for clothing. Thus, this cutting function facilitated the customization of shapes. In addition, since any shape can be cut out from a single-shaped sensor sheet, a low cost was achieved.

1.2.1.3 Communication networks and modularization

The major problem with commercial tactile sensors is that the wiring and connectors occupy substantial space. A typical tactile sensor requires M+N wires to measure M by N points, and the space occupied by them is significant, and it is difficult to store them near the body surface.

Therefore, we configured a sensor module that includes 32 detection points and a small micro-processing unit (MPU) with an analog-digital converter and serial communication functions. By connecting multiple of these modules, the connectors and wiring are distributed, resulting in a thin and slim sensor network. Specifically, we adopted the C8051 series of Silicon Laboratories as the MPU. Since SMBus was adopted for serial communication, the module was set up to be connected with four wires for power and communication. While SMBus is originally a bus-type network for intra-board communication, the communication in our application is prone to instability. Therefore, a small repeater IC is inserted to stabilize communication.

1.2.1.4 Pressure sensing by photo-reflector

Two types of pressure sensing devices are commonly used: the pressure-sensitive ink type, where a conductive filler is mixed with ink or other materials, and the capacitive type, where an elastic dielectric is inserted between two electrodes. However, both have the disadvantage of requiring a different manufacturing process than ordinal flexible printed circuits (FPCs) to be mounted on a module. Especially, the ink type requires considerable know-how.

Therefore, we decided to use a combination of a photo-reflector and a sponge as a pressure detection method. When the sponge is placed on top of the photo-reflector and a light emitting diode is illuminated, light is diffusely reflected by the sponge, and the reflected light is detected by a photo-transistor. As the sponge is pressurized, the density of the sponge increases, the intensity of reflected light increases, and the signal of the photo-transistor changes continuously. On this basis, pressure can be measured. This is an application of the technology used by Tactex Controls, Inc. [6]. They used optical fibers in their design, whereas we used an ultra-small photo-reflector IC. Photo-reflectors are common and inexpensive electronic components, and pressure sensing is also possible by reusing the sponges that cover the humanoid for safety. This made it possible to develop a sensor that is easy to implement and low cost.

1.2.1.5 Examples of implementation

Because of its modularity and customizability through cutting and folding, it is easy to use and has been used in various projects such as humanoids and sensor suits. Details of research using the adult humanoid robot (Figure 1.2) and how the sensors were used are described in section 1.3. The infant humanoid robot (Figure 1.3), which was developed for more active physical interaction, was able to perform the turn-over motion, exploratory behavior using visual and auditory information, and perceiving tactile information during human-robot physical interactions (Figure 1.4 (a)). A sensor suit for adults with both full-body tactile sensors and Inertial Measurement Unit (IMU)-based motion captures was used for research on measuring various daily behaviors [7]. We also developed a sensor suit that can be worn by both the baby and the mother and used it in a study to analyze the interactions when she holds the baby in her arms (Figure 1.4 (b)).

The Integrated Circuit (IC) is structurally prone to direct external forces, but the IC was rarely broken. Since external forces can cause the soldering of components to peel off from the FPC, the area around the IC is protected with epoxy. On the other hand, the 4-pin connectors for power and communication are prone to failure, and in some cases, soldering is used. Thus, the connection between the modules is not so well designed.

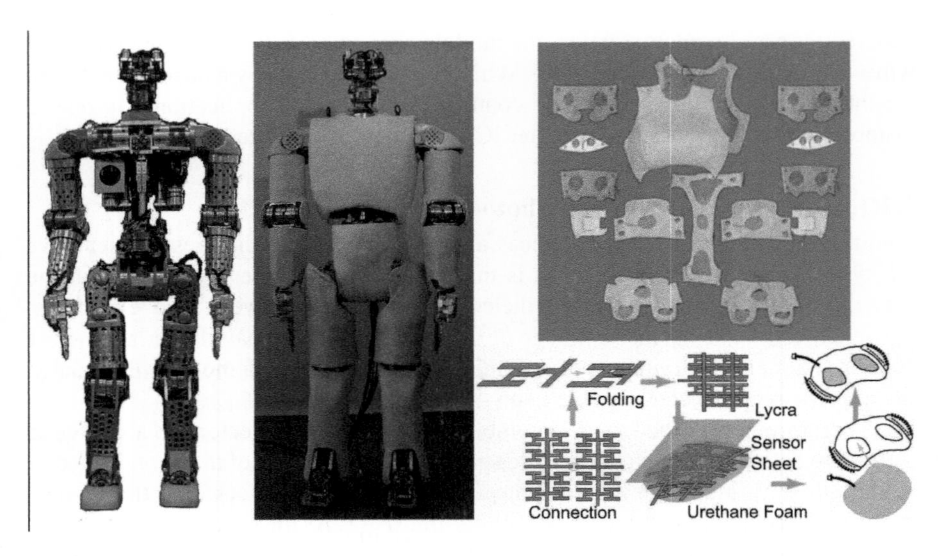

Figure 1.2 Developed adult humanoid robot and covers with tactile sensors

1.2.2 High-density tactile sensor for bare-hand-like sensor gloves

As the next step, we developed a tactile sensor for the human hand based on the sensor in the previous section [8, 9] (Figure 1.5). Although there have been many studies on tactile sensors for robot hands, very few tactile sensor gloves such as Grip System (Tekscan Inc.) and reference (e.g., Reference [10]) have been developed for the human hand. This is probably because human hands are bent and wrinkled, making it difficult to measure tactile data.

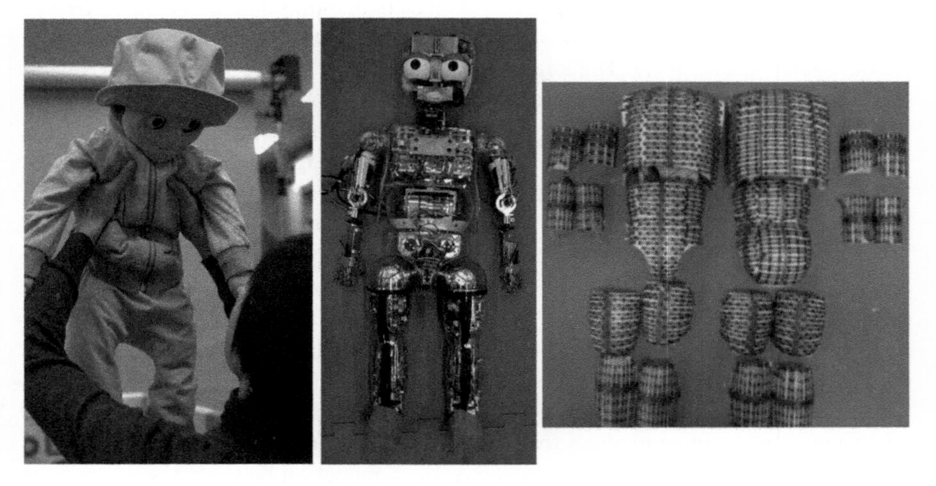

Figure 1.3 Developed infant humanoid robot and covers with tactile sensors

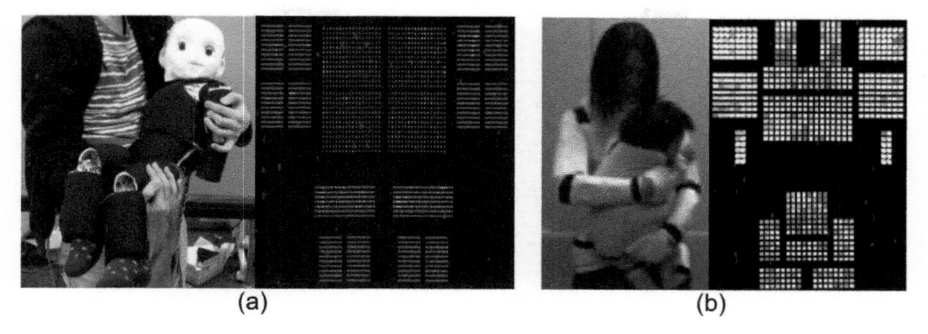

Figure 1.4 *The left shows an example of sensor pattern of the infant humanoid robot. The right shows an experimental scene of measuring interactions between a human baby and an adult. Both are wearing sensor suits.*

Despite the technical difficulties, we decided to develop a tactile sensor glove with relatively high resolution while adapting to joint bending and wrinkles. Furthermore, instead of attaching the sensor to a removable glove, we aimed to attach it directly to the skin to make the measurement as close to that of a bare hand as possible.

In addition, this project started as joint research with a major company that aims to market FPCs with built-in pressure sensors. Therefore, please note that this project envisions both the use of FPCs and the consistency with mass production and is not the result of a comparative study with different technologies such as stretchable wiring.

1.2.2.1 Design and implementation

This sensor is a thinner and more flexible version of the sensor in section 1.2.1. There are three important differences between the two sensors: the design of the shape to fit complex deformations, the thinness of the sensing element, which

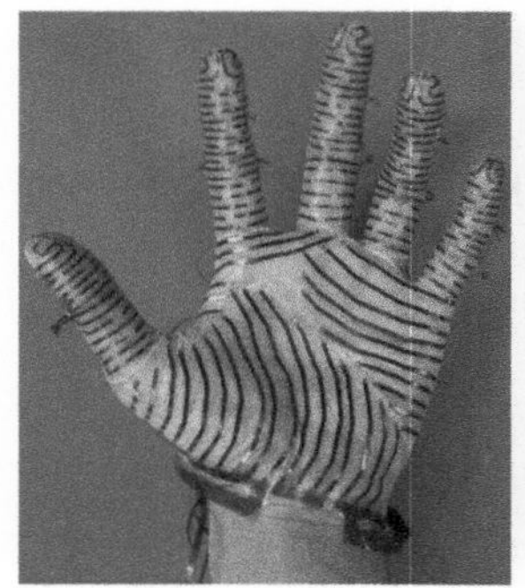
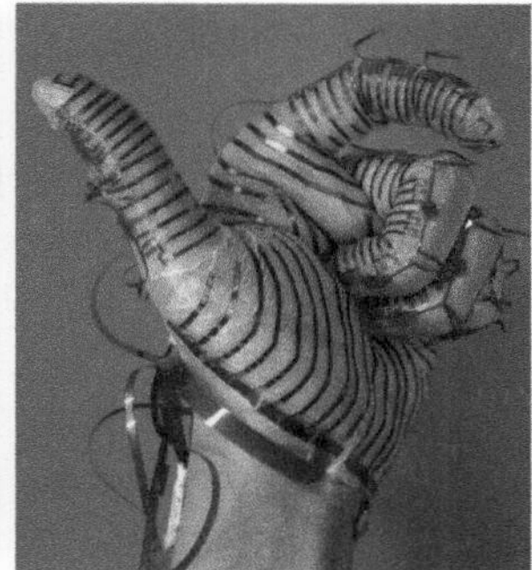
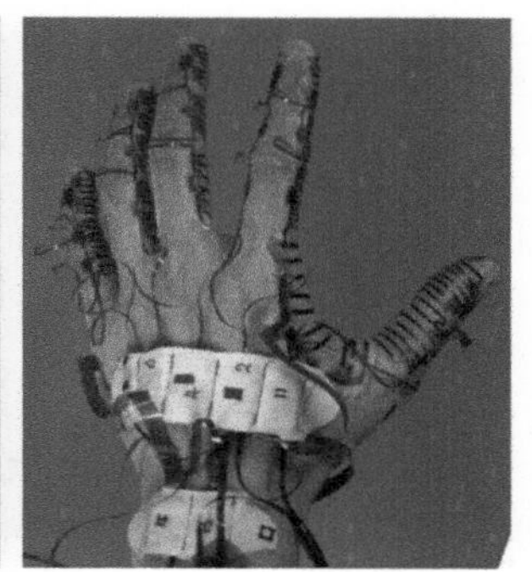

Figure 1.5 *Developed tactile sensor glove*

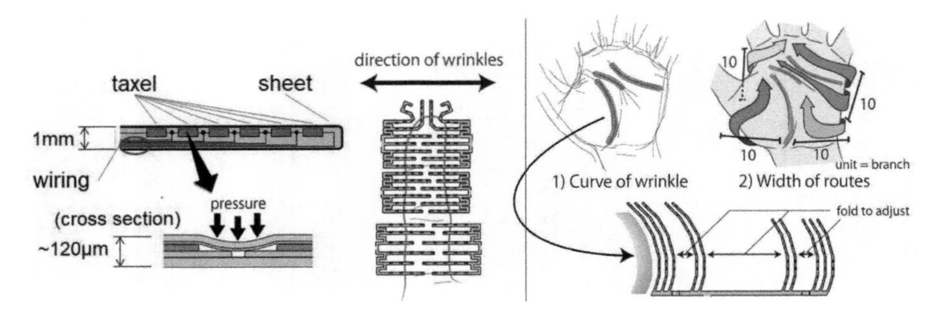

Figure 1.6 Structure and designing of the tactile sensor glove

is inserted between the layers of the FPC, and the consistency with the FPC manufacturing process. Specifications of the FPC required thin and flexible base sheets, coverlays and laminate adhesives, and durability against repeated bending of copper foil and through holes. The details of the sensing elements are described in section 1.2.2.3.

The fingers and palms of the hands have large wrinkles and raised skin due to the bending of the joints. The shape of the FPC was designed to fit this complex deformation (Figure 1.6). The skin tends to stretch more perpendicularly than parallel to the wrinkles. On the other hand, branch-shaped FPC cannot be stretched in the direction of the branches but can be stretched slightly between branches. Therefore, the branches and wrinkles were designed to be parallel. Furthermore, the shapes of the FPC were designed to minimize the number of patterns as much as possible, also the cutting and folding functions described in the section 1.2.1.2. The number of pressure sensing points is approximately 1,000 for one hand, and the distance between elements is 2 mm for fingertips, 3.5 mm for fingers, and 6 mm for palms. The electronic circuit board with the MPU is placed on the back of the hand, and the FPCs with built-in sensors are connected to the board with connectors.

The FPCs were fixed to the hand as follows. First, the hand was sprayed with silicone elastomer, then the FPCs were attached to it, and finally covered with medical polyurethane films. Those membranes have moisture permeability and stretchability. The total thickness of the sensor glove is approximately 0.2 mm, and it can also be removed and reused a few times.

Using this sensor glove, we could achieve unprecedented high-density measurements of pressure distribution on the fingers and palm when grasping and manipulating various objects in conditions close to those of bare hands (Figure 1.7). In addition, research on object identification by groping using the sense of touch was also conducted.

1.2.2.2 Consistency with mass production

In this project, the sensor was intended for mass production; thus, the pressure-sensitive element needed to be not only thin and flexible but also compatible with the general FPC manufacturing process. Specifically, it should be able to withstand

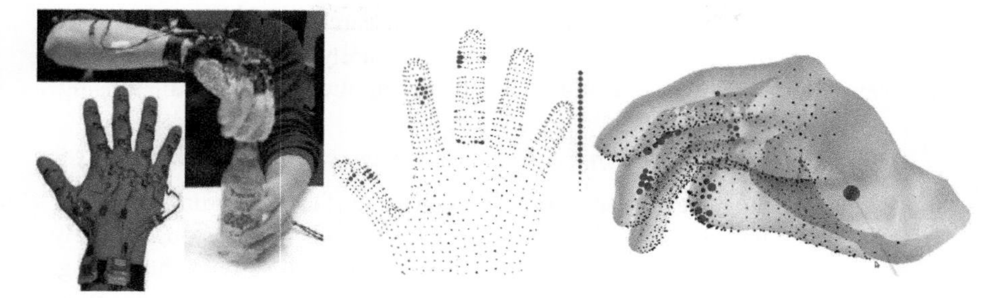

Figure 1.7 *Experimental scene of the sensor glove with a system including tactile sensor and small IMU-based motion capture when opening a plastic bottle lid*

high pressure and temperatures during interlayer adhesive, high temperatures during solder reflow, roll-to-roll process, and so on. As described in section 1.2.1.4, there are two common methods for pressure detection: the capacitance method and pressure-sensitive ink method. The former has problems with flexibility and error signals because the sheet is thicker due to the dielectric material and the two electrodes. The latter has drawbacks such as the printing of the ink being away from the normal manufacturing process of FPCs and the high temperature during reflow soldering.

1.2.2.3 Contact resistance method

Finally, we have adopted the contact resistance method. This method is based on the phenomenon that when a bad conductor and a good conductor come in contact, the electrical resistance changes continuously, not in an on-and-off manner, according to the pressure. Various bad conductors, such as oxidized metals and organic conductors, were deposited or coated on thin sheets, and several performance tests have been conducted, taking cost and manufacturability into consideration. Since it was necessary to separate the electrodes from the bad conductors without bonding them, the manufacturing process was devised so that the electrodes would be unaffected by the gas and molten liquid of the hot adhesive layer.

1.2.2.4 Toward new applications using tactile sensing

Finally, an FPC with a built-in pressure-sensitive element suitable for mass production has been developed. Various product manufacturers, including information equipment and home appliances, were interested in this device, and of course, many prototypes were fabricated. However, it has not yet been shipped in a product. The main problem is that many people are interested in high-resolution and dynamic pressure distribution images; but, how to extract useful information from data is a matter of software, not devices. For example, the real-time pressure distribution images of sensor insoles while walking or running give many people a sense of the potential for application in sports or rehabilitation; however, quantifying the "habit" of walking is difficult. It is sufficient to be a research topic in itself. A device that provides only the pressure distribution image is barely more than a measuring device. New software may be essential to invent new applications of tactile sensing.

1.2.3 Stretchable tactile sensor based on inverse problem analysis

Although the branch-shaped sensor can fit on curved surfaces, it is unsuitable for covering around joints such as shoulders and elbows, where the skin is highly stretched. It would be desirable if we could realize a sensor with stretchable wiring. However, we thought it would be technically difficult to achieve both stretchability and high reliability. After much deliberation, we devised a novel tactile sensor based on inverse problem analysis. For example, when measuring the resistance between two points on the boundary of a 10-cm^2 pressure-sensitive rubber sheet, changes in resistance are observed when the sheet is pressurized with a finger. The degree of the change depends on the positional relationship between the pressure and measurement points. Therefore, we thought that we could estimate the strength and position of the pressure point by measuring the resistance at various points and analyzing the results comprehensively. After substantive research, we found that there is a sophisticated technique called electrical impedance tomography (EIT).

1.2.3.1 Electrical impedance tomography

A well-known medical computed tomography (CT) scanner is one in which X-rays are irradiated from outside the human body from various directions, and the resulting data are analyzed in an integrated manner to calculate tomographic images of the human body. It is a type of inverse problem analysis.

EIT is a technique to estimate the resistance distribution inside a conductor by measuring potentials at the boundary of the conductor when injecting current, instead of X-rays, in various directions from the boundary (e.g., Reference [11]). A typical use of EIT is to estimate the shape and movement of the lungs and heart by attaching approximately 16 electrodes around the chest of the human. It has been researched as an inexpensive alternative technology to CT scans.

Based on the above, we thought that by applying EIT to stretchable pressure-sensitive conductive materials, we could realize a stretchable tactile sensor that can estimate pressure distributions from resistance distributions. Although stretchability near the boundary is limited by wires, there are no wires on the inside of the sheet, which allows for relatively free stretchability.

1.2.3.2 Conductive rubber sheet

The initial model [12, 13] was verified using commercially available pressure-sensitive rubber sheets. The effectiveness of this method was confirmed not only with simple squares and circles but also with various shapes and the arrangement of the electrodes, as shown in Figure 1.8.

However, with pressure in the thickness direction, the change in resistance to horizontal current was minute due to the thinness of the sheet, and it was necessary

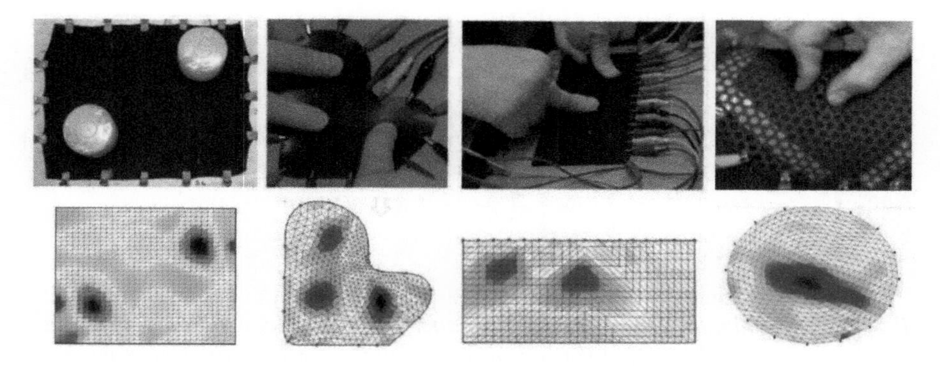

Figure 1.8 Examples of experimental data of tactile sensors based on EIT

to press hard over a wide area to obtain a sufficient signal change. On the other hand, since the pressure-sensitive rubber sheet changes its resistance, even in the horizontal direction, it works as a stretch distribution sensor as well. Therefore, we punched many holes in the sheet to increase its stretchability and placed it on a soft sponge sheet to convert pressure into a stretch, resulting in a more sensitive tactile sensor. At the same time, it also functioned as a new type of tactile sensor, capable of detecting not only pressure but also pinching and twisting.

1.2.3.3 Conductive knit fabric

Although ordinary pressure-sensitive rubber sheets are very soft, their stretchability is only a few percentages and they easily tear. Therefore, we have developed a conductive knit fabric with higher stretch and pressure sensitivity. This is made by impregnating cotton knit fabric with black ink in which carbon is dispersed, and then drying it. Without loading, a uniformly high resistance distribution is obtained. However, when pressurized, the density of the knitting increases, and the resistance value in that area decreases.

This sensor has been implemented on the movable elbow joint of a mannequin and could sufficiently follow the stretching of the surface around the joint (Figure 1.9 (a)). However, as with the pressure-sensitive rubber sheet, the resistance changed as it stretched, so it was necessary to distinguish between the resistance change caused by pressure and stretch. By recording the resistance distribution corresponding to the joint angle without pressure in advance and taking the difference between that data and the data obtained when pressure was applied, it was possible to extract only the pressure distribution while canceling out the effect of stretching [14]. Although this method can isolate the pressure to some extent, but if the stretch is large and the pressure is weak, it can happen that the error in the compensation data may exceed the pressure signal.

On the other hand, this conductive knit easily fits on complex curved surfaces such as the face of a mannequin (Figure 1.9 (b)). Since there is no dynamic stretching in this case, the pressure is stably detectable, and it works well.

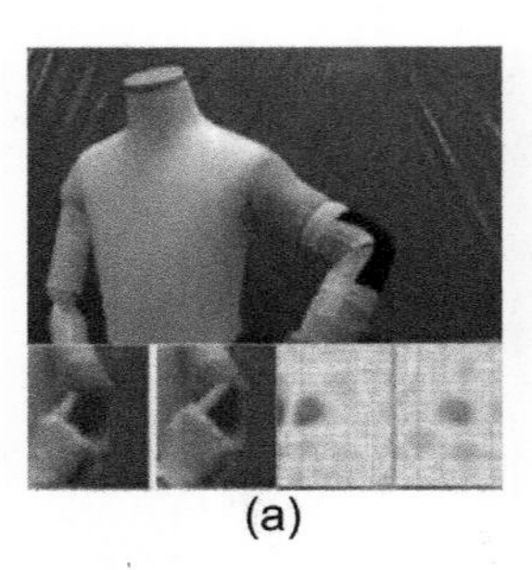
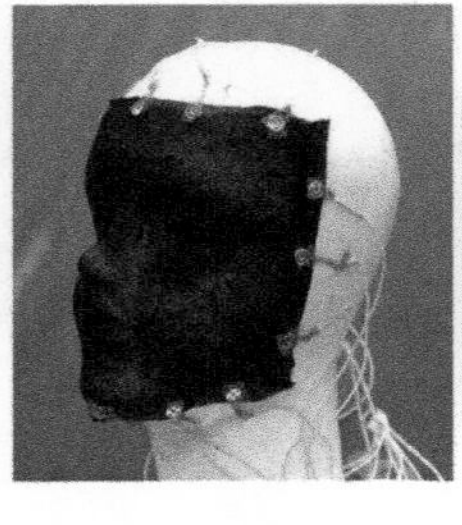
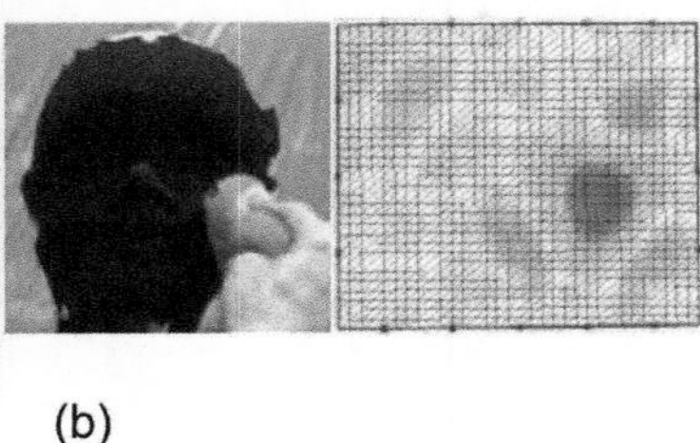

Figure 1.9 *Tactile sensors based on conductive knit fabric: experimental scene when mounting the sensor around the elbow joint (a) and example of implementation on a complex 3D-curved surface and its experimental scene (b)*

1.2.3.4 Pressure-sensitive but stretch-insensitive conductor

To solve the problem of the mixed effects of stretch and pressure, we developed a conductor that is insensitive to stretch and sensitive only to pressure [15]. In this case, we used the contact resistance method described in section 1.2.2.3 to detect pressure. First, we fabricate a stretchable bad-conductive sheet that is insensitive to both stretch and pressure and measure its resistance distribution via EIT. Then, a stretchable good conductor sheet is placed on top of it. If a certain area of this composite sheet is pressurized, the resistance of this area decreases continuously according to the pressure, and changes in resistance can be detected by EIT. Since the contact resistance is orthogonal to the stretching direction, a sensor that responds only to pressure will be realized.

Specifically, fibers coated with copper sulfide (CuS) were used as bad conductors, and these fibers were sewn into a mesh on ordinary knit fabric to make stretchable sheets of bad conductors. In addition, CuS-coated fibers were used to create knit fabrics, covering yarn with coiled fibers wrapped in spandex, and nets made from this covering yarn. As a good conductor, we used a conductive knit fabric made of silver (Ag)-coated fibers. However, in the case of a single large silver knit, when two points are pressurized, current flows between them through the silver knit, introducing errors in the data. To solve this problem, the silver knit was divided into many smaller areas and electrically insulated them. As a result, we have developed a tactile sensor that is sensitive only to pressure while having the same level of stretchability as regular knit fabric. It was confirmed that the system can measure stable pressure distributions against pressure and pinching, even when it is placed over an extremely deformable object such as a balloon and is greatly stretched or deformed (Figure 1.10). Furthermore, to evaluate the effectiveness of the sensor on the joints, the sensors were implemented around the hip (Figure 1.11) and the elbow joint (Figure 1.12), and it was confirmed that the data were almost unaffected by joint bending, and were stable even when measuring [16].

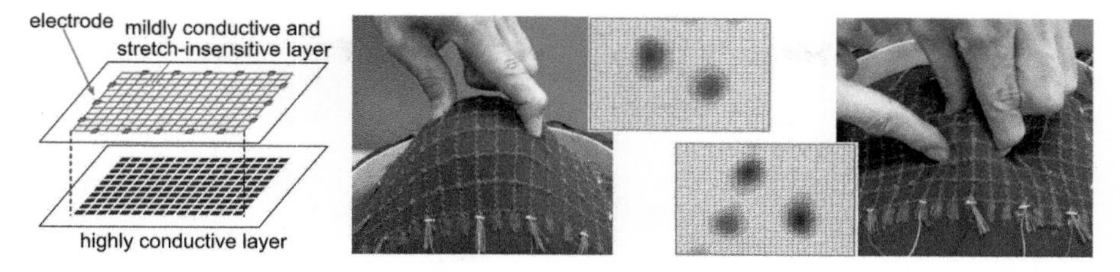

Figure 1.10 *Structure and experimental scenes of pressure-sensitive but stretch-insensitive tactile sensors based on EIT*

1.2.3.5 Issues and future prospects

This EIT-based sensor is very unique and distinctive, but it was not installed on our humanoids because it was inferior to the FPC type in terms of ease of use, fault tolerance, and complex signal processing. There were also some disadvantages in using this as a general-purpose tactile sensor: low sensitivity at a distance from the electrode, difficulty in detecting small pressure areas, rough shape and strength of the estimated pressure area, and difficulty in stably fixing the electrode to a soft material.

Nevertheless, the method based on inverse problem analysis is very interesting and has led to many related studies [17]. We believe that there are still many possibilities for sensors based on inverse problem analysis, such as the development of new conductors, improvement of the electrode arrangement, and application of tomography to other sensors than tactile sensors.

1.3 Tactile sensing and robotics: future direction

In this section, we introduce intelligent robotics research with tactile sensors in our team and discuss the importance of haptics in the intelligent robotics field and future

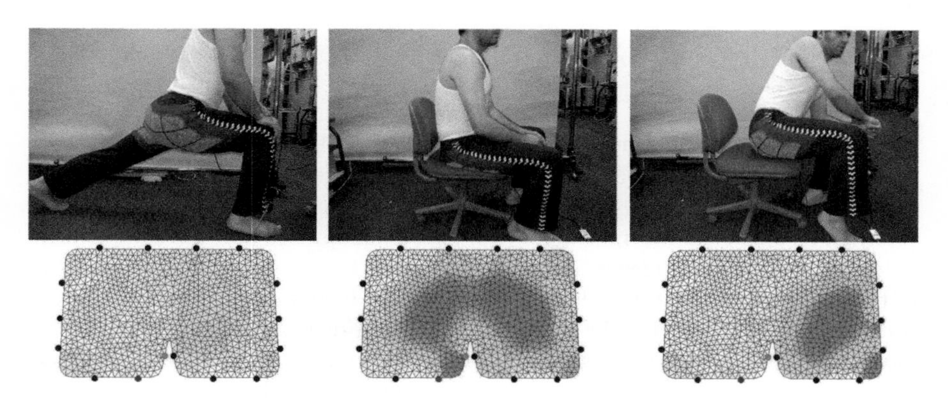

Figure 1.11 *Experimental scenes of the sensor based on EIT mounted around the hip joint*

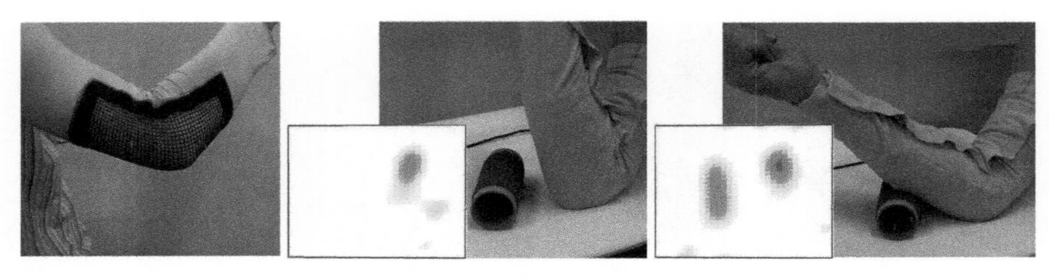

Figure 1.12 Experimental scenes of the sensor based on EIT mounted around the elbow joint

directions. We think haptics is essentially active sensing and meaningful information is not obtained without self-movement (except social communication). Therefore, somatosensory information often consists of both object/environmental information and self-motion information. How to extract useful information from such mixed information is a difficult but interesting problem for an intelligent system. First, we classified the robot application of tactile sensors into two categories based on differences in robot behavior models: automaton and state-action models. The automaton model was fully constructed by a human designer. We used the whole-body tactile sensor system described in the previous section. The tactile sensor element we used was only a pressure sensor, but the meaning of pressure can be changed by a predefined motion. It was one of the first applications of whole-body tactile sensors in humanoid robot behavior. The state-action model is typically used in reinforcement learning (RL) or imitation learning. Recently, we used the state-action model for learning grasping in the human-like five-finger robot hand. Finally, we show the future direction. We believe haptic exploration is the most difficult and interesting research field. Haptic exploration is active sensing to recognize object information from somatosensory information mixed with self-motion information.

1.3.1 Automaton model

We developed a humanoid robot that has a whole-body tactile sensor system. We used the system to realize whole-body motion, not for social communication, because we were interested in active sensing.

First, we introduce an application for heavy-object lifting [18]. In this behavior (Figure 1.13), the whole-body tactile sensor information was used to extract the environmental information. The control model was automaton, and somatosensory information was used for transition conditions and adjustment of the control variable. The movement sequence is shown as follows: 1. measurement of the position of the table side; 2. measurement of the width of the box; 3. pull the box toward the trunk, 4. push the box out of the table side; 5. exploration of the bottom of the box by the right hand, 6. hold the box and lift. We used the width of the box and the position of the table side to adjust the amount of push motion in state 4. Except for transition from state 4 to state 5, transition conditions were determined from somatosensory information.

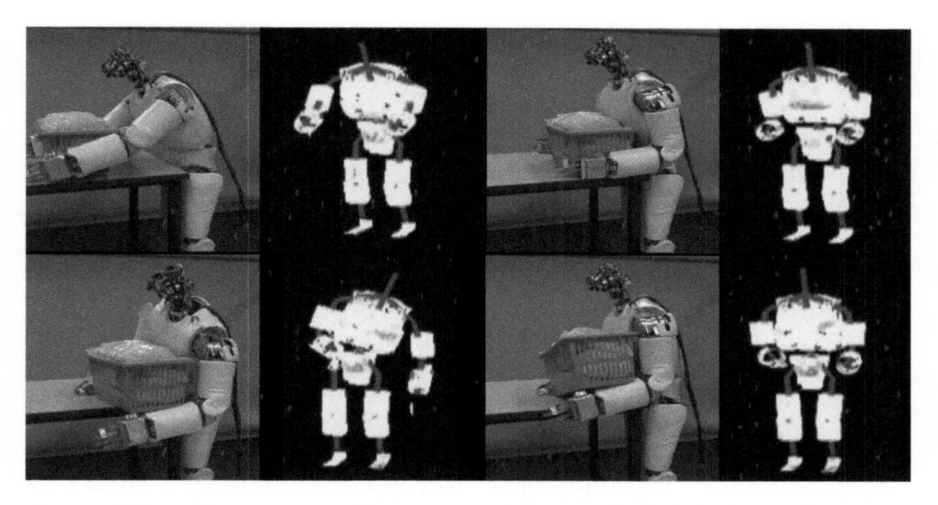

Figure 1.13 The sequence of heavy object lifting and whole-body tactile sensing

Next, we introduced an application for dynamic whole-body motion (Figure 1.14), which is a roll-and-rise motion [19, 20]. A sequence of roll-and-rise motion is shown as follows: 1. the flexion of the hip joint on the back; 2. the extension of the hip joint; 3. the flexion of the hip joint and knee joint simultaneously; 4. landing. This sequence is shown in Figure 1.14. This motion was also designed as an automaton. The tactile information was used to measure the posture and rolling velocity by calculating the movement of the center of pressure. All transitions were determined by tactile information. The start time of flexion of the hip and knee joint in state 3 is critical to realize this dynamic motion. If the start time is early, sufficient rotational momentum to lift the center of gravity after landing cannot be obtained. Conversely, if the start time is delayed, the duration is insufficient to complete the flexion of the hip and knee joint before landing because the distance between the landing position of the foot and the center of gravity is too far to lift the center of gravity. Therefore, the duration of flexion of the hip and knee joint in state 3 was set to the shortest time possible to achieve. To determine the start time of flexion

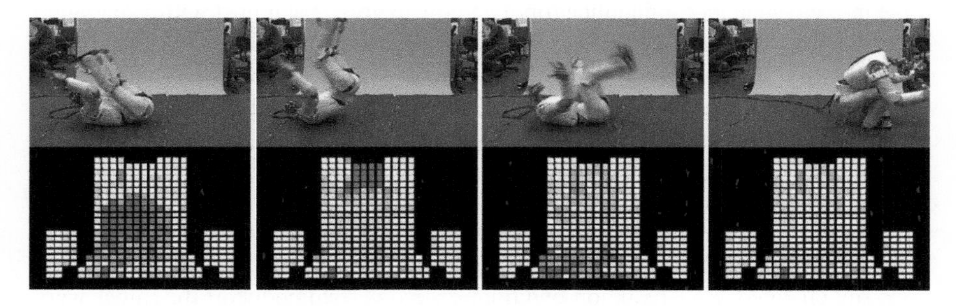

Figure 1.14 A roll-and-rise motion and tactile sensor pattern of back

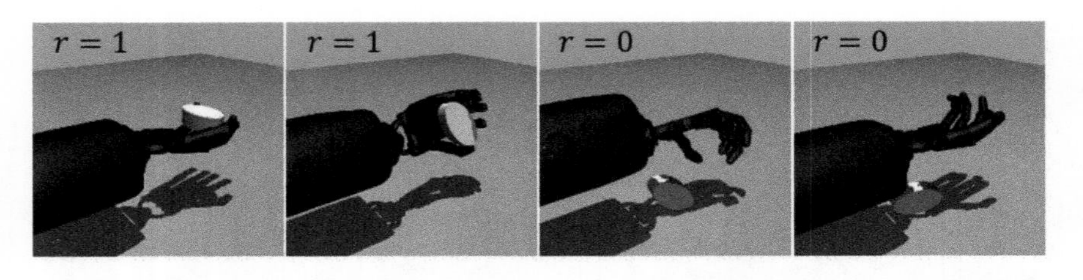

Figure 1.15 *Experiment setup. The wrist was rotated and a reward was given when the object was on the palm of the robotic hand.*

in state 3, the remaining time to landing was estimated from the center of pressure dynamics, which was a data-driven model. The start time of flexion was determined to match the timing of landing and the complete time of flexion. Importantly, in this dynamic motion, the meaning of tactile information was determined by the pre-defined automaton. The change of tactile information was caused by the movement. Furthermore, the movement was controlled by tactile information.

1.3.2 State-action model

Next, we introduce RL of grasping using tactile sensors. In RL or imitation learning, the state-action model is frequently used. Many studies have used RL to learn in-hand manipulation [21, 22]. We conducted experiments on learning in-hand grasping of multiple shapes of objects based on tactile recognition using an anthropomorphic robot hand [23]. We used a simulation model of Shadow Dexterous Hand in OpenAI Gym [24]. First, we attempted to train the robot to grasp an object on a table, but it was difficult to learn because the robot never succeeded in the task and did not gain a reward. To confirm that the robot can learn how to grasp objects, we changed the initial state. The object was placed with the palm facing up (Figure 1.15). Following the change in the initial state, the robot rotated its wrist to face the hand down using a predefined motion. If the robot learns how to grasp the object using five fingers, the object remains in the in-hand state. We defined a reward function as the duration of the in-hand state. We trained the robot on how to grasp the object with five fingers using Soft-Actor-Critic [25]. A learning step was set to 10^6 episodes.

The input was a set of 24 joint angles of robot fingers, one joint angle of the wrist, and their velocities. Furthermore, we used four conditions of distributed tactile sensors: (a) no tactile sensors, (b) 5 tactile sensors only in fingertips, (c) 44 tactile sensors, except for the palm, and (d) 92 tactile sensors on the entire surface of the hand (Figure 1.16). The variation of grasping objects was eight. The robot did not know which object was in its hand at the initial state. We tested the grasping performance 1,000 times (8 objects × 125, Figure 1.17). The positions of objects were randomly changed. As a result, the tactile sensors on the surface of the finger were required to grasp the multiple objects (success rate: (a) 9.5%, (b) 7.5%, (c) 95.7%,

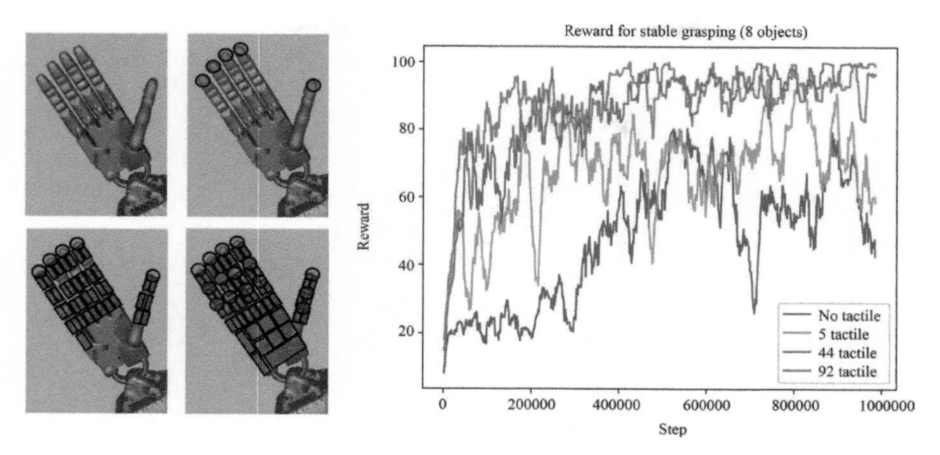

Figure 1.16 *The left shows the position of tactile sensors. Blue and red denote the positions of the tactile sensors. The right shows transitions of the reward for 4 conditions of tactile distributions.*

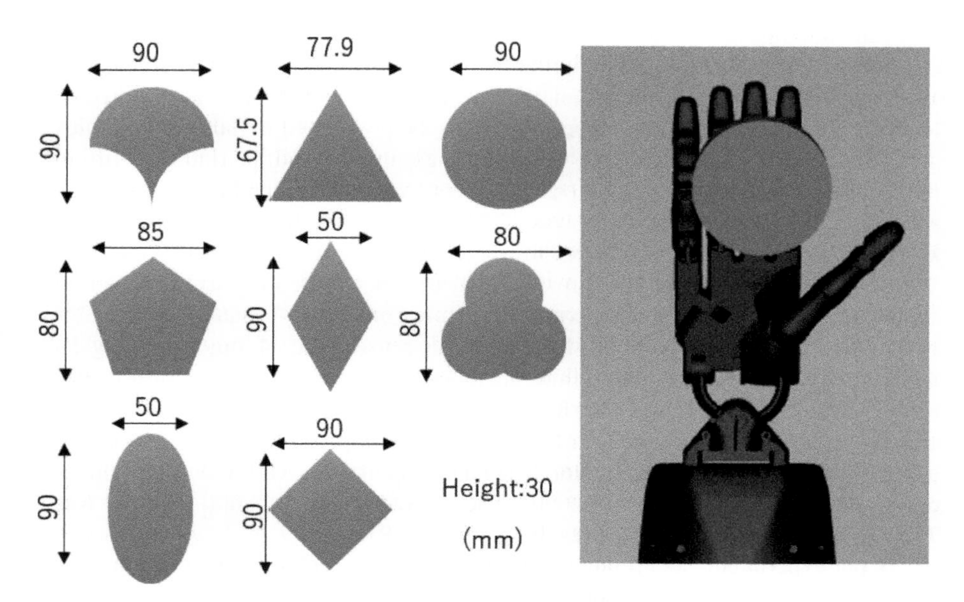

Figure 1.17 *Eight shapes of columnar objects, with a height of 30 mm. The left shows the shapes, which are ginkgo, triangle, circle, pentagon, diamond, flower, ellipse, and square from the upper-left. The right shows the scale of an object relative to the robot hand.*

Figure 1.18 The initial state for picking objects

and (d) 82.5%). Only the fingertip sensors were ineffective in learning the in-hand grasping of multiple objects. Furthermore, we confirmed that the trained policy was effective in picking up objects on a table if the initial posture of the finger was suitable for grasping.

Next, we attempted to train the robot how to pick up objects on the table again. We assumed that the suitable initial state of the finger and object was required to learn how to pick up multiple objects. We used the predefined initial posture of fingers (Figure 1.18). We trained the robot for finger motion and the timing of lifting the hand. We used the 44 tactile sensors except for the palm condition model. The success rate of the 8,000-time test was 42.2%. We hypothesize that if the robot can know the stability of grasping before lifting the hand, the success rate may increase. We trained a discriminator to estimate the success or failure of grasping based on somatosensory information to determine the timing of lifting. A linear discriminator was trained with 6,400 data points from 44 tactile sensors and 24 finger joint angles, including multiple success and failure grasping. Then, we tested the discriminator using 1,600 data points. The discrimination rate was 97.25%. We trained the discriminator for only the grasping motion, and the timing of lifting was determined using the output of the discriminator. The hand was lifted when the discriminator's output was "success" 10 times in a row. The success rate of the 8000-time test was 71.2%, which is greater than the result obtained by RL.

In our experiment, the initial state of picking up objects is critical in learning the policy. The initial state was a posture with the thumb and other fingers were facing each other, and the object was placed between the thumbs and other fingers. The robot could grasp the object only by bending its fingers. In this experiment, the robot cannot search for the object by haptic exploration. Haptic exploration cannot be used for random action. Humans can generate structured and intended motions for haptic

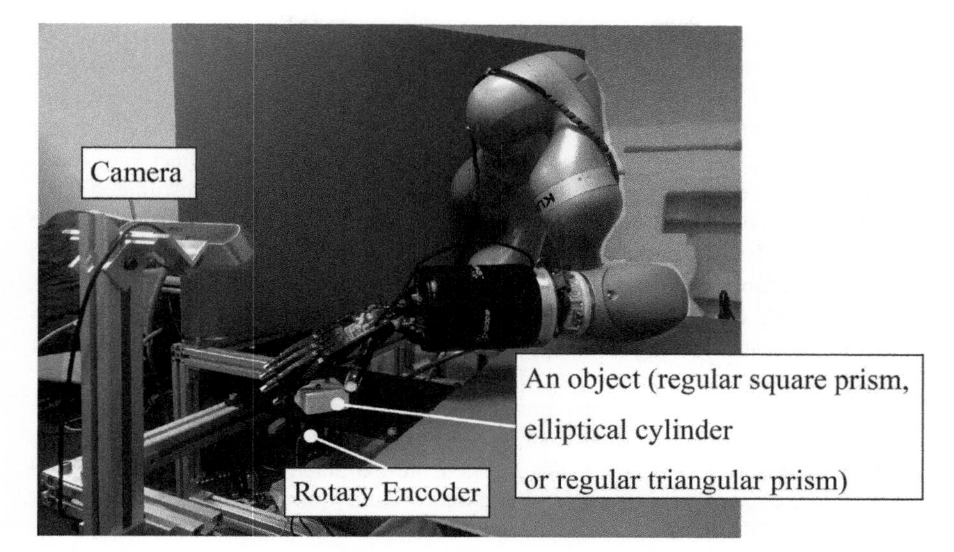

Figure 1.19 *Experimental setup. The robotic hand equipped with a fixed robot arm touching an object. The stereo camera captures images of the object. The rotary encoder measures the posture of the object.*

exploration. However, currently, we have no method for designing a reward function to learn such intelligent exploring motion.

1.3.3 Future direction

Finally, we introduce research on haptic exploration. Humans can recognize and manipulate objects in a situation, where visual information has been lost, e.g., in the dark, or when the object is in a pocket. While somatosensory information consists mainly of self-motion and posture-related information, humans frequently pay attention to the object's posture and shape rather than their hand pose. Because the object's posture and shape are more important than self-motion during manipulation, this attention bias is reasonable. However, the method used for extracting an object's information from somatosensory information is poorly understood.

The force of the hand can change the posture and location of an object during haptic exploration. To use haptic exploration for object manipulation, the explored objects should not be fixed in the environment. However, several studies have used immovable objects to research shape recognition by haptic exploration [26–28]. In our study, we used objects that can rotate around predefined hinge joints.

We developed a system, which can generate visual appearance from somatosensory information of a human-like five-finger robotic hand during haptic exploration [29]. In this experiment, we used a robotic hand (Shadow Dexterous Hand E Series, Shadow Robot Company), which has five pressure sensors at the fingertips (Figure 1.19). First, we measured somatosensory information and the posture of the object measured by the rotary encoder simultaneously during haptic exploration.

The finger motion was realized by teleoperation using a glove (CyberGlove II, CyberGlove Systems) because the motion of haptic exploration was difficult to generate autonomously. During haptic exploration, the appearance of an object cannot be seen from a camera. Therefore, we measured the appearance of the object recorded by a camera (ZED, Stereolabs) and the posture of the object simultaneously at other sessions. We show that the object's posture can be estimated from the time-series of the joint angles of the robot hand via regression analysis (Square prism: 89.9%, Elliptic prism: 92.3%, Triangular prism: 88.7%). Thus, we confirmed the somatosensory information has sufficient information to estimate the object's posture. In this study, we showed that conditional generative adversarial networks [30] can generate an image to show the appearance of invisible objects from their estimated postures.

Next, we examined whether unsupervised clustering can extract the object's posture information. The hand data generated when touching the objects at two or more points were extracted. We merged the extracted hand data with several steps before and after the extracted hand data were acquired to use the time-series information of the hand data. The duration of time-series data was 0.5 s. To extract the time-shift invariance feature, we used a fast Fourier transforms of the time-series data and performed k-means clustering. The number of clusters was set to 10. Finally, we created a histogram of the object's posture in each cluster (Figure 1.20). The histogram showed that diversity in the object's posture was large and the distribution had frequently overlapped. The clustering method is affected by the distance between somatosensory data points. Somatosensory information is determined not only by object information but the finger posture. Therefore, the similarity between the finger posture strongly affected the clustering results. Currently, it is difficult to extract object information from a fusion of object and finger information. The difficulty has to be decreased by decreasing the diversity in the finger posture. However, we did not have a suitable motion generation method for haptic exploration. Thus, we believe that a motion generation method for haptic exploration is an important research topic for the application of tactile sensors.

1.4 Conclusion

We introduced the development of tactile sensors and their application for detecting robot behavior. In these studies, we used pressure sensors. This type of sensor can detect only one-dimensional pressure. However, the distributed tactile sensor information has several meanings, depending on the application: the shape and size of objects, the position of objects, the stability of grasping, the posture of the self-body, and the rolling velocity of the body. The meaning of tactile sensor information and self-motion is strongly interdependent. Thus, the meaning of tactile sensor information cannot be determined before the motion and has to be constructed using motion data.

Our design of the tactile sensor system did not follow the conventional engineering design methodologies. Engineers typically determine the application of a sensor

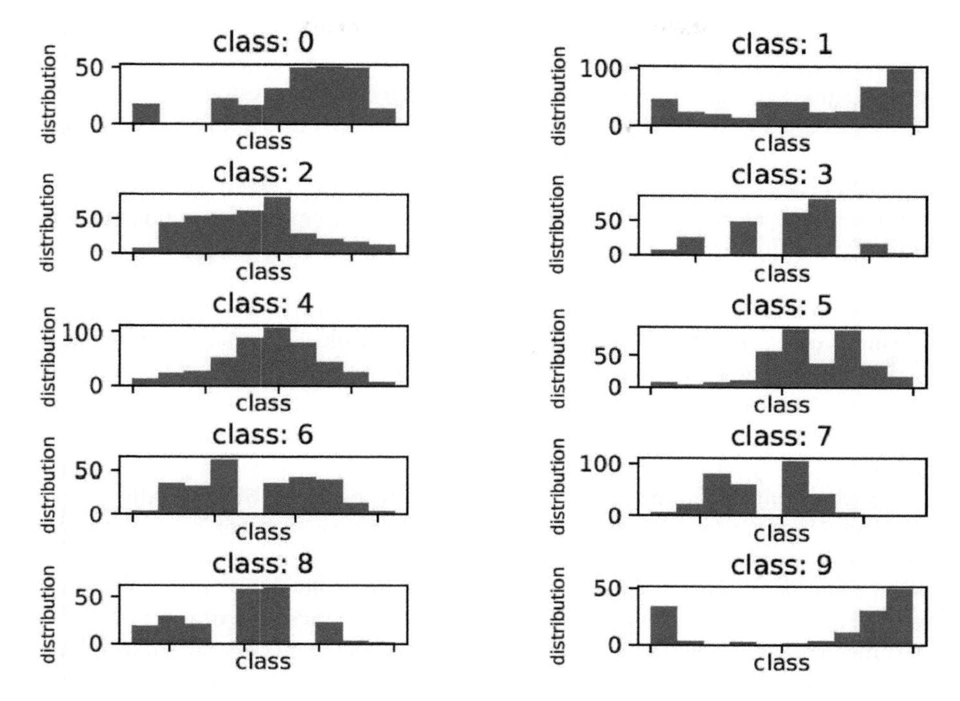

Figure 1.20 *Histogram of the posture of an object in each cluster calculated by k-means clustering of fast Fourier transforms of time-series of somatosensory information*

system before the design and optimize the specialized sensor for the predefined goal, e.g., a tactile sensor for slip detection. However, we believe such a one-to-one correspondence between sensor design and required functions is not suitable for intelligent robot research because the attractive feature of the tactile sensor is that the target motion can change the meaning of tactile information and the meaning cannot be determined without motion. The meaning of both motion and tactile information has to be constructed by cyclic causation between motion and sensing.

"*Anyone using a probe for the first time will feel its impact against his fingers and palm. But as we learn to use a probe, or to use a stick for feeling our way, our awareness of its impact on our hand is transformed into a sense of its point touching the objects we are exploring. This is how an interpretative effort transposes meaningless feelings into meaningful ones, and places these at some distance from the original feeling. We become aware of the feelings in our hand in terms of their meaning located at the tip of the probe or stick to which we are attending. This is so also when we use a tool.*", Michael Polanyi, "The tacit dimension," 1966.

Humans can attend to an object and the external world by removing our body or self-motion information. How can we extract the objective information from the fused information in somatosensory information? One plausible hypothesis is that humans can remove the self-related information because such information is caused by the self-generated motion command and can be estimated by the command. However, during haptic exploration, changes in the position and posture of the object are also caused by the motion command. Furthermore, the motion is also changed by contact with the object; therefore, only causation cannot explain the separation of object- and self-related information. During haptic exploration, the motion is not random [31]. However, the mechanism of motion generation suitable for haptic exploration is currently unknown.

Haptic exploration and whole-body motion development have an equivalent problem structure. The former is a simultaneous construction of motion and the meaning of the object or environment in contact with the body. The latter is a simultaneous construction of motion and the meaning of the motion. Because the meaning of an object or motion can be learned by even blind persons, such tacit knowledge has to be learned unsupervised. Therefore, we studied both [3, 29, 32].

Such simultaneous constructions are similar processes to general creative activities; novel and innovative concept creations in art, engineering, and natural science. We believe that these processes have to share a construction mechanism. Piaget also argued that development in humans is similar to the evolution of scientific knowledge [33]. He argued that the developmental mechanism consists of two processes: assimilation and accommodation [34]. Assimilation is the mechanism to assimilate the external environment into the internal structure. Accommodation is a mechanism to adjust the movement and change the internal structure. In creative activity research, a two-process model was proposed: coevolution of problem-discovering and problem-solving [35, 36]. Haptic exploration and development of behaviors also have two processes: meaning-discovering and behavior-generation. Piaget's assimilation, problem-discovering, and meaning-discovering are all related to representational structure mapping [37], including perception, inference, and analogical reasoning. Piaget's accommodation, problem-solving, and behavior-generation are all partly related to goal-directed control. The difference in these two-process models is which process can change the representational structure. However, a construction mechanism of representational structure suitable for these processes is currently unknown.

There is a difference between creative activities in adults and development in infants. An infant has only limited knowledge. We believe the initial knowledge, including embodiment, is crucial to realize the above coevolution of two processes. To clarify the initial knowledge and the mechanism, which is considered to be shared with general creative activity, is an important milestone in the research of human intelligence. During early development, the initial tacit knowledge has to be the minimum; therefore, the solution has to be simpler than creativity in adults, which requires a large amount of knowledge.

If creative activities in adults and development in infants have a shareable coevolution mechanism, research in creative activities should be valuable to elucidate the developmental principles. In the initial stage of creative activities, it is typical to use a relatively unstructured form of representation, which has to be reinterpreted and refined [38]. Importantly, the initial structure is not chaotic and requires minimum tacit knowledge.

In the initial stage of creative activity, discovery-oriented behavior is related to creativity [39]. In haptic exploration, when we use a stick, we expect the tactile input caused by the contact between the tip of the stick and the objects before exploration. When walking in a dark or a poorly visible room, we expect the wall and floors by touch between the foot/hand and the environment before the behavior. Attention is directed to discover external information before exploration. The discover-oriented behavior is also important for haptic exploration.

Humans always explore the unknown. The motivation to reveal the unknown is the driving force for development, haptic exploration, and creativity. To discover more "distal" concepts [40], humans must attend to not self but external objects or environments. To realize such a discover-oriented behavior, humans must have some tacit knowledge before the behavior. We believe that, in early development, such tacit knowledge has to be embedded in the embodiment integrating the physical body, brain, and sensory system.

References

[1] Kuniyoshi Y. 'Fusing autonomy and sociality via embodied emergence and development of behaviour and cognition from fetal period'. *Philosophical Transactions of the Royal Society of London. Series B, Biological Sciences.* 2019, vol. 374(1771), 20180031.

[2] Nagakubo A., Kuniyoshi Y., Cheng G. 'The ETL-humanoid system—a high-performance full-body humanoid system for versatile real-world interaction'. *Advanced Robotics*. 2019, vol. 17(2), pp. 64–149.

[3] Kuniyoshi Y., Sangawa S. 'Early motor development from partially ordered neural-body dynamics: experiments with a cortico-spinal-musculo-skeletal model'. *Biol Cybern.* 2006, vol. 95(6), pp. 589–605.

[4] Mori H., Kuniyoshi Y. 'A human fetus development simulation: self-organization of behaviors through tactile sensation'. Presented at Proceedings of the IEEE 9th International Conference on Development and Learning (ICDL 2010); Ann Arbor, Michigan, USA.

[5] Ohmura Y., Kuniyoshi Y. 'Nagakubo A. 'Conformable and scalable tactile sensor skin for curved surfaces'. *Proceedings of the IEEE International Conference on Robotics and Automation*; Robotic Society of Japan, 2006. pp. 53–1348.

[6] Reimer E.M., Danisch L. *Pressure sensor based on illumination of a deformable integrating cavity*. United States Patent; US5917180A. 1999

[7] Fujimori Y., Ohmura Y., Harada T., Kuniyoshi Y. 'IEEE international conference on robotics and automation (ICRA)'. Kobe, Japan, Kobe, 2009.

[8] Sagisaka T., Ohmura Y., Kuniyoshi Y., Nagakubo A., Ozaki K. '11th IEEE-RAS international conference on humanoid robots (humanoids 2011); bled,slovenia'. IEEE, 2011. pp. 42.

[9] Sagisaka T., Ohmura Y., K., Kuniyoshi Y., Nagakubo A., Ozaki K. 'Development and applications of high-density tactile sensing glove'. Presented at Proceedings of the EuroHaptics; Tampere, Finland, 2012.

[10] Bianchi M., Haschke R., Büscher G., Ciotti S., Carbonaro N., Tognetti A. 'A multi-modal sensing glove for human manual-interaction studies'. *Electronics*. 2016, vol. 5(4), p. 42.

[11] Harikumar R., Prabu R., Raghavan S. 'Electrical impedance tomography (EIT) and its medical applications: a review'. *International Journal of Soft Computing and Engineering*. 2013, vol. 3(4), pp. 98–193.

[12] Nagakubo A., Kuniyoshi Y. 'A tactile sensor based on inverse problem theory: the principle'. Presented at Proceedings of the24th Annual Conference of Robotics Society of Japan; Okayama, Japan, Sep 2006 (in Japanese): Robotic Society of Japan,

[13] Nagakubo A., Alirezaei H., Kuniyoshi Y. 'IEEE international conference on robotics and biomimetics (ROBIO)'; sanya, china'. IEEE, 2007. pp. 08–1301.

[14] Alirezaei H., Nagakubo A., Kuniyoshi Y. *7th IEEE-RAS International Conference on Humanoid Robots (Humanoids 2007); Pittsburgh, PA IEEE*. 2007.in press.

[15] Alirezaei H., Nagakubo A., Kuniyoshi Y. 'A tactile distribution sensor which enables stable measurement under high and dynamic stretch'. proceedings of the IEEE symposium on 3D user interfaces'. IEEE, 2009. pp. 87–93.

[16] Alirezaei H. 'Development of a highly stretchable and deformable fabric-based tactile distribution sensor'. [PhD Thesis]. Tokyo, Japan, University of Tokyo, 2011

[17] Silvera-Tawil D., Rye D., Soleimani M., Velonaki M. 'Electrical Impedance Tomography for Artificial Sensitive Robotic skin: a review'. *IEEE Sensors Journal*. 2015, vol. 15(4), pp. 16–2001.

[18] Ohmura Y., Kuniyoshi Y. 'IEEE/RSJ international conference on intelligent robots and systems' san diego, CA'. IEEE, 2007. pp. 41.

[19] Ohmura Y., Terada K., Kuniyoshi Y. 'Analysis and Control of Whole Body Dynamic Humanoid Motion – Experiments on a Roll-and-Rise Motion'. *Proceedings of the IEEE-RAS 3rd International Conference on Humanoid Robots*; Karlsruhe, Germany, 2003.

[20] Kuniyoshi Y., Ohmura Y., Terada K., Nagakubo A. 'Dynamic Roll-and-Rise Motion by an Adult-Size Humanoid Robot'. *International Journal of Humanoid Robotics*. 2004, vol. 1(3), pp. 497–516.

[21] Hellman R.B., Tekin C., van der Schaar M., Santos V.J. 'Functional contour-following via haptic perception and reinforcement learning'. *IEEE Transactions on Haptics*. 2018, vol. 11(1), pp. 61–72.

[22] van Hoof H., Hermans T., Neumann G., Peters J. 'Learning robot in-hand manipulation with tactile features'. *IEEE-RAS 15th International Conference on Humanoid Robots (Humanoids)*; Seoul, South Korea, IEEE, 2015. pp. 27–121.

[23] Sekiya K., Ohmura Y., Kuniyoshi Y. 'Learning to GRASP multiple objects with a robot hand using tactile information'. *Proceedings of the 26th Robotics Symposia*; in Japanese, Fukuoka, Japan, 2021. pp. 44–239.
[24] Brockman G., Cheung V., Pettersson L, *et al. OpenAI gym*. Available from ArXiv Preprint ArXiv 2016:1606.01540
[25] Haarnoja T., Zhuo A., Abbeel P., Levine S. *Soft actor-critic: off-policy maximum entropy deep reinforcement learning with a stochastic actor*. Available from arXiv preprint arXiv 2018:1801.01290
[26] Martinez-Hernandez U., Lepora N.F., Prescott T.J. ' Active haptic shape recognition by intrinsic motivation with a robot hand '. *IEEE World Haptics Conference (WHC)*; Evanston, IL, 2015.
[27] Jamali N., Ciliberto C., Rosasco L., Natale L. 'Active perception: building objects ' models using tactile exploration'. *IEEE-RAS 16th International Conference on Humanoid Robots (Humanoids)*; Cancun, Mexico, IEEE, 2016. pp. 85–179.
[28] Martinez-Hernandez U., Dodd T.J., Prescott T.J. 'Feeling the shape: active exploration behaviors for object recognition with a robotic hand'. *IEEE Transactions on Systems, Man, and Cybernetics*. 2018, vol. 48(12), pp. 48–2339.
[29] Sekiya K., Ohmura Y., Kuniyoshi Y. 'Generating an image of an object ' S appearance from somatosensory information during haptic exploration'. *IEEE/RSJ International Conference on Intelligent Robots and Systems (IROS)*; Macau, China, 2019. pp. 43–8138. Available from https://ieeexplore.ieee.org/xpl/mostRecentIssue.jsp?punumber=8957008
[30] Mirza M., Osindero S. *'Conditional Generative Adversarial Nets'*. Available from arXiv preprint arXiv 2014:1411.1784
[31] Lederman S.J., Klatzky R.L. 'Hand movements: a window into haptic object recognition'. *Cognitive Psychology*. 1987, vol. 19(3), pp. 342–68.
[32] Ohmura Y., Gima H., Watanabe H., Taga G., Kuniyoshi Y. 'Developmental changes in intralimb coordination during spontaneous movements of human infants from 2 to 3 months of age'. *Experimental Brain Research*. 2016, vol. 234(8), pp. 88–21.
[33] Piaget J., Garcia R. 'Psychogenese et historie des sciences'. *Frammarion*. 1983.
[34] Piaget J. *L épistémologiegénétique*. Paris:Press Universitaires de France. 1970.
[35] Poon J., Maher M.L. 'Co-evolution and emergence in design'. *Artificial Intelligence in Engineering*. 1997, vol. 11(3), pp. 27–319.
[36] Dorst K., Cross N. 'Creativity in the design process: co-evolution of problem–solution'. *Design Studies*. 2001, vol. 22(5), pp. 37–425.
[37] Gentner D. 'Structure-mapping: a theoretical framework for analogy*'. *Cognitive Science*. 1983, vol. 7(2), pp. 70–15.
[38] Purcell A.T., Gero J.S. 'Drawings and the design process'. *Design Studies*. 1998, vol. 19(4), pp. 389–430.

[39] Csikszentmihalyi M., Getzels J.W. 'Discovery-oriented behavior and the originality of creative products: a study with artists'. *Journal of Personality and Social Psychology*. 1971, vol. 19(1), pp. 47–52.
[40] Polanyi M. 'The tacit dimension' Garden City: N.Y. Doubleday; 1966.

Chapter 2
Developmental soft robotics

Luca Scimeca[1] *and Fumiya Iida*[2]

2.1 Introduction

Since the term artificial intelligence (AI) was coined in 1956, the research field of intelligence was initially dominated by the 'computational paradigm of intelligence' (traditional cognitivism). In this context, intelligence was regarded as a computational processes, where symbolic operations were of central interest without explicitly considering what the symbols actually meant. At the time, a strong connection was conjectured between the idea of 'intelligence', the power of symbolic representation (e.g., in the brain), and the possibility for a system to change from a state to another [1, 2]. In this context, an individual would create a symbolic representation of the world by means of sensory perception, then a process akin to rule-based symbol manipulation would allow them to exhibit intelligence [3, 4]. While the computational paradigm of intelligence has given significant impact mainly in cyberspace, there have been a number of aspects of intelligence that cannot be fully explained in this framework.

Because symbols must represent entities or concepts in the world, and because the entity representations must in some way be task independent, there is an issue with how exactly sensory perception can lead to a symbolic representation of the sensed entities in the world, whether these are physical entities, emotions or even concepts. This problem is also known as the symbol grounding problem. The physical grounding hypothesis arises to match the need for an agent to have its representation grounded in the physical world. In this context, the world becomes its own best model, and appropriate interactions modulate the behavioural intelligence, as eloquently explained in an early article titled 'Elephants Don't Play Chess' [5].

In contrast to the traditional cognitivistic view of intelligence, the 'embodied cognition' paradigm seems the mind and the body irrevocably linked in determining intelligence and its emergence. Biological systems, in fact, are not passively exposed to sensory perception, but instead actively interact with their surrounding

[1]Harvard University, Boston, United States
[2]Bio-Inspired Robotics Laboratory Department of Engineering, University of Cambridge, Cambridge, United Kingdom

environment [3]. Robotics systems, similarly, should strive for the action and motor control to contribute to the improvement of perceptual abilities. This change has brought on the characterization of frameworks that define the embodied cognition paradigm, below we will explore some such frameworks. The embodied view of intelligence has shown that physical system–environment interactions cannot be overlooked to understand intelligent adaptive systems especially in the physical real-world environment.

In robotics, many examples have shown the ability of robotics platform to exhibit complex 'intelligent' behaviour without the need of explicit high-level control. One famous case study is the passive dynamic walker, a robot with no motors or controller, which with low energy can achieve human-like walking locomotion down slopes [6]. The physical passive-dynamical characteristics of the robot, in fact, allow the hip joints and the legs to achieve the necessary swinging motion for a bipedal walking gate without the need of any explicit actuation; a valuable example of the concept of physical or mechanical intelligence.

In the book *Vehicles: Experiments in Synthetic Psychology* [7], Valentino Braitenberg describes a series of thought experiments in which vehicles can exhibit complex and meaningful behaviours through increasingly more complex sensory-motor interactions. In his thought experiments, the vehicles' steering is connected to sensor outputs in various ways by an increasingly more complex network of inhibitory or excitatory connections, and some analogue sensors and actuators, the vehicles can steer towards a light, they can avoid each other or group in various ways and exhibit basic behaviours of fear, aggression, liking and love, at least in the eye of the observer. The message in his book was key in showing how intelligence can be explained in a bottom-up fashion, and how intelligent behaviour can emerge from 'simple' sensory-motor interactions.

The advent of the field of developmental robotics (Chapter 7) has aided in the search for those elements, which determine the emergence of intelligence in living organisms, as well as those factors that are fundamental in the design and control of machines meant to achieve intelligence and adaptability. Likewise, the beginning of the era of 'soft robotics' has changed the robotics landscape, fuelling a revolution in the way robotics systems are thought of, designed, controlled and *evolved*. In this chapter, we will first introduce and review soft robotics research, with emphasis on how compliance and softness have changed the robotics landscape in the past two decades. We will then briefly discuss some key ideas in developmental robotics that are fundamental for understanding the relationship between biological systems and artificial systems and finally discuss how the developmental sciences and soft robotics are irrevocably linked, into what we have chosen to call 'developmental soft robotics'. Here, in fact, the two fields can be merged into one where the developmental sciences can aid in the design and make of soft robots, which can then be used as platforms to better understand biological systems. We will finally discuss how phylogenetic development, ontogenetic development and short-term adaptation are indeed naturally suited to be embedded within a 'soft' robotic context.

2.2 Bio-inspired soft robotics

Deformation is a fundamental characteristics of biological systems. Almost 90% of the human body is composed of soft tissue; many vital functions such as heart, lung, muscles, eye lenses, etc. depend on the deformation of materials. In bipedal walking, e.g., evidence has shown how the soft tissue of the body might not only cushion impacts on every stride but also both save muscles the effort of actively dissipating energy and perform a considerable amount of the total positive work per stride by soft tissue elastic rebound [8].

In the past few decades, there has been an unprecedented advancement in material sciences and manufacturing techniques, furthering our knowledge of functional materials and empowering artificial systems with new found abilities. These advancements, together with the better understanding of biological systems, gave rise to the era of soft robotics, where bio-inspired robotic platforms make use of soft and deformable materials to achieve more flexible, adaptable and robust behaviours [9, 10].

Since the dawn of soft robotics, the application of material science and soft body compliance has changed the robotics landscape. In manipulation, e.g., the 'universal gripper', a soft gripper capable of particle jamming through vacuum pressure control, has been shown to be able to grasp a large number of objects [11]. Other solutions for grasping and manipulation range from tentacle-like systems [12] to soft grippers [13] and human-inspired soft robotic hands [14] (Figure 2.1).

Animal-inspired soft robots are amongst the most developed sub-areas of soft robotics, where the robot platforms range from worms [15] or caterpillars [16], to octopuses [12], fish [17] and others besides (Figure 2.1). In worm-like soft robots, e.g., akin to their biological counterparts, the contraction of longitudinal muscles followed by the contraction of circumferential muscles simulates a travelling wave along the body, generating locomotion [19]. In caterpillars, motion is generated by coordinated control of the time and location of the prolegs attachment to the substrate, together with waves of muscular contraction [20].

The ability to mimic these unique systems makes soft robots an exciting new field, where the limits of the (rigid) robots of the past century can be overcome with new-found solutions.

2.2.1 Soft materials and soft actuation

The area of soft robotics is inevitably connected to the field of material science, where new discoveries in the latter facilitate progress in the former. For a soft robot to be able to use material compliance to aid in robotics tasks, it is necessary for the make of the robot to be, at least in part, deformable. Elastomeric (polymer) materials, like EcoFlex or DragonSkin [21], have been at the centre of researchers' attention for several years, with new materials composite materials being discovered every year. Moreover, the advent of 3D-printing technology has allowed for robot design and testing operations to be much faster than before, facilitating rapid and cheap prototyping in soft robotics.

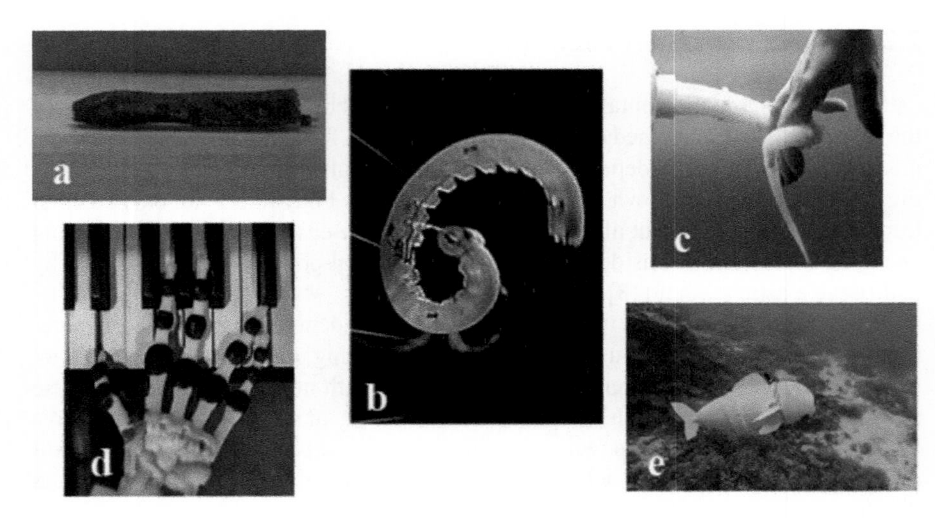

Figure 2.1 *Bio-inspired soft robot examples. (a) Worm-inspired soft robot [15]. (b) Caterpillar-inspired soft robot [16]. (c) Octopus-inspired tentacle [18] (d) Human-inspired soft passive hand [14]. (e) Fish-inspired soft robot [17].*

Actuation poses one of the biggest challenges in soft robotics. In many animals, the coaction of a large number of muscles distributed over their body is capable of generating relatively high forces, facilitating coordinated and robust action. Replicating this ability is no easy feat, as the majority of the robotics solution lack the ability to generate forces comparable to the industrial robots of the past.

Four main soft actuation techniques exist: tendon driven, pressurized air or fluids, dielectric elastomeric actuators (DEAs) and shape memory alloys (SMAs) [10]. Tendon-driven actuation mimics biological musculoskeletal systems, where actuation is achieved through the pull and release of tendons, via the appropriate control of motors (Figure 2.2a). Although a powerful and widespread actuation technique, a large number of tendons are usually necessary to achieve complex behaviours and control complexity increases with the number of motors necessary to control the tendons. For softer robots, like continuum soft robots, this type of actuation usually does not scale. Pressurized air and fluids are one of the most powerful of actuation techniques for soft robots, capable of generating high forces and displacements. The actuation usually consists of varying the pressure inside predesigned chambers within the body of the robot, to achieve their expansion and contraction and generate motion or morphological changes (Figure 2.2b). However, these actuation systems are usually bulky, heavy and require high power sources, making it unsuitable for untethered robotics systems [27]. DEAs are made of soft materials that can be actuated through electrostatic forces (Figure 2.2c). DEAs have been shown to have high strain/stress and mass-specific power, however, the need for DEAs to be pre-strained imposes rigid constraints on the robots' design [28]. Finally, SMAs with the most

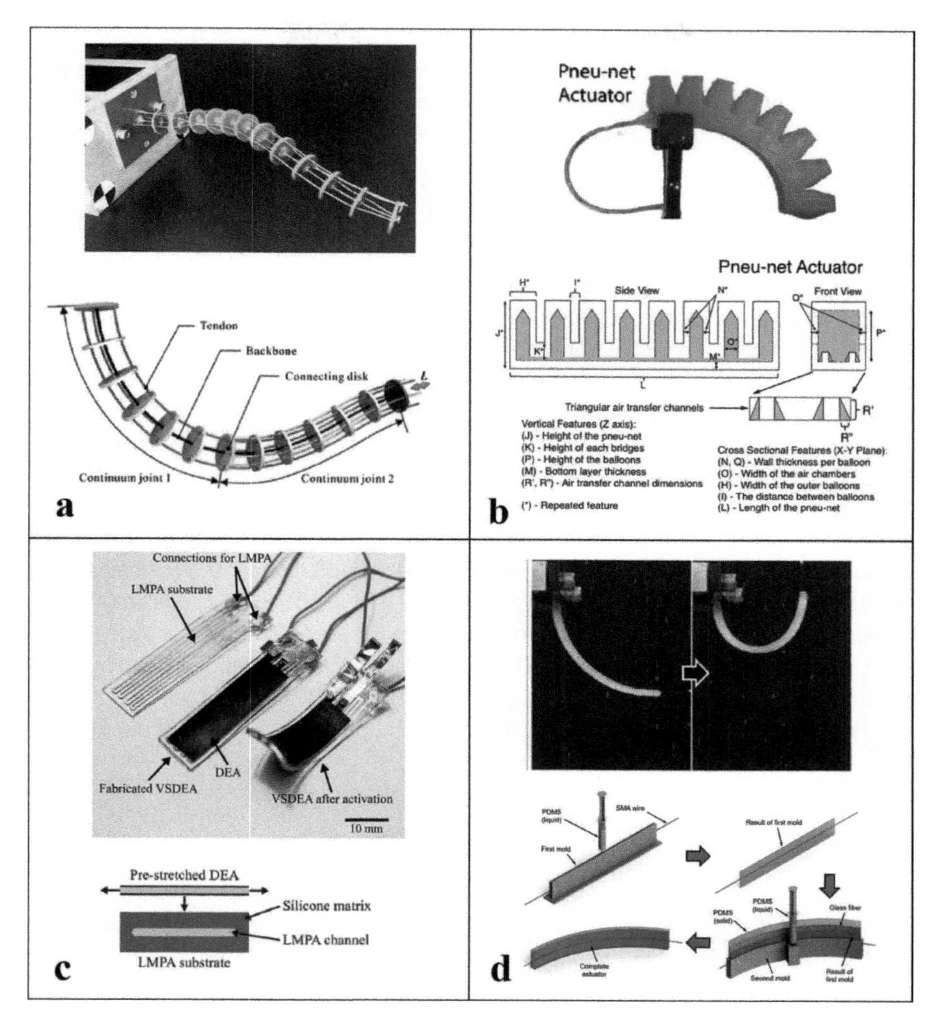

Figure 2.2 Examples of some of the main actuation mechanisms used for soft robotic systems. (a) Tendon-driven continuum robot [22] and model [23]. (b) Pneumatic soft actuator and [24]. (c) Variable stiffness dielectric elastomer actuator [25]. (d) Curved shape memory alloy-based soft actuator [26].

common nickel-titanium alloys can generate force through a change in shape due to a raise or fall in temperature of the material (Figure 2.2d). Temperature change control, however, is a challenge. High voltages are usually required to achieve temperature changes, and robustness over varying temperatures in the environment is still an issue to be overcome [29]. Other methods exist; it is possible, e.g., to induce pneumatic contraction by evaporating ethanol via restive heating [30], or achieve bending through combustion [31]. Other issues, like reduce output force or slow speed, however, come into play [32]. Soft robotics actuation and material sciences are still an ever changing field, with new solutions being expedited by fast prototyping and iteration.

2.2.2 Soft robot control, simulation and learning

Soft robotic control poses several challenges and opportunities. Here, the 'degree of softness' matters. Take, e.g., a rigid robotic hand, where the palms and fingertips are covered with an elastomeric material. The control of the hand may be achieved with classical methods (i.e., inverse kinematics), however, the complexity of the control may be reduced, as it may here be possible to make use of the mechanical passive dynamics of the soft-fingers to achieve a desired grasping behaviour, averting the need for submillimetre precision in the robot control [33, 34]. On the other hand, as the 'degree of softness' in the body increases, new challenges arise.

A robot made entirely of elastomeric materials, e.g., one simulating the tentacle of an octopus or the trunk of an elephant, cannot be controlled classically; moreover, proprioception and simulation become problematic. As opposed to the hard links with sliding or rotational joints in classical robots, the continuity and softness of the body make the control and simulation of continuous soft robots much harder. Novel actuation methods aid robotics researchers in their endeavours to achieve desired robot control (section 2.1), and new sensing and control methods are discovered on a daily basis [35]. Achieving a proprioceptive understanding of the robot's configuration is necessary to be able to control continuum soft robots accurately and repeatably, thus making the appropriate sensorization of soft-bodies fundamental.

Much effort has been put in the sensorization of soft robots. The most common soft sensors are perhaps strain sensors, which are soft deformable sensors capable of sensing body deformations through stretching. It is thus possible to embed such sensors into the (soft) body of a robot, without influencing its ability to deform. Other sensors have been used, based on restive [36] or capacitive [37] technologies. Recently, works in References [38] and [39] have shown how it is possible to achieve a high-fidelity proprioceptive understanding of a continuum soft body through its sensorization via fibre optic and capacitive tactile sensors, respectively.

In the context of control, and simulation, learning plays a fundamental role. With the infinite degrees of freedom posed by continuum soft body, e.g., precise control via classical methods is hard, and usually does not scale. Model-based solutions based on the piece-wise constant curvature assumption have been shown to work for small tentacle-like robots [40]. However, the error in the controller always increases with the increase in the number of soft segments within the robots. The models, in fact, are usually too simplistic to accurately capture the complexity of the make of continuum soft robots. Learning in this case has been shown to be useful in compensating for the lack of knowledge or model complexity [41].

2.3 Developmental soft robotics

Cognitive developmental robotics (CDR) is an area of research where the robotics and developmental sciences merge into a unique field; one that seeks to better robotics with insights from developmental sciences, and further our understanding of developmental sciences through the use of robotics platforms [3]. The need for CDR to be a research area on its own, arose at the dawn of the twenty-first century from the need to understand, not only the cognitive and social development of individuals as explored in the area of epigenetic robotics [42], but also how the acquisition and development of motor skills, as well as morphology, influence the development of higher order cognitive functions [3, 43, 44]. In this context, robots can be used as experimental subjects, where developmental models can be implemented in robotics platform and scientists can gain insights from behavioural analysis, an approach known and synthetic methodology [45, 46].

In stark contrast to the traditional cognitivistic approach, in developmental robotics, there is no clear separation between the physical body, the processes that determine reasoning and decision-making (cognitive structure) and the symbol representation of entities in the world. Rather, these processes influence each other and intelligence emerges from their interaction. Developmental robotics is treated in Chapter TO ADD: CHAPTER NUMBER; in the following section, we will briefly state some key concepts relevant within the context 'soft developmental robotics'.

2.3.1 Facets of development

In biology, ontogeny can be defined as the development of an organism, usually from the moment it is conceived and thereby throughout its lifespan. Ontogenetic development thus can be seen as the evolution of an organism throughout its life, as dictated by a coactive action of internal (endogenous) and external (environmental) factors to the organism itself [3, 47]. Within ontogeny, there is dissent on the role of physical development, through maturation and growth, and the role of learning. Although initially ontogenetic development was seen as tightly coupled to physical development, and learning would only occur as a consequence of development itself [48], several subsequent views thought to break the boundary between learning and development. One such view would see learning and development influencing each other bi-directionally, thus learning, modulated by physical developmental processes, could actively advance development itself [49]. Others, thought dynamics processes to be at the base of development and learning, breaking the boundaries between development and learning altogether in what is known as 'dynamic systems approach to development' [50]. Understanding ontogeny has the potential to give us invaluable insights to better understand biological system and thus build better artificial systems. A few components, or facets, of ontogenetic development are key to that end. Some of the main facets of development are defined and briefly reported below.

2.3.1.1 Incremental process and self-organization

Development, seen as a sequence of stages through which an organism advances, has been theorized to be an incremental process since the early days of developmental psychology [51, 52]. Development, moreover, is mainly a self-organized process. In infants, e.g., exploratory activity is not solely due to internal goal-oriented mechanisms, but is rather a result of a modulation between the body, mind and environment, whose complex, local, interactions can give rise to global order [53, 54].

2.3.1.2 Degrees of freedoms, freezing and freeing

The degree of freedom problem was first introduced by Russian physiologist N. Bernstein [55], and it refers to the ability of many biological organisms (humans in particular) to achieve highly controlled and coordinated behaviours, despite the non-linearity of their muscular-skeletal system and their high number of muscular degrees of freedom. Part of a solution to the degree of freedom problem was proposed by Bernstein himself, through what he called the principle of freezing and freeing degrees of freedom. By freezing, or tightly coupling the peripheral joins, a body can reduce its degrees of freedoms so that learning is possible. Freeing or weakening the coupling at the peripheral joins at a later stage would allow the body to reclaim its degrees of freedom and learn more complex motion patterns [56]. In this process, the emergence of coordinated motion from a dynamic interaction with the environment is fundamental for any organism to achieve controlled behaviours.

2.3.1.3 Self-exploration and spontaneous activity

At infancy, the inquiry of one's surroundings through physical exploratory action, and the perceptual consequences of said explorations, has been shown to play a crucial role in an infant's 'sense of bodily self' [57, 58]. The concept of self-exploratory action is tightly coupled with the concept of spontaneous activity. Infants, e.g., explore their physical constraints through coordinated behaviour emerging from spontaneous neural and motor activity (e.g., kicking or suckling), useful to create or reinforce joint coupling and motor synergies.

2.3.1.4 Intrinsic motivation

When infants explore the surrounding environment and its own physical constraints, not all their actions may be strictly goal oriented. An intrinsic value mechanism guides the motor action to achieve specific motion patters. Intrinsic motivation has a key role in learning and development, and, in infants, it has been linked to the concept of curiosity [59]. In artificial systems, such a mechanism could allow learning without explicit teaching or supervision [60, 61].

2.3.1.5 Categorization

Categorization is one of the most fundamental abilities for the majority of living organisms, without whom there would be no chance for an organism to distinguish any entity within its environment, including food sources, dangers, peers and so on [3]. The vast majority of the organisms, thus, are capable of categorizing and discriminating between a wide range of sensory stimuli [62]. Two factors have been shown to be tightly coupled with the categorization of percepts: motor skills and morphology. Active exploration is, in fact, fundamental in the categorization of stimuli and the formation of concepts [3, 52]52. The role of motor-coordinated behaviour is here that of purposefully influencing the sensory stimuli so that structure arises, and categorization is both possible and simpler [63]. Similarly, the morphology of an organisms inescapably influences the sensory perception of an organism, influencing the structure in the sensory stimuli via sensory-motor-coordinated behaviour [41].

2.3.1.6 Morphology and morphological computation

Akin to the role of sensor coordinated behaviour during action (section 2.1), the morphology of the body plays a fundamental role in the developmental process, and the acquisition of motor skills [64, 65]. Within the idea behind morphological computation, some complexity due to the interaction of an agent with its environment can now be outsourced to the body, leaving more canonical computational frameworks to serve higher level functions [66]. It is here that the role of morphology, materials and mechanical intelligence can be understood.

2.3.1.7 Sensory-motor coordination

In the context of embodied cognition, it becomes key to consider and understand the interactions arising when an individual learns and adapts its motor skills while interacting with entities in the world it lives in. The sensory-motor-coordinated behaviour, linking action and perception in a cohesive loop to better discriminate known or unknown entities in the world, gives rise to the concept of 'sensory-motor coordination' or 'sensory-motor contingencies' (SMC) [67]. Under the light of SMCs, the perceptual experience of an individual is no more thought of as a consequence of brain computations, but rather a synergic interaction of skilful actions and sensing, both influencing each other to improve the perceptive experience [67–69].

2.3.1.8 Body schema

Body schema, also known as forward models, are simulators of the musculoskeletal system and the environment [70–72]. These simulators confer the ability to an organism to understand both their body and the environment surrounding them, and thereby predict the consequences of their actions. Systems possessing body schema are capable of predicting future states, given their body configuration and other sensory inputs. In humans, instances of such systems have been previously hypothesized to exist in the cerebellum [73].

2.3.2 Soft robotics and developmental time scales

One of the most difficult tasks in modern day robotics is to achieve an appropriate robot design for a robot to perform certain tasks in the world. The advent of soft robotics, if anything, has increased the complexity of robots, revoking the rigidity constrains established in the earlier century, and bringing about a new era. In this new era, robot design is driven by factors much like biological systems, where functional morphology, coordinate sensory-motor action, physical adaptation and embodiment all contribute to the 'robot's survival' in the world, and to its ability to see a task to completion.

Developmental soft robotics aims at bringing together the areas of soft robotics with that of developmental robotics and the developmental sciences. These, in fact, are irrevocably linked, as we will later show. Within the developmental sciences, in its simplest form, the development of a biological organism can be distinguished on three different scales: phylogenetic, ontogenetic and short term.

In biological organisms, *Phylogenetic Development* has the largest timescale, where changes happen at the level of groups of organisms, over many generations, and where processes like natural selection are responsible for certain 'traits' to survive and evolve, while others to become extinct. Akin to phylogenetic development is soft robotics design, where the design of robots is adaptive and ever-changing, to comply and conform to the task the robot has to achieve. Currently, the majority of the adaptation is due to human design and biased by human skill and experience. However, new methodology for autonomous designed is a hot research topic, and processes like evolutionary algorithms have shown promise in the past [74, 75].

Ontogenetic development, like previously explained, concerns changes throughout and within the lifespan of an organisms, and includes growths and bodily adaptation. The ability or robots to 'morph' throughout their lifespan to achieve desired behaviour have been one of the key advantages of soft robots, as opposed to their rigid counterparts of the previous century. Robots navigating through growth, like fungal hyphae [76], elongating their bodies due to pressure, as well changing their stiffness to change their body dynamics and achieve different behaviours [77] are all examples of such adaptability.

Short-term adaptation refers to the shortest adaptive and developmental timescale of all, where adaptation needs to be achieved instantaneously. Short-term adaptation is perhaps the most naturally suited to be discussed in a soft setting. In the past, this type of adaptation needed to be actively achieved at the control level, where real-time control would allow short-time adaptive behaviour through mechanical or sensory feedback. Within the soft robotics framework, much like biological organisms, the short-time adaptation is just a consequence of the soft instantaneous deformation of the soft body itself. When we delicately slide our finger through a ridged surface, e.g., the need for complex and precise control is void by the ability of our dermis to deform and conform to the surface under touch. Much like the illustrated example, the compliance and softness of materials, in soft robots, can achieve short-term adaptation. The mechanical feedback becomes only a physical consequence of contact, and compliance can naturally suppress the need for complex controllers. Figure 2.3 illustrates the main idea behind the developmental soft robotics framework. For the remainder of the chapter, we will highlight some of the design principles to achieve short-term and ontogenetic adaptation, as well as briefly explain evolutionary algorithms on a phylogenetic timescale interest. Finally, we discuss some of the challenges and perspectives for the future.

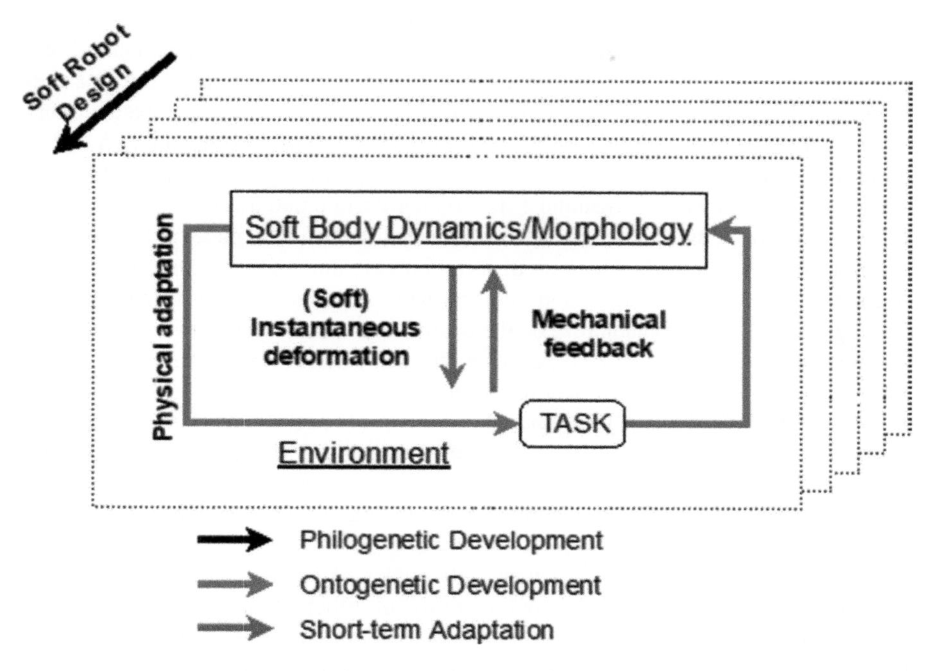

Figure 2.3 Developmental soft robotics

2.3.3 Design principles

As explained in section 2.1, the morphology of the body plays a fundamental role in living organisms, one that influences their learning and developmental process, and aids in the ability of said organisms to perform everyday tasks.

2.3.3.1 Functional morphology and morphological computation

When designing robotics systems, if shape was initially the most salient of morphological features, with the advent of soft robotics this may no longer be the case. Materials at different levels of elasticity have been shown to be able to perform 'computation' [41, 78]. Recent work in Scimeca *et al.* [41], e.g., shows how complex haptic information can be used to classify objects based on different properties, solely based on clustering analysis. The simplicity of the inference is possible due to a 'soft filter' or elastic layer between the tactile sensor and the object. When changing the properties of the elastic layer, the tactile information is appropriately influenced (spatially filtered) as to induce object similarities with respect to different object properties, like edges or elongation. The 'intelligence' is here in the body, since its ability to appropriately mould the sensory information to the agent allows for higher cognitive functions to solve the object classification problem with clustering methods, without prior training or supervision, an otherwise impossible feat.

A paradigm trying to make use of the complex body–environment interactions is the 'reservoir computing' framework of computation. The original idea behind reservoir computing begins with network computation, where an input is fed to a network, which computes a corresponding output. In reservoir computing, a fixed random dynamical system, also known as reservoir, is used to map input signals to a higher dimensional space. The 'readout' final part of the network, then, is trained to map the signals from the higher dimensional space to their desired output. As previously mentioned, soft robots, as well as biological organisms, are usually made, at least in part, of soft materials. The body dynamics of soft robots are thus very complex, highly non-linear and high dimensional, making control hard. Through the reservoir computing paradigm it is possible to capitalize on the complexity of such system by exploiting the soft body as a computational resource, using the body dynamics to emulate non-linear dynamical systems, and thus offloading some of the control to the body itself [79, 80]. Work in Nakajima *et al.* [81], e.g., has shown how it is possible to control a complex continuum soft arm, inspired by the tentacle of an octopus, in close loop without any external controller, by using the body of the robot as a computational resource. Under this light, high non-linearity and complexity may be a desirable property of the body, and design might have to be thought of accordingly.

An additional property that allows soft bodies to be used as a computational resource is memory. The soft-body dynamics of soft robots, in fact, can exhibit short-term memory, allowing robots to emulate functions that require embedded memory [81]. When under-actuating a continuum soft robot, e.g., it may be the case that control mechanism is not deterministic with respect to the behaviour of the robot. In these cases, the behaviour of the robot may depend, not only on the induced control and its current state, but also on the history of the previous robot states. This may be the case when actuating a soft tentacle arm via only moving one of its extremities.

2.3.3.2 Soft system–environment interactions

At the dawn of the twenty-first century, the concept of 'morpho-functional machines' was proposed. Morpho-functional machines were defined as machines that were adaptive by being able to change their morphology as they performed tasks in the real world [82]. In this context, changes at different timescales were already argued to be important, i.e., short term, ontogenetic and phylogenetic, or evolutionary. It is important to note that the adaptation and the resolution of the task are here achieved not at the control level, but at the morphological level.

As advocated by the developmental robotics, paradigm intelligence and coordinated action are the results of complex interactions between the body, the mind and the environment. The latter, in fact, plays an important role in determining the behaviours of the artificial or natural organisms living within it.

One of the most influential experiments of the last two decades was 'dead fish experiment', performed in a collaboration between Harvard and MIT in 2005 [83]. In the experiment, a dead fish was shown to be able to swim up-streams when no control impulse was clearly being sent by the brain. Upon further studies it was apparent how the streamlined body of the fish, passively oscillating, was capable of turning the surrounding energy into mechanical energy, and thus propel itself forward passively. Although the morphology and make of the body allowed the dead fish to transduce the surrounding energy, the environment was the enabling factor. The vortexes created by water streams were key in the experiment, as they generated the energy to be transduced and recreated the conditions for the body to manifest its propelling abilities. The interaction between the body and the environment was, in fact, the decisive factor in determining the observed behaviour. A similar influential experiment was the passive dynamic walker. The make of the robot, with knee caps, springs, pendulum-like leg swings and more, was capable of stable, human-like and low-energy, bipedal locomotion without any complex control. However, the walking locomotion was initiated and stabilized by the environment itself, as it manifested when the robot was placed on a downward slope [6], thus the potential energy could be skilfully be turned into kinetic energy.

In robot design, it is therefore always necessary to take the environment into account. Much like the examples previously mentioned, the body and the brain are often not enough to achieve useful objectives. Things in the world exist to affect and change their surroundings, and live within the environment they are situated in [84]. In this context, it is in the interplay of the body and the environment that intelligent, situated, behaviour can be observed, and that morphology can be empowered and purposefully adapted.

2.3.3.3 Sensor morphology and soft perception

In nature, morphology plays a fundamental role within the sensing landscape, mechanically converting, filtering and amplifying sensor stimuli from the outside world, to make sense of the surrounding environment, or internal states [85, 86]. In rats and mice, e.g., vibrissae, or sensitive tactile hairs, have been known to confer these mammals specialized tactile capabilities, aiding them in a number of sensory discrimination tasks [87]. In a similar light, most mammals have evolved to mediate vision through compound eyes, compromising resolution for larger fields of view and high temporal resolution, enabling fast panoramic perception [88]. Within the biomimetic robotics field attempts have been made to endow robotic systems with the capabilities of organisms observed in nature, haptic robot perception through whiskers [89] and compound vision [90] are two such examples (Figure 2.4).

Soft sensing is one of the most popular fields within the soft robotics landscapes. Augmenting soft robotics system with the ability of sensing the environment can enable robots to react to unknown events, to improve their control and morphology over time and capture information or reason about entities in the world. Sensorizing soft robots is no easy task. One of the goals within this field is to devise sensors which themselves exhibit some ‘soft’ behavioural characteristics, usually flexibility (i.e., can be bent) and stretchability [94] are desirable. Currently, approaches to achieve stretchable electronics include wavy circuits [95, 96] and conductive liquids [97]. One of the most widespread soft sensors is strain sensors, which have been

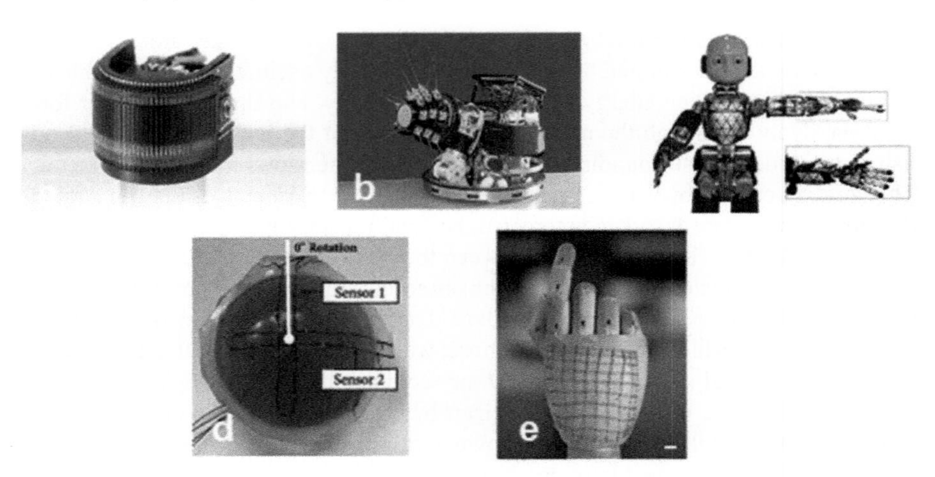

Figure 2.4 *Bio-inspired flexible and soft sensing examples. (a) Artificial compound eyes [90], (b) robotic tactile vibrissal sensing [89], (c) I-cub robot with large-area flexible capacitive tactile skin [91], (d) conductive thermoplastic elastomer's sensorized universal gripper [92], (e) stretchable and conformable sensor for multinational sensing [93].*

shown to be highly elastic [98]. New embedding methodologies have also shown the possibility of embedding strain sensors within elastomers through 3D-printing techniques. Other flexible sensing technologies, like capacitive tactile sensing [99] and optic fibres [38] have been used within soft robotics systems.

As previously mentioned, sensory-motor coordination and morphology can enhance the sensing capabilities of robotics systems. Sensors thus, should not be thought of simply as independent and self-sufficient technologies, but instead, it is fundamental to think of sensor technologies as apparatuses that reside within a body. The body dynamics derived from the sensor's morphological properties, coupled with the environment in which the robotic system is situated in, should all contribute to the sensor morphology, its characteristics and its perceptual capabilities. The appropriate coupling of these factors has been shown to be able to improve the sensing capabilities of robotic systems [100]. In Hughes and Iida [92], e.g., the sensorization of a universal gripper is achieved with a pair of conductive thermoplastic elastomer strain sensors Figure 2.4 (d). Differential sensing is then used to compute deformations within the soft body. Morphology, however, is key. By weaving the strain sensor in different patterns within the soft gripper, information regarding the magnitude, orientation or location of a deformation can be detected. Because the sensing is also inescapably linked to motor control, mechanical dynamics and the objectives of the robotic system, the concept of 'adaptive morphology' has recently been proposed [85], where the iterative design, assembly and evaluation of sensor mythologies attempt to explain the adaptive nature of the perceptual abilities of living organisms.

2.3.4 Ontogenetics and adaptivity

2.3.4.1 Adaptation and growth

The principles previously discussed encourage a different approach to design, in line with endowing robots with the ability to adapt to ever changing environments, and indeed make use of the environment as a means to solve the tasks given to them. Besides design principles at a phylogenetic scale, and instantaneous deformation on the short-term scale via material properties and design, another important factor is ontogenetic change and adaptation. Plants, e.g., are capable of continuously changing their morphology and physiology in response to variability within their environment, in order to survive [101]. Inspired by the unique abilities of plants to survive in diverse and extreme environments, a stream of researchers has more avidly tried to reproduce some of their adaptivity in robotics system. Plantoids, or robotic systems equipped with distributed sensing, actuation and intelligence to perform soil exploration and monitoring tasks, have started to gain traction in this direction [101]. Root-like artificial systems in References [102, 103], e.g., have been shown to be able to perform soil explorations through novel methodologies simulating growth via elongation of the tip. Other plant-inspired technologies in biomimicry and material sciences include Velcro, from the mechanisms behind the hooks of the plant burrs [104], bamboo-inspired fibres for structural engineering materials [105] or novel actuation mechanisms in Taccola *et al.* [106] based on reversible adsorption and desorption of environmental humidity, and in Mazzolai *et al.* [107] based on the osmotic principle in plants.

Another important factor in ontogenetic adaptivity is the ability for organisms to mend their own tissue over their lifespan. Endowing artificial systems with self-healing abilities has recently become of primary importance, setting the landscape for untethered robots to 'survive' for longer periods of time in more uncertain and dynamic task environments. Self-healing of soft materials is typically achieved through heat treatments of the damaged areas, which allow some polymers to reconnect and retrieve most of their structural properties. In Terryn *et al.* [108], e.g., a soft gripper, a soft hand and artificial muscles were developed with Diels–Alder materials [109]. In the developed systems, the Diels–Alder were shown to be reversible at temperatures of 80 °C, recovering up to 98–99% of the mechanical properties of the polymers post-damage.

2.3.4.2 Tool use and extended phenotype

In biology, the phenotype is known to be the set of observable traits of an organism, including its morphology, developmental process and physiological properties. The idea of extended phenotype was first introduced by Richard Dawkins in Reference [110], where he argued that the concept of phenotype might have been too restricted. In fact, the effects that a gene may have are not limited to the organism itself, but to the environment the organism is situated in, through that organism's behaviour. The coupling of an artificial agent and its environment was discussed in section 2.3.2. The extended phenotype notion, however, extends to even more radical concepts.

One of the most fascinating examples of this is found in primates, corvids and some fish, which have been found to be able to purposefully make and use 'tools' to achieve goals within their environments, such as acquiring food and water, defence, recreation or construction [111].

Extending the phenotype concept, the observable traits of the organisms in this case should be augmented to include the extended functionalities, behaviours and morphology derived from the tool under use. When a primate is holding a small branch, e.g., the physical characteristic of the primate is undeniably changed, its reach is longer, its weight and morphology are affected, as is the stance to keep balance on two or three limbs, or the ability to affect the environment around them. Under the extended phenotype concept, these and as well as many other changes need to be captured within the phenotypic traits of the organism.

In the context of soft developmental robotics, the ontogenetic development of robotics systems should include their ability to adapt to their environments over their lifespan (physical adaptation), and indeed the ability to augment their functionality by the active creation and use of tools, initially excluded from their phenotypic traits. This ability was previously investigated in References [112, 113] where it was obvious that at the foundation of the idea of tool use there was the concept of body schema, previously mentioned in section 2.1. The body schema in this scenario requires adaptability and alterability throughout ontogenetic development, to cope with the changes in one's body, including growth, as well as with the extended capabilities conferred by the use of tools. In Nabeshima *et al.* [113], the temporal integration of multisensory information was argued to be a plausible candidate mechanism to explain tool use incorporation within the body schema. Another core component in this context is proprioceptive sensing, or the ability to sense self-movement and body position. Previously discussed in section 2.2 to be important in soft robotics, proprioception also plays a significant role in the perception/action model of body representations [114].

2.4 Challenges and perspectives

Through this chapter, we have been treating the various aspects bio-inspired robotics with emphasis on soft robotics and the idea that intelligence exhibits as an interplay, and reciprocal dynamical coupling, of the brain, the body and the environment. The concept of developmental soft robotics was introduced in this context, where some design principles can be established on three different timescales, aiding and enabling romanticists and researchers to develop systems for the new generation or robotics. Many enabling technologies for sensing and actuation have driven progress in the past few decades, and have allowed robots to pass from rigid, and industrial, to soft and human-friendly. These robots have been shown to achieve locomotion, to pick and manipulate objects, to be able to safely interact with humans and much more. However, many challenges still await this field, as the road to the ultimate goal of creating machines with abilities akin to those of organisms in the animal world is only at its early stages.

2.4.1 Evolutionary robotics

On the phylogenetic timescale, the question of how to achieve complex embodied behaviour has been answered by nature for a very long time. The concept of evolution in biological organisms is fairly straightforward, where evolution is thought of as the change in inheritable characteristics of populations over successive generations [115]. Due to various sources of genetic variation, new generations have increasingly different traits, and by a mediating process like that of natural selection some traits will ensure higher or lower chances of survival [116]. Eventually, the surviving population has all the different traits that we can now see in the immense variety of living organisms in our planet, which have adapted to use a plethora of different methodologies and techniques to ensure their survival.

The field of phylogenetics is tightly coupled with this concept, and consequently this field has a major impact in emergent design and control in robotics. In the area of 'evolutionary robotics', evolutionary computation is used to develop physical design or controllers for robots (cf. Chapter 6). Evolutionary computation takes inspiration from biological evolution. In robotics, e.g., it is possible to create an initial set of candidate robots, and encode their physical and or control characteristics numerically. By testing the robot population against a specific task, it is then possible to identify which combination of morphology and control performed better. The encoded characteristics of the best performing robots can then be perturbed and used to create a new generation of robots that can now be tested again. The iteration of this process for thousands of iterations has been shown to be able to achieve robust controls [117, 118] and designs [119–121].

One of the biggest limitations of evolutionary algorithms lies with the resources and time necessary to be able to achieve good controllers or designs. Because iteration of robot design or robot control and robot evaluation are very time consuming, it is generally not feasible to apply evolutionary algorithms in very complex problems, by starting from a generic, non-bounded, robot characteristics' encoding. The world of simulation has historically been more suited for evolutionary algorithms [118, 119, 122] given the ease with which populations can be created, tested and iteration achieved. The controllers and designs found, however, are usually not robust real-world solutions, as simulation environments are still very limited, and the solutions found within them do not necessarily correspond to solutions in the real world [123]. Moreover, depending on the complexity of the problem, computational resources are still an issue.

In soft robotics, given the complexity of the bodies, and the interactions emerging from them, design and control pose one of the biggest problems. Evolutionary algorithms find themselves suited as a candidate solution, but the limitations previously mentioned still apply. Further advancements in virtual reality engines, new manufacturing methods for fast prototyping, advancements in material science and the ever increasing computing power, however, may solve some of the mentioned limitations in the near future.

2.4.2 Complexity and scalability

As of today, the robots we see still 'feel' unnatural, they move slowly and sluggishly, humanoid robots still do not possess the ability to walk, run or move the way humans do, they cannot reason about the world the same way we do and get confused when unknown events occur [124]. One of the several reasons contributing to this fact is complexity. The amount of actuators and distributed sensors present in humans is much too high to be replicated by motors and standard sensors in machines. This complexity poses a problem, as it does controlling the coupling of a high number of motors and sensors. Even when dealing with sub-problems, like humanoid hands, the complexity may very well be already too high to try and tackle with standard methods. Some attempts to replicate complexity have already been made, for example, by replicating in a robotic manipulator many of the degrees of freedom present in a human hand [125]. This approach, however, did not give the results many were hoping for, as complexity in the body was coupled with complexity in the control, and achieving adaptable, smooth grasp and manipulation behavior was no easy task. Recent advances have shown how under-actuated or even passive hand can achieve complex behaviors if the interactions with the environment is appropriately exploited [9, 14]. It is here that complexity can be avoided, since complex behaviour can emerge from simple design, when appropriate interaction takes place.

Within this framework, many questions still remain. It is, in fact, unclear how design should be achieved so that complexity can be avoided or exploited. Exploiting environmental constraints is no easy feat, as the constraints to be exploited are also tightly linked to the task under hand. In soft robotics the make of the robot themselves makes for highly non-linear behaviours and robots with complex dynamics. Paradigms like that of reservoir computing can capitalize on the complexity of such structures, using them as a computational resources and thus making complexity a desirable feature. Control, however, is still hard to achieve, and mathematical models fail to comprehensively account for dynamical interactions when the complexity of the body becomes too high. The complexity challenge presents infinite challenges and opportunities, and the ever changing landscape or robotics will have to face many of them in the near future.

2.4.3 Learning through the body

The advancements in AI in the last two decades have begun a scientific revolution, endowing machines with the possibility of achieving superhuman performance levels in several different fields, like image-based object detection [126], virtual agent control [127] or haptic texture identification [128]. In robotics, machine learning has extensively been used both on the perceptual side, like object detection and recognition, and on the control side, like robot trajectory planning and motor control.

The most powerful machine learning algorithms make use of supervision, or the knowledge of target labels, to improve performance over time or trials. Broadly speaking, from the machine learning point of view, it is common to try and fit the best function to some collected data, to be able to achieve good behaviour in future instances of similar data. The data could, e.g., be streaming images from a camera mounted on an indoor mobile robotic platform, and the supervised machine learning module could have learned when and how to turn the wheels left and right, based on collected and labelled visual feed in a similar indoor environment. Throughout the sections in this chapter, we have treated the concepts of morphology, with the repercussions of what is known as morphological processing, sensory-motor-coordinated behaviour and environment. In cases such as the one mentioned above, it is common that this interconnection of mind, body and environment is neglected. In fact, in robotics, the data are usually perceptual information collected by the robot itself. As such, the perceptual information is subject to influences from both the way in which the robot interacts with entity in the world (sensory-motor-coordinated action), and the morphology of the robot's body itself. The robot can be thus be seen as a reality filter, which can act in its environment and affect the information the way that is most appropriate for learning.

Previous research has shown robots to be capable to purposefully affect the information gathered from its environment through both morphological processing, and sensory-motor coordination [66, 67]. In this context, not only the information can be structured so to be rendered suitable for learning, but the structure information itself can guide both the morphology and the control of the robot, creating a sensory-motor and morphological adaptation loop capable of intrinsically drive the robot's behaviour. Learning through the body refers to the ability for the robot to understand how its own body and actions filter the information retrieved from the world, and change its configuration and interactions so to optimize information retrieval. This simplification can then drive learning and further the adaptive capabilities of autonomous robotics systems. In Scimeca *et al.* [41], e.g., the morphology of the robot is shown to be able to achieve the cluster separation of stimuli belonging to different object types. Learning, thus, can be achieved with unsupervised methods, as the 'labels' or classes come from the skilful body–environment interaction, which induces the sensory separation.

The ability of robotics systems to purposefully shape the sensory information through their actions, or morphology, and learn from the induced structure, has the potential to change the learning landscape within robotics systems. In this context, learning may not be thought of as a process that starts in the information world, but rather one that exists in the world, where 'learning' the actions and interactions appropriate for sensory perception is the first step to appropriate learning in of the sensory stimuli at a later stage.

References

[1] Fodor J.A. 'Representations: philosophical essays on the foundations of cognitive science'. *British Journal for the Philosophy of Science*. 1983, vol. 34(2), pp. 175–82.

[2] Newell A., Simon H.A. *Computer science as empirical inquiry: symbols and search*. ACM, New York City: Philosophy of Psychology; 1975. p. 407.

[3] Lungarella M., Metta G., Pfeifer R., Sandini G. 'Developmental robotics: a survey'. *Connection Science*. 2003, vol. 15(4), pp. 151–90.

[4] Pfeifer R., Scheier C. *Understanding Intelligence*. Cambridge, MA: The MIT Press; 1999.

[5] Brooks R.A. 'Elephants Don't play chess'. *Robotics and Autonomous Systems*. 1990, vol. 6(1–2), pp. 3–15.

[6] Collins S., Ruina A., Tedrake R., Wisse M. 'Efficient bipedal robots based on passive-dynamic walkers'. *Science*. 2005, vol. 307(5712), pp. 1082–85.

[7] Braitenberg V. Vehicles: experiments in synthetic psychology. Cambridge: MIT Press; 1986.

[8] Zelik K.E., Kuo A.D. 'Human walking isn't all hard work: evidence of soft tissue contributions to energy dissipation and return'. *The Journal of Experimental Biology*. 2010, vol. 213(Pt 24), pp. 4257–64.

[9] Hughes J., Culha U., Giardina F., Guenther F., Rosendo A., Iida F. 'Soft manipulators and grippers: a review'. *Frontiers in Robotics and AI*. 2016, vol. 3, p. 69.

[10] Kim S., Laschi C., Trimmer B. 'Soft robotics: a bioinspired evolution in robotics'. *Trends in Biotechnology*. 2013, vol. 31(5), pp. 287–94.

[11] Brown E., Rodenberg N., Amend J., *et al.* 'Universal robotic gripper based on the jamming of granular material'. *Proceedings of the National Academy of Sciences*. 2010, vol. 107(44), pp. 18809–14.

[12] Laschi C., Cianchetti M., Mazzolai B., Margheri L., Follador M., Dario P. 'Soft robot arm inspired by the octopus'. *Advanced Robotics*. 2012, vol. 26(7), pp. 709–27.

[13] Yap H.K., Ng H.Y., Yeow C.H. 'High-force soft printable pneumatics for soft robotic applications'. *Soft Robotics*. 2016, vol. 3(3), pp. 144–58.

[14] Hughes J.A.E., Maiolino P., Iida F. 'An anthropomorphic soft skeleton hand exploiting conditional models for piano playing'. *Science Robotics*. 2018, vol. 3(25), eaau3098.

[15] Seok S., Onal C.D., Wood R., Rus D., Kim S. 'Peristaltic locomotion with antagonistic actuators in soft robotics'. *IEEE International Conference on Robotics and Automation*; Anchorage, AK, 2010. pp. 1228–33.

[16] Lin H.-T., Leisk G.G., Trimmer B. 'GoQBot: a caterpillar-inspired soft-bodied rolling robot'. *Bioinspiration & Biomimetics*. 2011, vol. 6(2), p. 026007.

[17] Katzschmann R.K., DelPreto J., MacCurdy R., Rus D. 'Exploration of underwater life with an acoustically controlled soft robotic fish'. *Science Robotics*. 2018, vol. 3(16), eaar3449.

[18] Cianchetti M., Arienti A., Follador M., Mazzolai B., Dario P., Laschi C. 'Design concept and validation of a robotic arm inspired by the octopus'. *Materials Science and Engineering*. 2011, vol. 31(6), pp. 1230–39.

[19] Trueman E.R. Locomotion of soft-bodied animals. London: Edward Arnold; 1975.

[20] Belanger J.H., Trimmer B.A. 'Combined kinematic and electromyographic analyses of proleg function during crawling by the caterpillar Manduca

sexta'. *Journal of Comparative Physiology. A, Sensory, Neural, and Behavioral Physiology*. 2000, vol. 186(11), pp. 1031–39.

[21] Siegenthaler K., Künkel A., Skupin G., Yamamoto M. 'Ecoflex R and ecovio R: biodegradable, performance-enabling plastics' in Rieger B., Künkel A., Coates G.W., *et al* (eds.). *Synthetic Biodegradable Polymers*. Berlin, Heidelberg: Springer; 2011. pp. 91–136.

[22] Rucker D.C., Webster R.J. 'Mechanics of continuum robots with external loading and general tendon routing' in Khatib O., Kumar V., Sukhatme G. (eds.). *Experimental Robotics*. Berlin, Heidelberg: Springer-Verlag; 2014. pp. 645–54.

[23] Camarillo D.B., Milne C.F., Carlson C.R., Zinn M.R., Salisbury J.K. 'Mechanics modeling of tendon-driven continuum manipulators'. *IEEE Transactions on Robotics*. 2008, vol. 24(6), pp. 1262–73.

[24] Yirmibesoglu O.D., Morrow J., Walker S. 'Direct 3D printing of silicone elastomer soft robots and their performance comparison with molded counterparts'. *IEEE International Conference on Soft Robotics (RoboSoft)*; Livorno, 2018. pp. 295–302.

[25] Shintake J., Schubert B., Rosset S., Shea H., Floreano D. 'Variable stiffness actuator for soft robotics using dielectric elastomer and low-melting-point alloy'. *IEEE/RSJ International Conference on Intelligent Robots and Systems (IROS)*; Hamburg, Germany, 2015. pp. 1097–102.

[26] Rodrigue H., Wang W., Han M.-W., Kim T.J.Y., Ahn S.-H. 'An overview of shape memory alloy-coupled actuators and robots'. *Soft Robotics*. 2017, vol. 4(1), pp. 3–15.

[27] Laschi C., Cianchetti M. 'Soft robotics: new perspectives for robot bodyware and control'. *Frontiers in Bioengineering and Biotechnology*. 2014, vol. 2(2), p. 3.

[28] O'Halloran A., O'Malley F., McHugh P. 'A review on dielectric elastomer actuators, technology, applications, and challenges'. *Journal of Applied Physics*. 2008, vol. 104(7), 071101.

[29] Rodrigue H., Wang W., Kim D.-R., Ahn S.H. 'Curved shape memory alloy-based soft actuators and application to soft gripper'. *Compos Struct*. 2017, vol. 176(176), pp. 398–406.

[30] Miriyev A., Stack K., Lipson H. 'soft material for soft actuators'. *Nature Communications*. 2017, vol. 8(1), 596.

[31] Tolley M.T., Shepherd R.F., Karpelson M. 'An untethered jumping soft robot'. *IEEE/RSJ International Conference on Intelligent Robots and Systems (IROS 2014)*; Chicago, IL, 2014. pp. 561–66.

[32] Rich S.I., Wood R.J., Majidi C. 'Untethered soft robotics'. *Nature Electronics*. 2018, vol. 1(2), pp. 102–12.

[33] Iida F., Laschi C. 'Soft robotics: challenges and perspectives'. *Procedia Computer Science*. 2011, vol. 7(7), pp. 99–102.

[34] Pfeifer R., Lungarella M., Iida F. 'Self-organization, embodiment, and biologically inspired robotics'. *Science*. 2007, vol. 318(5853), pp. 1088–93.

[35] Rus D., Tolley M.T. 'Design, fabrication and control of soft robots'. *Nature*. 2015, vol. 521(7553), pp. 467–75.

[36] Homberg B.S., Katzschmann R.K., Dogar M.R., Rus D. 'Haptic identification of objects using a modular soft robotic gripper'. *IEEE/RSJ International Conference on Intelligent Robots and Systems (IROS)*; Hamburg, Germany, 2015. pp. 1698–705.

[37] Maiolino P., Galantini F., Mastrogiovanni F., Gallone G., Cannata G., Carpi F. 'Soft dielectrics for capacitive sensing in robot skins: performance of different elastomer types'. *Sensors and Actuators A*. 2015, vol. 226(226), pp. 37–47.

[38] Galloway K.C., Chen Y., Templeton E., Rife B., Godage I.S., Barth E.J. 'Fiber optic shape sensing for soft robotics'. *Soft Robotics*. 2019, vol. 6(5), pp. 671–84.

[39] Scimeca L., Hughes J., Maiolino P., Iida F. 'Model-free soft-structure reconstruction for proprioception using tactile arrays'. *IEEE Robotics and Automation Letters*. 2019, vol. 4(3), pp. 2479–84.

[40] Della Santina C., Katzschmann R.K., Biechi A., Rus D. 'Dynamic control of soft robots interacting with the environment'. *IEEE International Conference on Soft Robotics (RoboSoft)*; Livorno, 2018. pp. 46–53.

[41] Scimeca L., Maiolino P., Iida F. 'Soft morphological processing of tactile stimuli for autonomous category formation'. *IEEE International Conference on Soft Robotics (RoboSoft)*; Livorno, Italy, 2018. pp. 356–61.

[42] Zlatev J., Balkenius C. *Introduction: why'epigenetic robotics'?. in proceedings of the first international workshop on epigenetic robotics: modeling cognitive development in robotic systems*. Vol. 85; 2001. pp. 1–4.

[43] Asada M., Hosoda K., Kuniyoshi Y, *et al*. 'Cognitive developmental robotics: a survey'. *IEEE Transactions on Autonomous Mental Development*. 2009, vol. 1(1), pp. 12–34.

[44] Asada M., MacDorman K.F., Ishiguro H., Kuniyoshi Y. 'Cognitive developmental robotics as a new paradigm for the design of humanoid robots'. *Robotics and Autonomous Systems*. 2001, vol. 37(2–3), pp. 185–93.

[45] Scheier C., Pfeifer R. 'The embodied cognitive science approach' in *Dynamics, Synergetics, Autonomous aagents: Nonlinear Systems a Approaches to Cognitive Psychology and Cognitive Science*. Singapore: World Scientific; 1999. pp. 159–79.

[46] Sporns O. 'Embodied cognition' in Arbib M.A. (ed.). Handbook of brain theory and neural networks. Cambridge: MIT Press; 2003. pp. 395–98.

[47] Johnson M.H. Developmental cognitive neuroscience. Cambridge, MA: Blackwell; 1997.

[48] Parker S.T., McKinney M.L. Origins of Intelligence: The Evolution of Cognitive Development in Monkeys, apes, and humans. Baltimore, MD: JHU Press; 2012.

[49] Kuhl P.K. 'Language, mind, and brain: experience alters perception'. *The New Cognitive Neurosciences*. 2000, vol. 2(2), pp. 99–115.

[50] Fogel A. 'Theoretical and applied dynamic systems research in developmental science' [online]'. *Child Development Perspectives*. 2011, vol. 5(4), pp. 267–72. Available from http://doi.wiley.com/10.1111/cdep.2011.5.issue-4
[51] Piaget J. The psychology of intelligence. London, New York: Routledge; 2003. Available from https://www.taylorfrancis.com/books/9781134524693
[52] Piaget J., Cook M. *The Origins of Intelligence in Children*. Vol. 8. New York: International Universities Press; 1952. p. 5. Available from http://content.apa.org/books/11494-000
[53] Goldfield E.C. Emergent forms: origins and early development of human action and perception. New York: Oxford University Press; 1995.
[54] Turvey M.T. 'Coordination'. *The American Psychologist*. 1990, vol. 45(8), pp. 938–53.
[55] Bernstein N. *The co-ordination and regulation of movements*. Oxford: Pergamon Press; 1967.
[56] Taga G., Takaya R., Konishi Y. 'Analysis of general movements of infants towards understanding of developmental principle for motor control'. *IEEE SMC'99 Conference Proceedings. IEEE International Conference on Systems, Man, and Cybernetics*; Tokyo, Japan, 1999. pp. 678–83.
[57] Rochat P. 'Self-perception and action in infancy'. *Experimental Brain Research*. 1998, vol. 123(1–2), pp. 102–09.
[58] Tsakiris M. 'My body in the brain: a neurocognitive model of body-ownership'. *Neuropsychologia*. 2010, vol. 48(3), pp. 703–12.
[59] Oudeyer P.-Y., Kaplan F., Hafner V.V. 'Intrinsic motivation systems for autonomous mental development'. *IEEE Transactions on Evolutionary Computation*. 2007, vol. 11(2), pp. 265–86.
[60] Baldassarre G., Mirolli M. (eds.) 'Intrinsically motivated learning in natural and artificial systems' in *Springer*; 2013. pp. 1–14.
[61] Barto A.G. 'Intrinsic motivation and reinforcement learning' in *Intrinsically motivated learning in natural and artificial systems*. Berlin, Heidelberg: Springer; 2013. pp. 17–47.
[62] Edelman G.M. *Neural Darwinism: the theory of neuronal group selection*. New York: Basic books; 1987.
[63] Thelen E., Smith L.B. *A Dynamic Systems Approach to the Development of Cognition and Action*. Cambridge: MIT Press; 1996.
[64] Berthouze L., Lungarella M. 'Motor skill acquisition under environmental perturbations: on the necessity of alternate freezing and freeing of degrees of freedom'. *Adaptive Behavior*. 2004, vol. 12(1), pp. 47–64.
[65] Pfeifer R. 'On the role of morphology and materials in adaptive behavior'. *From Animals to Animats*. 2000, vol. 6, pp. 23–32.
[66] Pfeifer R., Iida F., Gómez G. 'Morphological computation for adaptive behavior and cognition'. *International Congress Series*. 2006, vol. 1291, pp. 22–29.
[67] Pfeifer R., Scheier C. 'Sensory—motor coordination: the metaphor and beyond'. *Robotics and Autonomous Systems*. 1997, vol. 20(2–4), pp. 157–78.

[68] Buhrmann T., Di Paolo E.A., Barandiaran X. 'A dynamical systems account of sensorimotor contingencies'. *Frontiers in Psychology*. 2013, vol. 4(4), 285.

[69] O'Regan J.K., Noë A. 'A sensorimotor account of vision and visual consciousness'. *The Behavioral and Brain Sciences*. 2001, vol. 24(5), pp. 939–73.

[70] Clark A., Grush R. 'Towards a cognitive robotics'. *Adaptive Behavior*. 1999, vol. 7(1), pp. 5–16.

[71] Grush R. 'In defense of some 'Cartesian' assumptions concerning the brain and its operation'. *Biology & Philosophy*. 2003, vol. 18(1), pp. 53–93.

[72] Wolpert D.M., Doya K., Kawato M. 'A unifying computational framework for motor control and social interaction'. *Philosophical Transactions of the Royal Society of London. Series B, Biological Sciences*. 2003, vol. 358(1431), pp. 593–602.

[73] Miall R.C., Weir D.J., Wolpert D.M., Stein J.F. 'Is the cerebellum a Smith predictor?'. *Journal of Motor Behavior*. 1993, vol. 25(3), pp. 203–16.

[74] Doncieux S., Bredeche N., Mouret J.-B., Eiben A.E.G. 'Evolutionary robotics: what, why, and where to'. *Frontiers in Robotics and AI*. 2015, vol. 2(2), p. 4.

[75] Nolfi S., Floreano D., Floreano D.D. *Evolutionary Robotics: The Biology, Intelligence, and Technology of Self-Organizing Machines*. Cambridge: MIT Press; 2000.

[76] Hawkes E.W., Blumenschein L.H., Greer J.D., Okamura A.M. 'A soft robot that navigates its environment through growth'. *Science Robotics*. 2017, vol. 2(8), eaan3028.

[77] Cianchetti M., Ranzani T., Gerboni G., De Falco I., Laschi C., Menciassi A. 'Stiff-flop surgical manipulator: mechanical design and experimentalcharacterization of the single module'. *IEEE/RSJ International Conference on Intelligent Robots and Systems*; Tokyo, 2013. pp. 3576–81.

[78] Eder M., Hisch F., Hauser H. 'Morphological computation-based control of a modular, pneumatically driven, soft robotic arm'. *Advanced Robotics*. 2018, vol. 32(7), pp. 375–85.

[79] Nakajima K., Hauser H., Kang R., Guglielmino E., Caldwell D.G., Pfeifer R. 'A soft body as a reservoir: case studies in a dynamic model of octopus-inspired soft robotic arm'. *Frontiers in Computational Neuroscience*. 2013, vol. 7(7), 91.

[80] Nakajima K., Hauser H., Li T., Pfeifer R. 'Information processing via physical soft body'. *Scientific Reports*. 2015, vol. 5(5), 10487.

[81] Nakajima K., Li T., Hauser H., Pfeifer R. 'Exploiting short-term memory in soft body dynamics as a computational resource'. *Journal of the Royal Society, Interface*. 2014, vol. 11(100), p. 20140437.

[82] Hara F., Pfeifer R. Morpho-Functional machines: the new species. Tokyo: Springer Science & Business Media; 2003. Available from http://link.springer.com/10.1007/978-4-431-67869-4

[83] Beal D.N., Hover F.S., Triantafyllou M.S., Liao J.C., Lauder G.V. 'Passive propulsion in vortex wakes'. *Journal of Fluid Mechanics*. 2006, vol. 549(1), 385.

[84] Matarić M. J. 'Situated robotics' in Encyclopedia of cognitive science. Philadelphia, PA: John Wiley & Sons; 2006.

[85] Iida F., Nurzaman S.G. 'Adaptation of sensor morphology: an integrative view of perception from biologically inspired robotics perspective'. *Interface Focus*. 2016, vol. 6(4), 20160016.

[86] Towal R.B., Quist B.W., Gopal V., Solomon J.H., Hartmann M.J.Z., Friston K.J. 'The morphology of the rat vibrissal array: a model for quantifying spatiotemporal patterns of whisker-object contact'. *PLoS Computational Biology*. 2011, vol. 7(4), e1001120.

[87] Prescott T.J., Pearson M.J., Mitchinson B., Sullivan J.C.W., Pipe A.G. 'Whisking with robots'. *IEEE Robotics & Automation Magazine*. 2009, vol. 16(3), pp. 42–50.

[88] Land M.F., Nilsson D.-E. *Animal eyes* [online]. New York: Oxford University Press; 2012. Available from https://academic.oup.com/book/10639

[89] Pearson M.J., Mitchinson B., Sullivan J.C., Pipe A.G., Prescott T.J. 'Biomimetic vibrissal sensing for robots'. *Philosophical Transactions of the Royal Society of London. Series B, Biological Sciences*. 2011, vol. 366(1581), pp. 3085–96.

[90] Floreano D., Pericet-Camara R., Viollet S, *et al*. 'Miniature curved artificial compound eyes'. *Proceedings of the National Academy of Sciences of the United States of America*. 2013, vol. 110(23), pp. 9267–72.

[91] Hoffmann M., Straka Z., Farkaš I., Vavrečka M., Metta G. 'Robotic homunculus: learning of artificial skin representation in a humanoid robot motivated by primary somatosensory cortex'. *IEEE Transactions on Cognitive and Developmental Systems*. 2017, vol. 10(2), pp. 163–76.

[92] Hughes J., Iida F. 'Localized differential sensing of soft deformable surfaces'. *2017 IEEE International Conference on Robotics and Automation (ICRA)*; Singapore, 2017. pp. 4959–64.

[93] Hua Q., Sun J., Liu H, *et al*. 'Skin-inspired highly stretchable and conformable matrix networks for multifunctional sensing'. *Nature Communications*. 2018, vol. 9(1), 244.

[94] Lu N., Kim D.-H. 'Flexible and stretchable electronics paving the way for soft robotics'. *Soft Robotics*. 2014, vol. 1(1), pp. 53–62.

[95] Majidi C. 'Soft robotics: a perspective—current trends and prospects for the future'. *Soft Robotics*. 2014, vol. 1(1), pp. 5–11.

[96] Rogers J.A., Someya T., Huang Y. 'Materials and mechanics for stretchable electronics'. *Science*. 2010, vol. 327(5973), pp. 1603–07.

[97] Cheng S., Wu Z. 'Microfluidic electronics'. *Lab on a Chip*. 2012, vol. 12(16), pp. 2782–91.

[98] Muth J.T., Vogt D.M., Truby R.L, *et al*. 'Embedded 3D printing of strain sensors within highly stretchable elastomers'. *Advanced Materials (Deerfield Beach, Fla.)*. 2014, vol. 26(36), pp. 6307–12.

[99] Maiolino P., Maggiali M., Cannata G., Metta G., Natale L. 'A flexible and robust large scale capacitive tactile system for robots'. *IEEE Sensors Journal*. 2013, vol. 13(10), pp. 3910–17.

[100] Iida F., Pfeifer R. 'sensing through body dynamics'. *Robotics and Autonomous Systems*. 2006, vol. 54(8), pp. 631–40.

[101] Mazzolai B., Beccai L., Mattoli V. 'Plants as model in biomimetics and biorobotics: new perspectives'. *Frontiers in Bioengineering and Biotechnology*. 2014, vol. 2(2), p. 2.

[102] Sadeghi A., Tonazzini A., Popova L., Mazzolai B. *'IEEE International Conference on robotics and automation (ICRA)'* Karlsruhe, Germany'. 2013. pp. 3457–62.

[103] Sadeghi A., Tonazzini A., Popova L., Mazzolai B. 'A novel growing device inspired by plant root soil penetration behaviors'. *PloS One*. 2014, vol. 9(2), e90139.

[104] Velcro S. *Improvements in or relating to a method and a device for producing a velvet type fabric. 721338*. 1955.

[105] Li S.H., Zeng Q.Y., Xiao Y.L., Fu S.Y., Zhou B.L. 'Biomimicry of bamboo Bast fiber with engineering composite materials'. *Materials Science and Engineering*. 1995, vol. 3(2), pp. 125–30.

[106] Taccola S., Zucca A., Greco F., Mazzolai B., Mattoli V. 'Electrically driven dry state actuators based on PEDOT: PSS nanofilms'. *In EuroEAP 2013, International Conference on Electromechanically Active Polymer (EAPeap) Transducers & Artificial Muscles*; 2013.

[107] Mazzolai B., Mondini A., Corradi P., *et al.* 'A miniaturized mechatronic system inspired by plant roots for soil exploration'. *IEEE/ASME Transactions on Mechatronics*. 2011, vol. 16(2), pp. 201–12.

[108] Terryn S., Brancart J., Lefeber D., Van Assche G., Vanderborght B. 'Self-healing soft pneumatic robots'. *Science Robotics*. 2017, vol. 2(9), eaan4268.

[109] Scheltjens G., Diaz M.M., Brancart J., Van Assche G., Van Mele B. 'A self-healing polymer network based on reversible covalent bonding'. *Reactive and Functional Polymers*. 2013, vol. 73(2), pp. 413–20.

[110] Dawkins R. The extended phenotype. Vol. 8. Oxford: Oxford University Press; 1982.

[111] Shumaker R.W., Walkup K.R., Beck B.B. *Animal Tool Behavior: The Use and Manufacture of Tools by Animals*. Baltimore, MD: JHU Press; 2011.

[112] Hoffmann M., Marques H., Arieta A., Sumioka H., Lungarella M., Pfeifer R. 'Body schema in robotics: a review'. *IEEE Transactions on Autonomous Mental Development*. 2010, vol. 2(4), pp. 304–24.

[113] Nabeshima C., Kuniyoshi Y., Lungarella M. 'Adaptive body schema for robotic tool-use'. *Advanced Robotics*. 2006, vol. 20(10), pp. 1105–26.

[114] de Vignemont F. 'Body schema and body image– pros and cons'. *Neuropsychologia*. 2010, vol. 48(3), pp. 669–80.

[115] Hall B., Hallgrimsson B. Strickberger ' S evolution. Burlington, MA: Jones & Bartlett Learning; 2008.

[116] Scott-Phillips T.C., Laland K.N., Shuker D.M., Dickins T.E., West S.A. 'The niche construction perspective: a critical appraisal'. *Evolution; International Journal of Organic Evolution*. 2014, vol. 68(5), pp. 1231–43.
[117] Fleming P.J., Purshouse R.C. 'Evolutionary algorithms in control systems engineering: a survey'. *Control Engineering Practice*. 2002, vol. 10(11), pp. 1223–41.
[118] Mautner C., Belew R.K. 'Evolving robot morphology and control'. *Artificial Life and Robotics*. 2000, vol. 4(3), pp. 130–36.
[119] Lipson H., Pollack J.B. 'Automatic design and manufacture of robotic life-forms'. *Nature*. 2000, vol. 406(6799), pp. 974–78.
[120] Lund H.H., Hallam J., Lee W.-P. 'Evolving robot morphology'. *Proceedings of 1997 IEEE International Conference on Evolutionary Computation (ICEC'97)*; Indianapolis, IN, 1997. pp. 197–202.
[121] Pfeifer R., Iida F., Bongard J. 'New robotics: design principles for intelligent systems'. *Artificial Life*. 2005, vol. 11(1–2), pp. 99–120.
[122] Nolfi S., Floreano D., Miglino O., Mondada F. 'How to evolve autonomous robots: different approaches in evolutionary robotics'. *Artificial life IV: Proceedings of the Fourth International Workshop on the Synthesis and Simulation of Living Mystems*; Cambridge, MA, The MIT Press, 1994. pp. 190–97.
[123] Jakobi N., Husbands P., Harvey I. 'Noise and the reality gap: the use of simulation in evolutionary robotics'. *European Conference on Artificial Life*; Granada, Spain, 1995. pp. 704–20.
[124] Pfeifer R., Lungarella M., Iida F. 'The challenges ahead for bio-inspired " soft " robotics'. *Communications of the ACM*. 2012, vol. 55(11), pp. 76–87.
[125] Tuffield P., Elias H. 'The shadow robot mimics human actions'. *Industrial Robot*. 2003, vol. 30(1), pp. 56–60.
[126] Schmidhuber J. 'Deep learning in neural networks: an overview'. *Neural Networks*. 2015, vol. 61(61), pp. 85–117.
[127] Mnih V., Kavukcuoglu K., Silver D, *et al*. 'Human-level control through deep reinforcement learning'. *Nature*. 2015, vol. 518(7540), pp. 529–33.
[128] Fishel J.A., Loeb G.E. 'Bayesian exploration for intelligent identification of textures'. *Frontiers in Neurorobotics*. 2012, vol. 6, article 4.

Chapter 3

Three-axis tactile sensor using optical transduction mechanism

Masahiro Ohka[1] *and Hanafiah Yussof*[2]

3.1 Introduction

Tactile information plays a major role in our daily life and is crucial when we handle objects such as food, tools, books, or clothing. It is well known that we cannot button a shirt when we lose tactile sensation due to disease. For robots, tactile information is important to fulfill handling tasks. Many engineers and researchers in robotics have developed tactile sensors for robots that apply various physics phenomena, such as electric resistance variation, magnetic field variation, piezoelectric effect, piezoresistance variation, and optical variation. Recently, new technologies have been utilized, such as microelectromechanical systems (MEMS), metal-oxide-semiconductor field-effect transistors (MOSFET), micro-chips, and 3D printing. As some authors have previously reported on these [1–4], we focus on optical tactile sensors in this chapter.

The optical tactile sensor is a candidate for actual use in tactile sensors because of its attractive characteristics of impact toughness and electromagnetic noise. Optical tactile sensors are grouped into two categories, specifically, extrinsic and intrinsic optical sensors [3]. With the former sensors, applied force interacts with light external to the primary light path. With the latter sensors, the applied force modulates the transmitted light without interrupting the optical path.

We will begin with a discussion of some extrinsic optical tactile sensors. Rebman and Morris [5] produced an extrinsic optical tactile sensor that modulates light transmission between a light-emitting diode and a photodetector via the applied force. Kampmann and Kirchner [6] produced fiber-optic sensor arrays in which several pairs of optical fibers are inserted into foam. Of the pair, one fiber is an emitter and the other is a detector. If the applied force compresses the foam, more light is collected by the detector because light scattering is reduced. Later, a GelForce sensor was developed at the University of Tokyo Tachi Lab [7]. This sensor consists of a transparent elastic body and two layers of colored markers within the body; when

[1]Graduate School of Information Science, Nagoya University Furo-cho, Nagoya, Japan
[2]Faculty of Mechanical Engineering, Universiti Teknologi MARA, Selangor, Malaysia

force is applied to the body's surface, the internal strain on the body is optically measured through the movement of the markers to calculate force vectors from the strain using elasticity. A GelSight sensor was developed at the MIT laboratory that uses gel sight to precisely measure high-resolution surface topography; tactile data are obtained by image data processing [8]. Ito *et al.* produced a tactile sensor using a balloon filled with red-colored water. A dot pattern is drawn on the inside surface of the balloon, and this tactile sensor detects several data, such as applied force, object shape, and slippage, from changes in the dot pattern configuration [9].

Recently, a commercial three-axis tactile sensor of this type was produced (http://optoforce.com/) that is based on the reflection design. It is composed of a hollow hemispherical elastomer and a circular substrate. The inside surface of the hemisphere is coated with a reflection layer that reflects infrared light; the circular substrate is covered with the hemisphere elastomer and has an infrared emitter at the center and four detecting elements. The emitted infrared beam is reflected on the reflection layer, and the reflection angle is inclined according to a combination of the three-axis force components caused by applied force on the hemispherical outer surface. Since this tactile sensor does not detect the distribution of the force on the sensing surface, it should be considered a three-axis force sensor.

In this chapter, we focus on intrinsic optical sensors, which modulate the transmitted light without interrupting the optical path, as mentioned above. Furthermore, intrinsic optical tactile sensors are mainly divided into two categories: the absorption type and the reflection type. Kawashima and Aoki [10] produced an absorption type optical tactile sensor using a computer tomography (CT) reconstruction method; a transparent circular elastomer plate is sandwiched between two black circular rubber sheets having an array of hemispherical projections on the transparent circular plate side. A circular scanner composed of infrared (IR) diodes and phototransistors is installed around the transparent plate. If force is applied to the black circular rubber sheet, some hemispherical projections on the back side of the black circular rubber sheet collapse to generate contact areas. At that time, IR-light emitted from the IR-diodes is absorbed on the contact areas. If the applied force increases, the absorption coefficient of IR increases. Since the IR-diodes and phototransistors surround the transparent circular plate, the 2D absorption coefficient distribution is obtained through the multidimensional projection provided by circular scanning and a CT reconstruction algorithm. However, usage of Kawashima and Aoki's method is limited to the circular region because it is based on circular scanning and radon transformation. After approximately two decades, Ohka *et al.* [11, 12] deduced a new CT algorithm for square regions. In their formulation, they assumed virtual meshes in the square region; algebraic equations are derived from the relationship between the input and output light intensities on the assumed light projections. The absorption coefficient distribution is obtained by solving a series of algebraic equations using lower-upper decomposition methods. However, in the CT algorithm, ghost images sometimes appeared near the receivers. This problem seems to be caused by a small voltage difference via applied force; since the CT reconstruction is calculated based on this difference, photo-electronic characteristics of the IR-diodes and phototransistors are uneven compared to this difference. We expect a new generation of progression in IR-diodes and phototransistors to solve this problem.

On the other hand, there are several tactile sensors of the reflection type. During the earlier development, Tanie *et al.* [13] produced a sandwiched structure composed of a rubber sheet, an acrylic plate, and a phototransistor array in which light is introduced into the acrylic plate from the end of the plate. The light totally reflects on the inside of the acrylic plate. If force is applied to the rubber sheet, the acrylic plate contacts the back surface of the rubber sheet. Since this acrylic plate acts as an optical waveguide, evanescent light goes out around a wavelength at the reflecting point. On the contact area, the reflected light on the rubber surface is observed as a blight and is retrieved through the phototransistor array. Tanie came up with this idea when he saw a measurement system for flat feet at a hospital in the United Kingdom. Similar sensors were developed by other researchers [14, 15], while Tanie *et al.* progressed this tactile sensor to mount it onto a multi-articulated robotic hand [16].

In this chapter, we introduce an optical three-axis tactile sensor that is an advancement of Tanie's sensors to measure normal and tangential force distributions. An advantage of this tactile sensor is that each sensing element's three-axis force sensibility can be calibrated one by one. Although this sensor has the disadvantages of difficulties in miniaturization, creating a thin structure, and high sampling rate due to image data processing, its high impact resistance and noise robustness are very attractive to users who are attempting to perform actual robotic tasks. We performed several tasks using a two-hand-arm robot equipped with an optical three-axis tactile sensor.

The three-axis tactile sensor was first examined in the national project "Advanced Robot Technology Research" of Japan, supervised by the Japanese Ministry of International Trade and Industry (MITI) from 1986 to 1992. At that time, the three-axis tactile sensor was not based on the optical tactile sensor, but on piezoresistance, and was made of crystal silicon wafers. Since the three-axis tactile sensors developed in this project were previously described in a different survey paper [17], we provide just a brief introduction here. Fuji Electric Company provided two types of silicon semiconductors based on the three-axis tactile sensor using MEMS in the project; one was a 3-by-3-mm cell type and the other a 1-by-1-mm cell type. The 3-by-3-mm cell was evaluated through a parallel-fingered robotic hand to perform inference of the grasped object using an if-then rule base [18] and Fuzzy rules [19]. However, since the supply of the silicon-semiconductor-based three-axis tactile sensors stopped with the project's termination, we conceived a new three-axis tactile sensor based on a different mechanism.

Ohka *et al.* [20] produced a tactile sensor based on Tanie's tactile sensor [13], as mentioned in the preceding paragraph. When Ohka inspected the tactile sensor in his laboratory, it sparked an idea for a new three-axis tactile sensor; if an additional projection array was formed on the outer surface of Tanie's tactile sensor, it would work as a three-axis tactile sensor. If the outer additional projection accepts three-axial force, the tips of inner surface projections will collapse in a nonuniform fashion, and three-components of applied force can be estimated from the distribution of the collapsing inner projections. First, we developed a one-columnar feeler and four-conical feeler type [21–23] because a position sensitive detector (PSD) algorithm is available. We progressed to a one-columnar feeler and eight-conical feeler type [23] to more precisely measure the tangential force direction. Simultaneously, we attempted to miniaturize it. We first performed experiments using a table-sized sensor system and then reduced its size to that of a lunch box. Finally, the flat sensing surface was changed to a hemispherical surface and then we reduced it to Φ35 × 50 mm to mount it onto a microactuator produced as a motor for robotic fingers (microactuator, Yasukawa, Co.).

While developing the abovementioned three-axis tactile sensor, we further developed another type of three-axis tactile sensor to perform miniaturization. Since the columnar-conical-feeler-type tactile sensor has a complicated structure, it is not suitable for miniaturization. We therefore made the structure simpler and utilized image data processing to compensate for the simplification of the hardware. The outer projections were removed and the inner conical projections remained; the normal force was calculated from an integrated grayscale value and the tangential force was calculated from movement of the contact area. We achieved a micro-three-axis tactile sensor having conical feelers of Φ300 × 100 μm on the back surface of a rubber sheet. The distance between the conical feelers was 600 μm and array size was 10 × 12 [24]. Furthermore, we applied this idea to a normal size three-axis tactile sensor because the columnar-and-conical feeler type is expensive and several tasks do not require precious three-axis tactile data. After the micro-three-axis tactile sensor, we progressed the three-axis tactile sensor based on optical flow and a human-finger-sized three-axis tactile sensor. Although these tactile sensors need further improvement, they obtained satisfactory results in the initial trials [25, 26].

In the remainder of this chapter, we will precisely explain the optical three-axis tactile sensors. In section 3.2, we explain the design concept of the optical three-axis tactile sensor and some advanced ideas based on the basic concept. In section 3.3, we present the actual designs of the optical three-axis tactile sensor, such as the aluminum-dome and columnar-conical feeler type, rubber-dome and columnar-conical feeler type, and tracking-contact-area-movement type. In the next, we describe applications of the optical three-axis tactile sensor, such as slippage judgment, cap twisting, paper counting, and communication between a robot and a person. Finally, we conclude this chapter and describe future related to this tactile sensor.

3.2 Design concept of the optical three-axis tactile sensor

3.2.1 Basic principle

Figure 3.1 shows the basic configuration of the optical tactile sensor. It is composed of an acrylic plate, a rubber sheet, and a light source. Since planar light is preferred for inside total reflection, the light is usually introduced through optical fibers. The introduced light is totally reflected on the inside surface of the acrylic plate because the refraction index of the acrylic plate is different from that of the air. At the reflection point, the light goes out slightly from inside the acrylic plate. This phenomenon

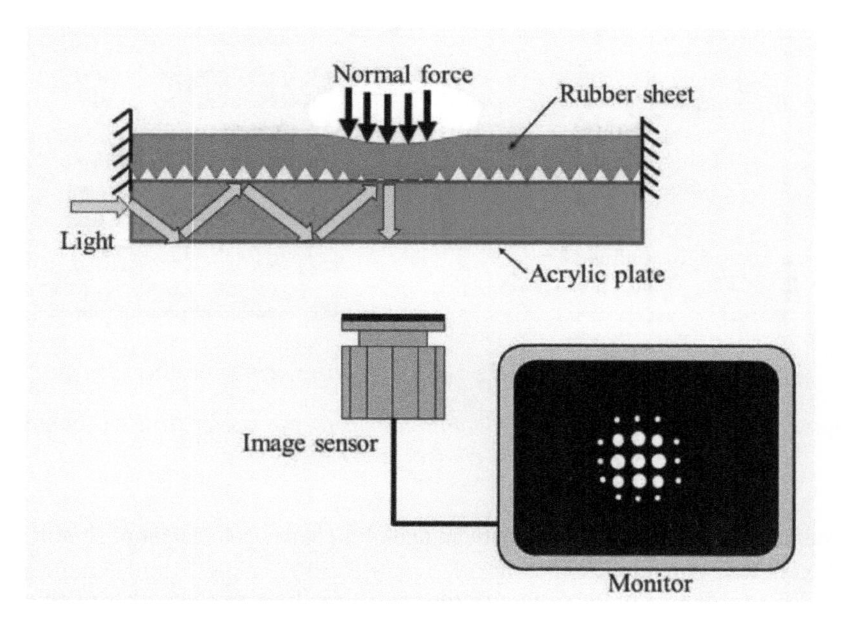

Figure 3.1 *Basic configuration of the optical tactile sensor*

is called evanescent reflection, in which light goes out around the wavelength of visible light (0.4–0.8 μm). Thus, if an object approaches within this region, the light reflects on the object's surface. This reflected light appears as a bright image from beneath the acrylic plate. If we create a texture of small cones array on the rubber sheet's back surface, we observe bright spots, as shown in Figure 3.1.

On the other hand, according to finite element analysis [19, 27], when the tip of a conical elastic body makes contact with a flat object and force is applied to the base of the cone, the relationship between the contact area and the applied force is linear. Since the contact area is proportional to brightness, the applied force is also proportional to brightness. Since the relationship between applied force and brightness becomes linear, we can obtain the applied force value from the brightness of the spots. In order to measure distributed force, the brightness distribution should be obtained through an array of phototransistors and image sensors. In Tanie's tactile sensor, described in section 3.1, they used a phototransistor array. Since image sensors, such as charge-coupled device and complementary metal oxide semiconductor (CMOS) image sensors, have decreased in size, it is convenient to use these image sensors.

3.2.2 Conical-columnar feeler-type optical three-axis tactile sensor

We have two approaches to progress the optical tactile sensor described in the preceding section to a three-axis sensor: one uses a special structure, while the other uses image data processing. The advantage of the former is to obtain precious

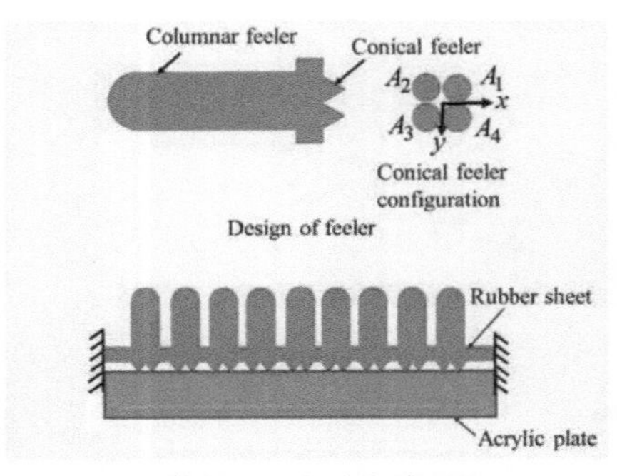

(a) Four-conical feelers type

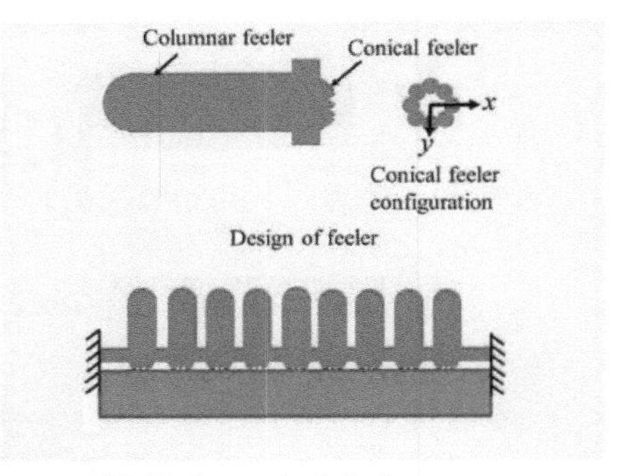

(b) Eight-conical feelers type

Figure 3.2 Feeler design of optical three-axis tactile sensor (a) Four-conical feeler type (b) Eight-conical feeler type

three-axis force components; the advantage of the latter is to accomplish miniaturization in terms of simple structure.

As the structure of the former tactile sensor, we form another array of columnar feelers on the outer surface of the rubber sheet, as shown in Figure 3.2(a). First, we used four-conical feelers and a one-columnar feeler type because we applied a PSD algorithm for tactile sensing. If we obtain contact areas as shown in Figure 3.2(a), we derive the following formulas to obtain a three-component force vector, F_x, F_y, and F_z:

$$F_x = K_xA_x,\ F_y = K_yA_y, F_z = K_zA_z \tag{3.1}$$

$$A_x = A_1 + A_4 - A_2 - A_3 \tag{3.2}$$

$$A_y = A_3 + A_4 - A_1 - A_2 \tag{3.3}$$

$$A_z = A_1 + A_2 + A_3 + A_4 \tag{3.4}$$

where Kx, Ky, Kz are coefficients for transformation from the contact area to force.

In subsequent experiments, we discovered that the four-conical feeler type did not have enough sensing force direction precision, except for the x–y axes. Thus, we developed an eight-conical feeler type, as shown in Figure 3.2(b). In order to calculate three components of force, we assumed an integral area A, which surrounds the contact areas of the eight conical feelers. In area A, we measured the distributed grayscale value $G(x, y)$ to derive the following formulas:

$$F_x = \widehat{Kx}xG, F_y = \widehat{Ky}yG, Fx_z = \widehat{Kz}\int G(x,y)dAA \tag{3.5}$$

$$xG = \int xG(x,y)dAA \int G(x,y)dAA \tag{3.6}$$

$$yG = \int yG(x,y)dAAA \int G(x,y)dA \tag{3.7}$$

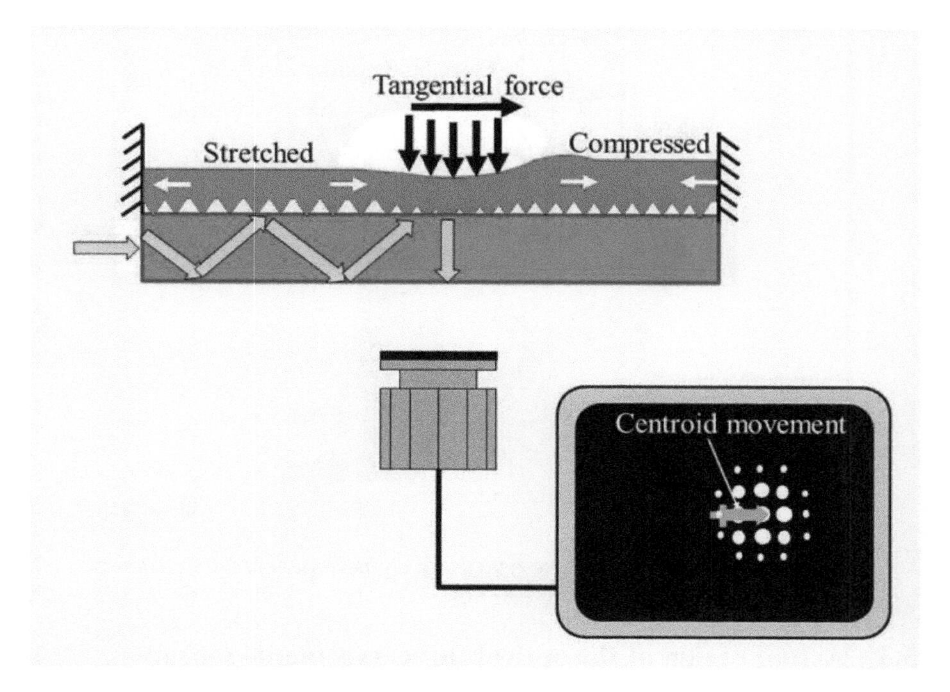

Figure 3.3 Tracking-centroid-movement type

where $\hat{K}_x$, $\hat{K}_y$, $\hat{K}_z$ are coefficients for transformation from centroid movement or grayscale value to force.

3.2.3 Tracking-centroid-movement-type optical three-axis tactile sensor

Since the conical-columnar feeler type has a complicated structure, it is not suitable for miniaturization. Thus, we made the sensor structure as simple as possible. To compensate for this simplification, we enhanced the image data processing to obtain three-axis force data. This type of tactile sensor basically has the same structure as the basic one shown in Figure 3.1. Only the usual integration of grayscale value, as described in section 3.2.1, is required for the normal force sensing, but the tangential force sensing requires the following special procedure. As shown in Figure 3.3, tangential force causes the rubber sheet to stretch and compress, which maintains equilibrium of the applied tangential force. Thus, the tangential force is proportional to the movement of the contact areas. To measure the movement, we programmed tracking of the contact area movement, for which we use several program modules in OpenCV. For example, calculating the image's gravity center, the optical flow, and particle filter are effective to calculate the contact area movement.

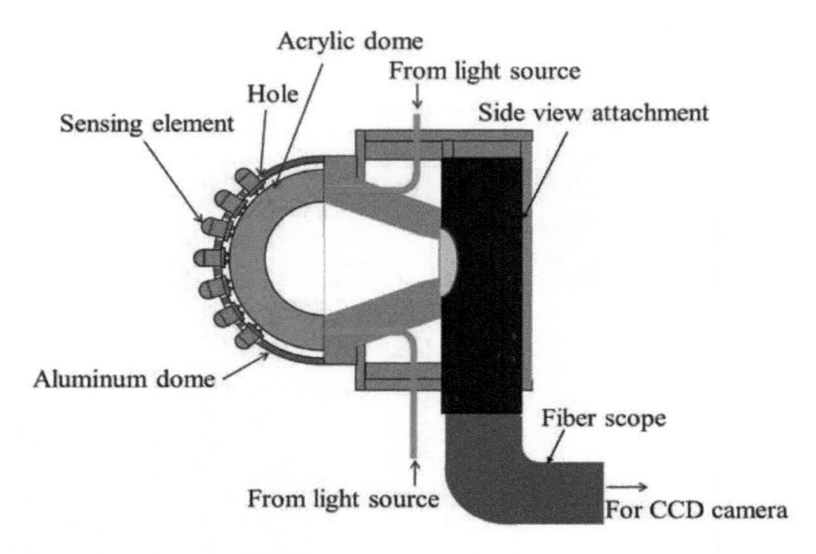

Figure 3.4 Aluminum-dome type using eight-conical feelers elements

3.3 Actual design of the optical three-axis tactile sensor

3.3.1 Aluminum-dome type

In order to apply the conical-columnar-feeler-type optical three-axis tactile sensor to a fingertip of a multi-fingered robotic hand, we developed the hemispherical three-axis tactile sensor, as shown in Figure 3.4 [24]. This sensor features an aluminum dome with holes arranged on concentric circles and a sensing element inserted in each hole. As shown in Figure 3.5, if the tangential component of force is applied to the sensing element tip, the element rotates with regard to the rotational axis, which is assumed at the hole. Since the hole inhibits the element's horizontal movement, the tangential component is effectively transformed to rotation. Due to this rotation, the eight-conical feelers nonuniformly collapse to obtain three components of force according to (3.5) to (3.7).

In this tactile sensor, horizontal movement of the sensing element is blocked by the hole in the aluminum dome; therefore, image processing for identifying the sensing element is not needed and calibration of a specific element is available. Thus, this type of three-axis tactile sensor is suitable for tasks requiring precise tactile sensing, such as texture sensing and surface fine unevenness detection.

3.3.2 Rubber-dome type

While the aluminum-dome type provides some advantages, as described in the last section, it is expensive and creates insensible zones. Since the aluminum dome is made with a five-axis machining center for making holes perpendicular to the hemispherical surface, the production cost increases. Since we want to mount this

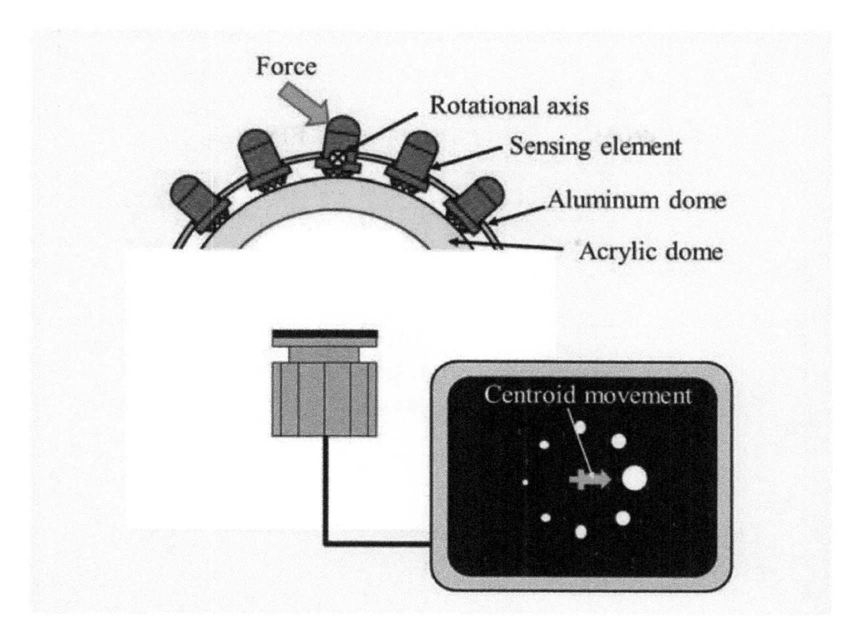

Figure 3.5 *Aluminum-dome type using eight-conical feelers elements*

three-axis tactile sensor on each finger of multi-fingered robotic hands, this cost increase is not a trivial issue. Furthermore, the insensible zone causes problems between sensing elements; even if an object touches the portion between the sensing elements, we appreciate any change in image data, as shown in Figure 3.5.

In order to overcome the above-mentioned problems, we exclude the aluminum dome and introduce a new rubber dome to design a new structure, as shown in Figure 3.6. Each sensing element is embedded in a hole of the rubber dome [28]. In this design, a CMOS camera and LEDs are installed in the tactile sensor body. Image data are retrieved via a USB interface and sent to a computer. While the aluminum-dome type is equipped with a light source and an image fiber scope on the outside of the sensor, this tactile sensor is equipped with all devices on the inside. Therefore, when the rubber-dome-type tactile sensor is optimized, its all-in-one structure effectively performs several robot tasks.

We evaluated the precision of these three-axis tactile sensors. Figure 3.7(a) and (b) shows normal force sensing and tangential force sensing, respectively, when force is applied to the summit of the sensors. As shown in Figure 3.7(a), variation in relationship between grayscale value and normal force of the aluminum-dome type is quite different than that of the rubber-dome type. Since these types use a different camera and light source, comparison of the value of the ordinate between these two types is irrelevant. However, the graph shapes of the two types differ; while the graph shape of the rubber-dome type shows an “s” shape variation, the aluminum-dome type approximates a bilinear variation. The latter characteristic is preferable because the regression of the aluminum-dome type is easier than that of

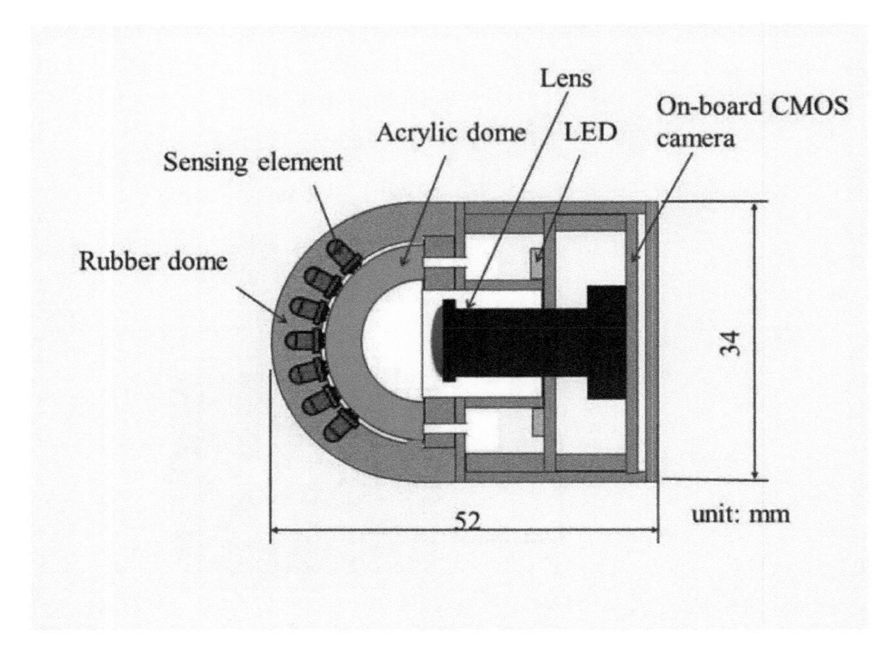

Figure 3.6 Rubber-dome type

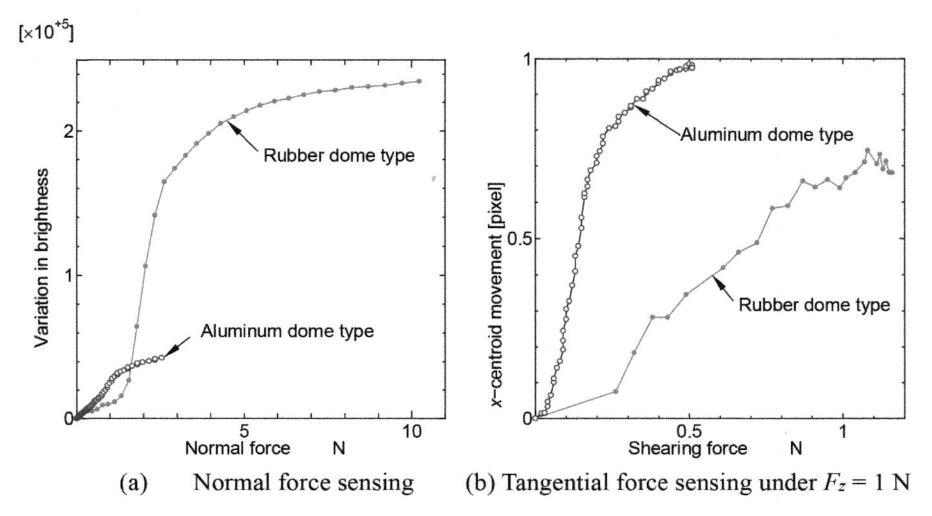

(a) Normal force sensing (b) Tangential force sensing under $F_z = 1$ N

Figure 3.7 Comparison of rubber dome and aluminum dome

the rubber-dome type. This difference is caused by contact between the rubber dome and the acrylic dome at a higher applied force region.

For tangential force sensing, looking at Figure 3.7(b), we see that the sensitivity of the aluminum-dome type is greater than that of the rubber-dome type, while the sensible region of the rubber-dome type is larger than that of the aluminum dome. This difference is caused by the support methods for the sensing element; the support method of the aluminum dome is looser than that of the rubber dome in the low tangential force region, but becomes tighter in the high force region. Consequently, the sensible tangential force region of the rubber-dome type is larger than that of the aluminum-dome type.

As described above, these two types have both advantages and disadvantages. Although the aluminum-dome type seems better if we focus on measuring precision in the low force region, we favor the rubber-dome type because there is room for improvement in sensing the characteristics through optimal design of the rubber dome.

3.3.3 Tracking-contact-area-movement type

As described in section 3.2.3, the tracking-contact-area-movement type is suitable for miniaturization. We provided three tactile sensors based on this type. First, we created a micro three-axis tactile sensor with conical feelers of Φ300 × 100 μm on the back surface of a rubber sheet. The distance between the conical feelers was 600 μm and the array size was 10 × 12 [24]. Figure 3.8(a) and (b) shows the normal and tangential force sensing characteristics, respectively. As shown in Figure 3.8(a), there is linearity between the grayscale value and applied force. On the other hand, in tangential force

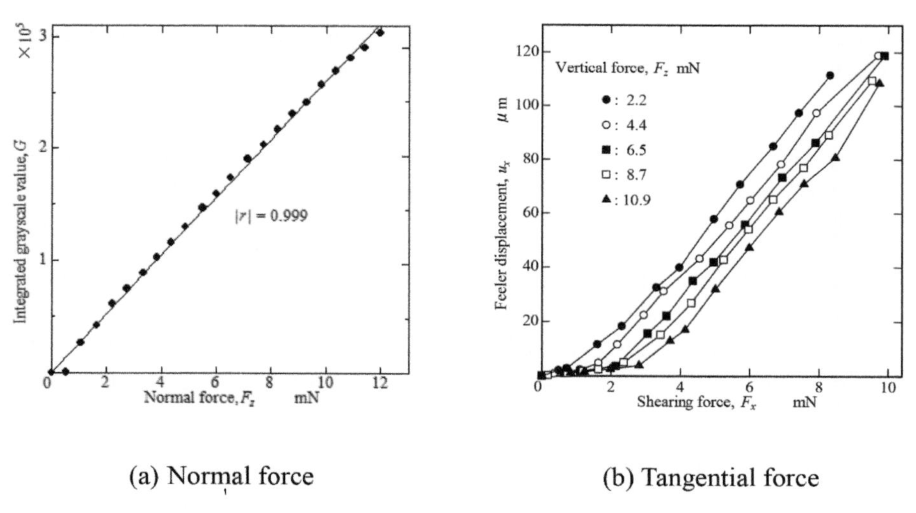

(a) Normal force (b) Tangential force

Figure 3.8 Normal and tangential force sensing characteristics of the micro-optical three-axis tactile sensor

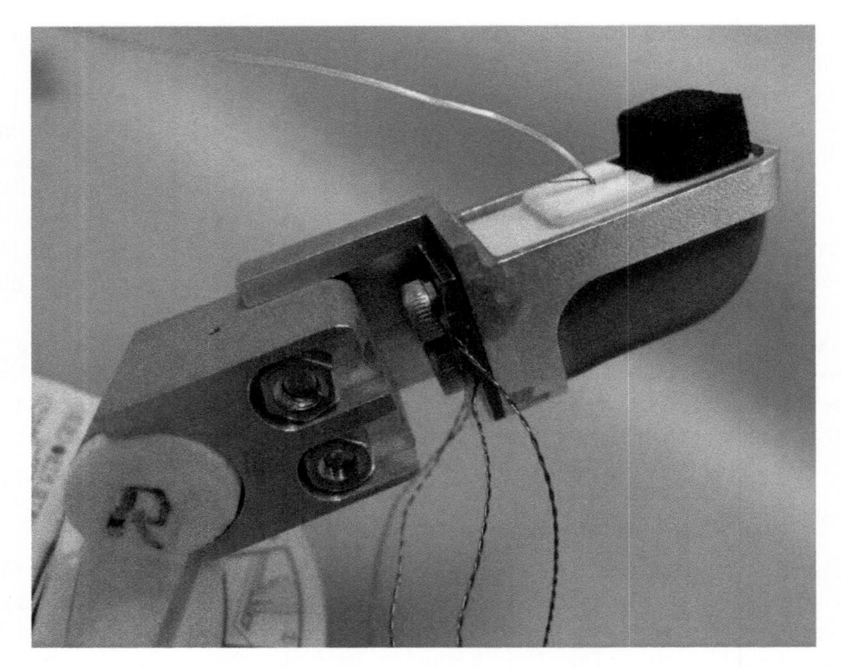

Figure 3.9 Fingertip three-axis tactile sensor is based on tracking-contact-area-movement type; this tactile sensor size is similar to an adult human's fingertip size

sensing, although the relationship between the centroid movement and tangential force is approximated as bilinear, the relationship shifts according to the applied normal force level. Since this type of tactile sensor uses contact area displacement of the rubber sheet on the acrylic plate, it is affected by basic physics, i.e., the maximum friction force increases with an increase in normal force. Although the relationship between the displacement and tangential force is not one curve, the applied tangential force value is uniquely determined because the relationship between the grayscale value and normal force is not affected by tangential force; if the applied normal force is determined by the grayscale value according to Figure 3.8 (a), then tangential force is determined by the displacement and obtained normal force according to Figure 3.8(b).

In the subsequent study, we used optical flow for the contact area movement analysis, while contact area movement was determined by labeling and momentum calculation for the former tactile sensor. We adopted the Lucas-Kanade method for contact movement analysis [25]. The sensing characteristics of this sensor were almost the same as the abovementioned tactile sensor. We tried to make this type of sensor resemble human fingers, as shown in Figure 3.9, because we want to apply to it a humanoid hand [26]. At present, we are improving this tactile sensor based on the experimental results of the preceding studies.

3.4 Applications

3.4.1 Tasks achieved by three-axis tactile sensing

Using three-axis tactile sensors, we conquered various tasks, such as fine step height recognition, cap twisting, object passing, and assembling using two hands. These tasks require three-axis force information. For example, a robot should notice variations in tangential force to determine several-microns step height because no significant difference occurs in the normal force variation for such a small step height. Other tasks, specifically, the object passing task, cap twisting task, and assembling task using two hands, need slippage force, which is calculated from the time derivative of the tangential force. Since we introduced these tasks in our other study [17], we will explain them briefly here. Recent new achievements are introduced in subsequent sections.

Fine step height recognition: based on the human tactile sensing mechanism, we formulated a neuron model incorporating a temporal summation effect, which was formulated by convolution of the step response [23]. In the evaluation experiments, a robotic manipulator equipped with the four-conical-feeler-type optical three-axis tactile sensor was able to find a 50-μm step height on a flat surface.

Cap twisting task: using two fingers equipped with the eight-conical-feeler-type optical three-axis tactile sensors, the robot twisted a cap on a PET-bottle to fasten it [28]. To perform this twisting, the fingertips moved along square trajectories, which were provided in the initial programing. During this task, if there was slippage on the fingertips, the fingertip positions of two fingers were moved to the inside of the cap to grasp it more tightly. Through repeated slippage and tighter grasping, the robot performed a circular trajectory along the cap's contour.

Object passing task: in order to prevent dropping grasped objects, if there is slippage on the fingers, the robot should increase the grasping force. On the other hand, when the robot hands the object over for delivery, it should reduce the grasping force. This means that increasing or decreasing should be decided according to the applied external force direction. The robot equipped with two fingers was able to release the grasped object to place it at a specific position if upward slippage occurred [29].

Assembling task using two hands: through a combination of the abovementioned cap twisting task and an object passing task, the robot completed assembling a bottle and a cap in the air [29]. First, the robot grasps a cap and a bottle in its right and left hands, respectively, and moves the cap toward the bottle. When the cap collides with the bottle, a tangential impact force is applied along the longitudinal bottle axis. The robot then begins to fasten the cap onto the bottle according to the aforementioned procedure because it notices that the tangential force signals the cap twisting task. When the cap twist becomes tight, the tangential force along the tangent of the cap's contour exceeds a threshold. The robot then pinches the cap to retrieve the assembled bottle. The fingers of the left hand feel the tangential force along the longitudinal bottle axis in the opposite direction of the former, and when the robot senses this tangential force on its left hand, the left hand releases the bottle.

3.4.2 Picking-up and counting paper

Turning flexible and thin objects such as paper is a well-known challenge because it is difficult for robots. Several studies on this issue have been conducted [30, 31]. One solution uses a special machine for paper turning [31]. However, the idea that a specific task requires a specific machine necessitates too many machines in the work place. In daily life, we prefer versatile machines such as humanoid robots. Therefore, our aim is to have an articulated-fingered hand handle papers based on three-axis tactile information.

In order to perform the abovementioned task, we used a two-hand-and-arm robot equipped with optical three-axis tactile sensors on each fingertip, as shown in Figure 3.10 [30]. The maximum tangential force generated on the paper is controlled by normal force because the Coulomb friction is low. We adopted force control based on position control. In Figure 3.10, the robot tries to turn and pick up one sheet from sheets stacked on a table. Since the bill's mechanical properties and size

Figure 3.10 A robot trying to pick up a bill from a stack on a table

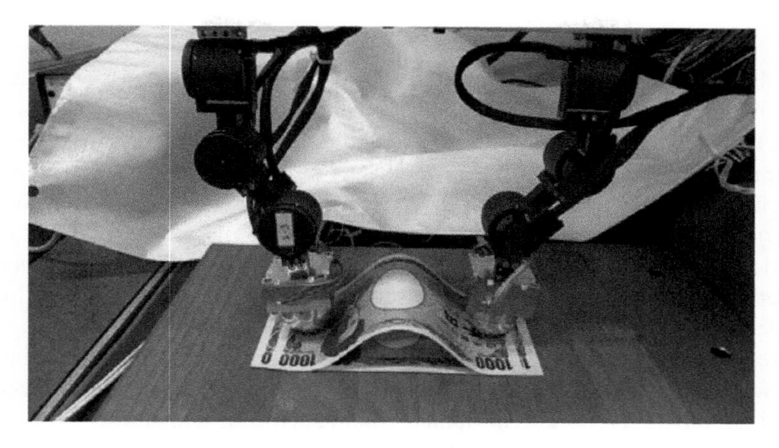

Figure 3.11 A bill turned from stacked bills

are very stable, we adopted ¥1 000 bill as specimen in this experiment. It was able to successively pick up one sheet, as shown in Figure 3.11.

After removing one bill from the stacked bills, the robot checked that it was really one. In order to count paper numbers, we built an algorithm for checking whether one sheet was picked up. Although the normal force distribution is changed according to the number of papers, when the robot slides fingers as shown in Figure 3.12, the difference between the maximum tangential force and the minimum tangential force measured at the start of sliding depends on whether one sheet or more than one sheet is picked up.

Figure 3.12 Evaluation of paper number using three-axis tactile sensing

3.4.3 Human-robot communication

It has been suggested that human orders to robots vary according to the human's needs in life. However, it is impossible for robot companies to provide robot programs that address each individual's need and circumstance. Furthermore, in order to command a robot, it provides stress to users to learn specific skills such as gestures: in a noisy environment, utterances are not always effective. To address these concerns, researchers strive to develop robots capable of recognizing multimodal sensation. As utterance recognition is being studied by many researchers, we focused our attention on nonverbal instruction. Research in nonverbal instruction includes the following: Kurita *et al.* achieved high-speed image tracking using hand gestures [32]; Ong and Ranganath surveyed recent automatic sign language analysis [33]; Shotton developed real-time 3D pose recognition [34]; Ito and Tsuji developed a mounted tactile pad to accept human commands [35]; Quintero *et al.* used the human body as a pointing device with Kinect to enable a robot to comprehend commands [36]; and Admoni and Scassellati found that nonverbal social cues, such as eye movement and gesture, are effective for socially assistive robots [37].

Taking this information into account, we progress to a new interaction between humans and robots using visual and tactile sensations. The robot introduced in the preceding section (Figure 3.10) performed a retrieving and passing task without requiring special training or specific gestures or utterances by users [38]. We assumed that a user stood in front of the robot and indicated his/her desire to make the robot place an object into his/her hand. In this case, the user naturally points out the object with his/her index finger. We therefore developed a finger direction recognition (FDR) system [39]. In order to accomplish the communication via tactile sensing, we combined the FDR system and the aforementioned object-passing task skill. Thus, if the operator wants to get a specific object and points it out, the robot recognizes it and takes it. The operator then opens his/her hand to receive the object, and the robot places it in their hand. When the object's bottom touches the hand, the robot releases the grasped object to accomplish a soft transfer. The sequential photographs in Figure 3.13 show the passing of the object.

3.5 Conclusion

Since unexpected events frequently occur in daily life, the demand for a tactile sensor will increase in order to avoid danger. Furthermore, people expect robots to be dexterous, which requires the use of tactile sensors. Robotics researchers have therefore sought the ideal tactile sensor, as demonstrated by Harmon [1], for over three decades. Although the optical tactile sensor explained in this chapter is not always ideal, it is suitable for actual use in robotics because of its high impact resistance. This is proven by the various applications of the optical three-axis tactile sensor that we described in the last chapter.

In this chapter, we explained various optical three-axis tactile sensors, specifically, the aluminum-dome type, rubber-dome type, and tracking-contact-area-movement type.

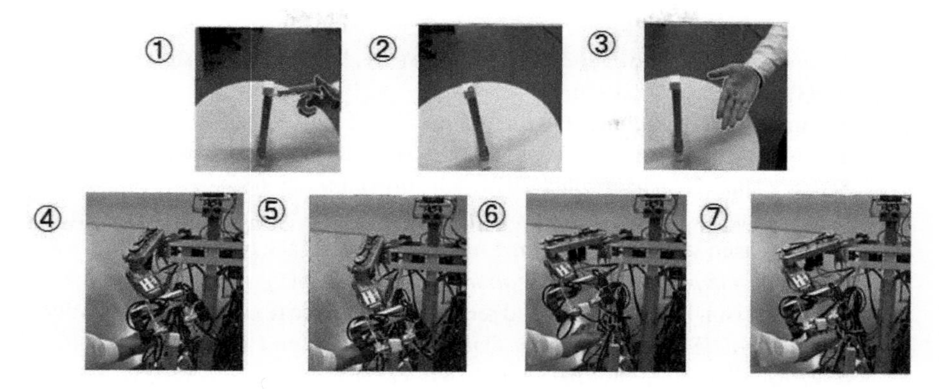

Figure 3.13 A robot handing over a cube to a hand. The person points out a cube using their index finger and opens their hand to signal the robot put the object on it (No. 1 – 3); the robot releases the cube when it accepts upward slippage caused by touching the cube bottom on the hand (No. 4 – 7).

We described these tactile sensors to help readers understand that a variety of sensing characteristics are caused by differences in the design of the rubber sheet that serves as the robot's skin. In other tactile sensor designs, the objectives of the rubber sheet covering the tactile sensors were impact resistance and adhesion ability. In contrast to these tactile sensors, the objectives of the rubber sheet for our optical tactile sensors are generating tactile sensation and the abovementioned objectives. In human skin, the skin has a similar function related to sensitivity as in Maeno's research [40]; the papilla between the epidermis and dermis of the skin plays a role in sensitivity enhancement. Therefore, these tactile sensor designs can be applied to other tactile sensor designs to enhance their characteristics in the future. We hope that our optical three-axis tactile sensor will progress with new technology and be useful for other tactile sensor designs.

References

[1] Harmon L.D. 'Automated tactile sensing'. *International Journal of Robotics Research*. 1982, vol. 1(2), pp. 3–32.

[2] Nicholls H.R., Lee M.H. 'A survey of robot tactile sensing technology'. *International Journal of Robotics Research*. 1982, vol. 8(3), pp. 3–30.

[3] Dahiya R.S., Valle M. *Robotic tactile sensing* [online]. Dordrecht: Springer; 1982. Available from http://link.springer.com/10.1007/978-94-007-0579-1

[4] Girão P.S., Ramos P.M.P., Postolache O., Miguel Dias Pereira J. ' Tactile sensors for robotic applications'. *Measurement*. 1982, vol. 46(3), pp. 71–1257.

[5] Rabman J., Morris K.A. 'Tactile and non-vision' in *A tactile sensor with electro-optical transduction, robot sensors*. Vol. 2. Kempston, Bedford: IFS(Pubs), Springer; 1986. pp. 55–145.

[6] Kampmann P., Kirchner F. 'Integration of fiber-optic sensor arrays into a multi-modal tactile sensor processing system for robotic end-effectors'. *Sensors (Basel, Switzerland)*. 2014, vol. 14(4), pp. 76–6854.

[7] Kamiyama K., Vlack K., Mizota T., Kajimoto H., Kawakami N., Tachi S. 'vision-based sensor for real-time measuring of surface traction fields'. *IEEE Computer Graphics and Applications*. 2005, vol. 25(1), pp. 68–75.

[8] Li R., Adelson E.H. 'Sensing and recognizing surface textures using a gelsight sensor'. *'IEEE Conference On Computer Vsiona And Pattern Recognition (CVPR)';* Portland, OR, 2013. pp. 1241–47.

[9] Ito Y., Kim Y., Obinata G. 'Contact region estimation based on a vision-based tactile sensor using a deformable touchpad'. *Sensors (Basel, Switzerland)*. 2014, vol. 14(4), pp. 5805–22.

[10] Kawashima T., Aoki Y. 'Electronics and communications in Japan (Part II)' in *An Optical Tactile Sensor Using the CT Reconstruction Method*. Vol. 70; 1987. pp. 35–43.

[11] Ohka M., Sawamoto Y., Zhu N. 'simulations of an optical tactile sensor based on computer tomography'. *Journal of Advanced Mechanical Design, Systems, and Manufacturing*. 2007, vol. 1(3), pp. 378–86.

[12] Sawamoto Y., Ohka M., Zhu N. 'sensing characteristics of an experimental CT tactile sensor'. *Journal of Advanced Mechanical Design, Systems, and Manufacturing*. 2008, vol. 2(3), pp. 454–62.

[13] Tanie K., Komoriya K., Kaneko M., Tachi S., Fujiwara A. 'Tactile and non-vision' in *A High Resolution Tactile Sensor, Robot Sensors*. Vol. 2. Kempston: IFS; 1986. pp. 190–98.

[14] Mott H., Lee M.H., Nicholls H.R. 'An experimental very-high-resolution tactile sensor array'. *Proceedings of the 4th International Conference on Robot Vision and Sensory Controls*; London, U.K, 1984. pp. 50–241.

[15] Nicholls H.R. *'Tactile sensing using an optical transduction method, traditional and non-traditional robot sensors'* in Springer-verlag; 1990.

[16] Maekawa H., Tanie K., Komoriya K., Kaneko M., Horiguchi C., Sugawara T. 'Development of a finger-shaped tactile sensor and its evaluation by active touch'. *Proceedings of the IEEE International Conference on Robotics and Automation*; 1992.

[17] Ohka, M., Yussof, H.B., Abdullah, S.C. *Three-Axis Tactile Sensor, Introduction to Modern Robotics II*. Concept Press Ltd; 2012. pp. 26–113.

[18] Ohka M., Kobayashi M., Shinokura T., Sagisawa S. 'Tactile expert system using a parallel-fingered hand fitted with three-axis tactile sensors'. *JSME International Journal. Ser. C, Dynamics, Control, Robotics, Design and Manufacturing*. 1994, vol. 37(1), pp. 46–138.

[19] Takeuchi S., Ohka M., Mitsuya Y. 'Tactile recognition using fuzzy production rules and fuzzy relations for processing image data from three-dimensional

tactile sensors mounted on a robot hand'. *Proceedings of the Asian Control Conference*; 1994. pp. 631–34.

[20] Ohka M., Morisawa N., Yussof H.B. 'Trajectory generation of robotic fingers based on tri-axial tactile data for cap screwing task'. *Presented at IEEE International Conference on Robotics and Automation (ICRA)*; Kobe, Japan, 2009.

[21] Ohka M., Mitsuya Y., Takeuchi S., Ishihara H., Kamekawa O. 'A three-axis optical tactile sensor (Fem contact analyses and sensing experiments using a large-size tactile sensor)'. *IEEE International Conference on Robotics and Automation*; 1995. pp. 24–817.

[22] Ohka M., Mitsuya Y., Matsunaga Y., Takeuchi S. 'sensing characteristics of an optical three-axis tactile sensor under combined loading'. *Robotica*. 2004, vol. 22(2), pp. 213–21.

[23] Ohka M., Takayanagi J., Kawamura T., Mitsuya Y. 'A surface-shape recognition system mimicking human mechanism for tactile sensation'. *Robotica*. 2006, vol. 24(5), pp. 595–602.

[24] Ohka M., Mitsuya Y., Higashioka I., Kabeshita H. 'An experimental optical three-axis tactile sensor for micro-robots'. *Robotica*. 2005, vol. 23(4), pp. 65–457.

[25] Ohka M., Matsunaga T., Nojima Y., Noda D., Hattori T. 'Basic experiments of three-axis tactile sensor using optical flow'. *IEEE International Conference on Robotics and Automation (ICRA)*; St Paul, MN, 2012. pp. 09–1404.

[26] Ohka M., Nomura R., Yussof H., Zahari N.I. 'Development of human-fingertip-sized three-axis tactile sensor based on image data processing'. *9th International Conference on Sensing Technology (ICST)*; Auckland, New Zealand: University of Auckland, 2015.

[27] Ohka M., Kobayashi H., Takata J., Mitsuya Y. 'An experimental optical three-axis tactile sensor featured with hemispherical surface'. *Journal of Advanced Mechanical Design, Systems, and Manufacturing*. 2008, vol. 2(5), pp. 73–860.

[28] Takagi A., Yamamoto Y., Ohka M., Yussof H., Abdullah S.C. 'Sensitivity-enhancing all-in-type optical three-axis tactile sensor mounted on articulated robotic fingers'. *Procedia Computer Science*. 2015, vol. 76, pp. 95–100.

[29] Ohka M., Abdullah S.C., Wada J., Yussof H.B. 'Two-hand-arm manipulation based on tri-axial tactile data'. *International Journal of Social Robotics*. 2012, vol. 4(1), pp. 97–105.

[30] Murakami K., Hasegawa T. 'Novel fingertip equipped with soft skin and hard nail for dexterous multi-fingered robotic manipulation'. *IEEE International Conference on Robotics & Automation*; Taipei, Taiwan, IEEE, 2003. pp. 13–708.

[31] Watanabe Y., Tamei M., Yamada M., Ishikawa M. 'Automatic page Turner machine for high-speed book digitization'. *IEEE/RSJ International Conference on Intelligent Robots and Systems (IROS 2013)*; Tokyo, 2013. pp. 79.

[32] Kurata T., Kato T., Kourogi M., Keechul J., Endo K. 'A functionally-distributed hand track-ing method for wearable visual interfaces and its applications' in IAPR *Workshop On Machine Vision Applications*; 2002. pp. 84–89.

[33] Ong S.C.W., Ranganath S. 'Automatic sign language analysis: a survey and the future beyond lexical meaning'. *IEEE Transactions on Pattern Analysis and Machine Intelligence.* 2005, vol. 27(6), pp. 91–873.

[34] Shotton J., Fitzgibbon A., Cook M, *et al.* 'Real-time human pose recognition in parts from single depth images'. *IEEE Conference on Computer Vision and Pattern Recognition (CVPR)*; Colorado Springs, CO, 2011. pp. 1297–304.

[35] Ito T., Tsuji T. 'command recognition of robot with low dimension whole-body haptic sensor'. *IEEJ Transactions on Industry Applications.* 2010, vol. 130(3), pp. 99–293.

[36] Quintero C.P., Fomena R.T., Shademan A., Wolleb N., Dick T., Jagersand M. 'SEPO: selecting by pointing as an intuitive human-robot command interface'. *IEEE International Conference on Robotics and Automation (ICRA)*; Karlsruhe, Germany, 2012. pp. 1158–63.

[37] Admoni H., Scassellati B. 'Data-driven model of nonverbal behavior for socially assistive human-robot interactions' [online]. Istanbul Turkey, New York. 2014. Available from https://dl.acm.org/doi/proceedings/10.1145/2663204

[38] Ikai T., Kamiya S., Ohka M. 'Robot control using natural instructions via visual and tactile sensations'. *Journal of Computer Science.* 2012, vol. 12(5), pp. 54–246.

[39] Ikai T., Ohka M., Yussof H. 'Behavior control of robot by human finger direction'. *Procedia Engineering.* 2012, vol. 41, pp. 91–784.

[40] Maeno T., Kobayashi K., Yamazaki N. 'Relationship between the structure of human finger tissue and the location of tactile receptors'. *JSME International Journal Series C.* 1998, vol. 41(1), pp. 94–100.

Chapter 4

Strain sensors for soft robotic applications

Oliver Ozioko[1] and Ravinder Dahiya[2]

4.1 Introduction

Strain sensors have recently received tremendous research interest due to their ability to support real-time monitoring of strain, which enables the detection of the movements of humans and robots body parts [1–5]. This capability makes them very useful for applications such as wearable systems [6], health monitoring [7–9], prosthesis [8, 10–12], human-robot interaction [13], gaming, and soft robotics [14]. In principle, strain sensors generally transduce mechanical stimuli into a proportional electrical signals that represents the strain, and interestingly, the recent advances in material science and nanotechnology have reformed the conventional brittle strain gauges into soft, stretchable strain sensors [15]. These soft and stretchable strain sensors have shown significant potential to use as sensory skin for soft robots, which has improved the granularity of information obtained from robot's interaction with its environment as well as human. To date, strain sensors with different materials [16], gauge factor (GF) [15], stretchability [17] and transduction mechanism, including piezoresistive [18, 19], capacitive [20], inductive [21], and optical [22], have been reported. Popular among these are the piezoresistive-type strain sensors, which are commonly realized using an active material (e.g., graphene [23], carbon nanotubes (CNTs) [24], Ag nanowires (NWs) [25], ZnO [26], carbon black [27], graphite [5, 28]) that provides the conductive network and a stretchable polymeric material (e.g., Ecoflex and polydimethylsiloxane (PDMS)) for stretchability [29–32].

The requirements for strain sensors may vary for different applications; for instance, wearable applications require strain sensors with high stretchability (>50%), high sensitivity, and high durability [3]. This is because such properties enable these sensors to accommodate the multiscale and dynamic deformations that

[1]Department of Electrical and Electronic Engineering, University of Derby, Derby, UK
[2]Bendable Electronics and Sensing Technologies (BEST) group, University of Glasgow, Glasgow, UK

may occur during use. A high GF, good mechanical robustness, large linear range, fast transient response, low hysteresis, and high stability are desired in an ideal stretchable strain sensor. To fabricate flexible and stretchable strain sensors with the desired properties, the nanofillers and polymer matrices are carefully chosen and are treated with various physical and chemical processing techniques. However, to obtain a certain set of desired characteristics, certain performance parameters are compromised, i.e., a trade-off always exists [33]. Other desirable properties of strain sensors for such wearable applications include high mechanical compliance and conformability to curved surfaces, such as the human skin. The sensitivity of a strain sensor plays a very big role in terms of its performance and is evaluated with GF, which is the ratio of the percentage change of sensor response to the applied strain [15]. Reportedly, the traditional silicon and metal oxide-based strain sensors have moderate GFs of about 1 800 but generally exhibit poor stretchability [9]. Today, the advances in material science and nanotechnology have enabled the realization of strain sensor with both high stretchability and GF. For instance, strain sensors with stretchability of the order of several hundred per cent of the linear strain [7, 14, 31, 32] and unprecedented GF of 1 834 140 [15] and 1.1×10^9 [34] have been reported. These sensors are also able to detect small- to large-scale strains, making them very attractive for soft robotics and human motion detection for monitoring small to large body movements such as facial expression and joint and limb movements.

4.2 Mechanisms for strain sensors

This section focuses on the mechanism that has been utilized for different types of strain sensors. In general, the resistive-type strain sensors depend on mechanisms such as intrinsic resistive properties of materials, tunneling effect, the disconnection of micromaterials/nanomaterials, and controlled propagation of microcracks in the sensing material itself [3, 33, 35, 36]. However, capacitive- and optical-type strain sensors mainly depend on the geometrical changes that occur during deformation.

4.2.1 Strain sensing based on intrinsic properties of materials and tunneling effect

Strain sensors based on the intrinsic properties of the active material are the fundamental property of semiconductor (e.g., germanium and silicon) materials and are referred to as the change in the electrical properties of a material as a result of external deformations [3, 33]. This change in resistance is a result of the corresponding change that occurs in the bandgap on the interatomic spacing of these semiconductor materials [3, 37].

Strain sensors also depend on the tunneling effect, which is a phenomenon of current that flows through adjacent conductive particles as long as their gaps are very small [38]. So, stain sensor realized using nanocomposites, for instance, have more than one conduction path. One is the electrical path that exists because of

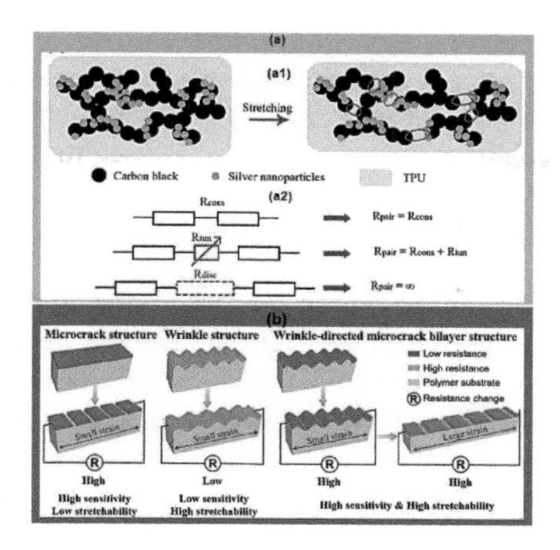

Figure 4.1 *(a) Disconnection mechanism reproduced with permission under the terms of the creative commons (CC-BY 4.0) [48]. Copyright 2018 (b) Microcrack propagation mechanism reproduced with permission [34] Copyright 2022, Elsevier.*

connected nanomaterials, and the other is the current flow because of the tunneling effect that occurs between adjacent nanomaterials [38–40].

4.2.2 Disconnection and microcrack propagation mechanism

As mentioned in the introduction, the strain sensing capability of the resistive-type strain sensors could be due to the changes that occur in the conducting network during deformation. Meanwhile, for there to be a conducting network for a nanomaterial-based resistive strain sensor, there is a threshold (percolation threshold) for the minimum number of nanomaterials required to establish a conducting path [41–44]. Upon deformation (stretching, bending, or twisting), some of the nanomaterials that form the conducting path experience disconnection from their overlapping area with neighboring particles due to stiffness mismatch between nanomaterials, thereby causing an increase in electrical resistance [3, 9, 38, 45]. This mechanism has been observed in sensors based on NWs [46, 47] and composites of carbon black and silver nanoparticles (CB/AgNPs) [48]. Figure 4.1(a) shows the structure and disconnection mechanism for a strain sensor realized using composites of CB/AgNPs as sensing layer and thermoplastic polyurethane for stretchability [48].

Figure 4.1(a1) shows the sensor structure before and after stretching with the red circles showing separation between pairs of nanoparticles in the material during stretching. Figure 1.1(b) shows the different resistance (R_{con}, R_{tun}, and R_{disc}) that plays the role of sensing in the material. If R_{pair} is regarded as the overall resistance between pairs of nanoparticles, it will vary depending on the state of the sensor as shown in Figure 1.1(a). R_{cons} is the constant resistance of the composite material which does not change during stretching, R_{disc} is the resistance due to the total disconnection that occurs between pair of nanoparticles during stretching, and R_{tun}

is the resistance due to tunneling effect. So, when the sensor is stretched, there is a disconnection between pairs of nanoparticles which increases the overall electrical resistance of the sensor (R_{pair}). Hence, R_{pair} becomes the sum of R_{tun} and R_{cons}. R_{tun}, which is a variable, continues to change in accordance to the strain. This change continues until electrons are no longer able to flow in the conducting network; at this point, R_{pair} becomes infinite (Figure 1.1(b)) [48].

Recently, the introduction of microcracks into strain sensing materials has been explored as a way to enhance the sensitivity of strain sensors. Microcracks occur when there is a mechanical mismatch between the sensing layer and the soft substrate [3, 34, 49, 50]. Popularly, this kind of strain sensors is realized by coating flexible substrates with thin brittle sensing films which can crack under minute deformations. However, a serious drawback of strain sensors based on microcracks mechanism is the narrow strain sensing range, often lower than 20%, due to the intrinsic lack of stretchability of the brittle thin sensing films coated on the soft substrates [34]. The GF strain sensors have been enhanced by using mechanically heterogeneous substrates with heterogeneous mechanical properties [51], introducing thickness gradient [52] and customized curvatures [34]. Figure 1.1(b) shows the microcrack and wrinkle mechanism used to realize a stretchable strain sensor [34]. The strain sensor was realized using a prewrinkled graphene oxide layer and a highly susceptible layer of conformally deposited AgNPs. When stretched, the prewrinkled interlayer guides preferential microcracking at the wavy troughs of the conformal Ag surface layer, which led to high sensitivity and stretchability [34].

4.3 Classification of strain sensors

Strain sensors could be classified based on the underlying transduction mechanism such as capacitive, piezoresistive, piezoelectric, optical, and triboelectric [53]. Among these, the capacitive and piezoresistive-type sensors are the most common, due to their high sensitivity, ease of fabrication, diverse choice of materials, designs, and use in diverse environments and applications [11], while the piezoelectric and triboelectric are rarely explored as they are not able to capture static strain because they operate at high frequencies.

4.3.1 Piezoresistive strain sensors

Piezoresistive strain sensors work by converting mechanical strain as a proportional change in resistance of the sensing material. They are primarily composed of conductive nanocomposites whose electrical resistance changes when stretched, bent, or twisted as a result of the reversible changes that occur in their conducting network [33]. When released, the resistance of these sensors is able to return to their initial state. The common active materials used for these type of strain sensors are often in the form of conducting micromaterials/nanomaterials–polymers composites [54], thin films, or conductive yarns/fabrics [37, 55]. These types of strain sensors are quite common and have been explored for decades. In 1940, it was initially based on metal foils where it could only detect limited strains up to 5%, such as small

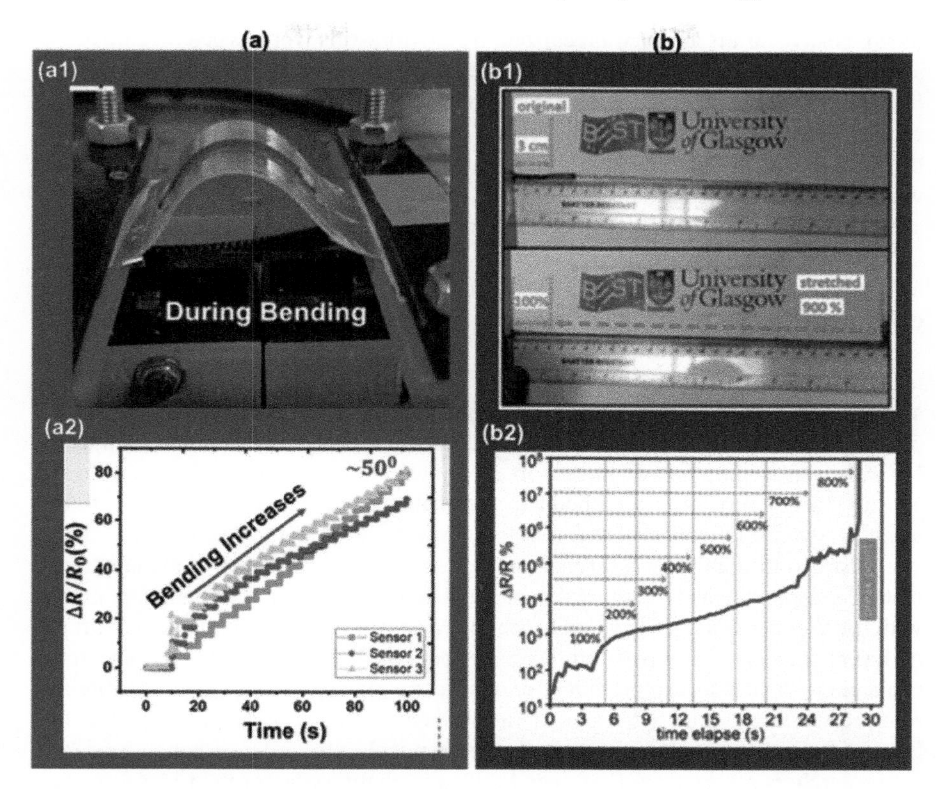

Figure 4.2 *(a) Flexible piezoresistive bend sensor for soft crawling robot reproduced with permission [57] Copyright 2022, IEEE.*
(b) Bioinspired strain sensor for earthworm- and inchworm-type soft robot. Reproduced with permission under the terms of the creative commons (CC BY 4.0) [4] Copyright 2021.

deformations in composites and rigid bodies [3]. Today, piezoresistive-type strain sensors have significantly advanced from brittle to flexible and stretchable format and have been explored for numerous stretchable and wearable applications. The stretchable strain sensor is mainly composed of active sensing material, such as graphene, CNTs [56], etc., and a stretchable material, such as Ecoflex and PDMS, etc., for stretchability [29–32].

A number of piezoresistive strain sensors with different materials have been explored. Figure 4.2(a) shows a strain sensor realized on a flexible (but not stretchable) material.

This bend sensor was proposed for the unique measurement of undulations of a crawling robot whose locomotion is controlled using magnetic force coming from the interaction of a flexible coil and a flexible magnet. The strain sensor was fabricated using ~80 μm-thick polyvinylchloride substrate and a graphite solution prepared using Polyvinylidene fluoride (PVDF) as a binder. In this work a 5 × 5 cm-wide piezoresistive strain sensor was realized [57]. Unlike this work, when piezoresistive strain sensors are fabricated using stretchable substrates, a number of them have the limitation of distinguishing stretchability and bendability as both conditions lead to change in resistance. So, the work reported in Reference [57] is focused mainly on the measurement of bending angle and so is primarily advantageous for applications where stretchability is not necessary. However, if stretchability and conformability are key, these nonstretchable type strain sensors are limited.

To achieve stretchability, researchers have fabricated these sensors on flexible and stretchable materials [4, 9, 38]. Figure 1.2(b) shows a highly stretchable and conformable bioinspired strain sensor for soft robotic applications [4]. The strain sensor reported a high stretchability of up to 900%. The developed strain sensor was used to realize a bioinspired closed-loop controlled inchworm- and earthworm-like soft structures with intrinsic strain sensing capability. The movement of the demonstrated inchworm and earthworm was controlled using tiny permanent magnets incorporated at the ends of these soft structures. This movement is based on the sensory-feedback from the embedded stretchable strain sensor. This type of self-sensing advances the field of soft robotics towards cognitive soft robotics.

Apart from graphite, researchers have also explored other active materials. For instance, a piezoresistive strain sensor has been realized by impregnating PDMS with graphene nanopaper [58], and recently, a combination of graphene and graphite has also been used to realized strain sensors [59]. Instead of PDMS as substrate, researchers have also explored the use of silicone sealant as a flexible substrate and reduced graphene oxide (rGO) as the active sensing material [9]. In this case, few layers of rGO were synthesized by electrochemical exfoliation of pencil lead, and the researchers reported a GF of over 4 000, stretchability larger than 100%, as well as durability longer than 1 600 cycles at 100% strain. As presented in this work, the use of silicone sealant as the substrate helped to enhance the overall stretchability of the strain sensor. A stretchable strain sensor based on overlapped CNT bundles coupled with a silicone elastomer has also been explored [45]. The presented strain sensor was prepared by synthesizing line-patterned vertically aligned CNT bundles and rolling and transferring them to the silicone elastomer. When stretched, there will be a sliding and disconnection of the overlapped CNTs which cause a change in the electrical resistance. The sensor has a sensing range above 45% strain and a GF of 42 300 at a strain of 125–145%.

Recently, an ultra-high GF strain sensor with wide-range stretchability has been reported as shown in Figure 1.3(a) [15]. The stretchable strain sensors are capable of operating over multistrain range and have an excellent GF (up to 1 834 140). In this work, different combinations of elastomer and filler particles were studied. The result presented for the different combinations of the elastomer, conductive filler and graphene–carbon paste (GCP) showed an excellent stretchability when only filler particles are present in the elastomer and high sensitivity when filler particles are mixed with GCP. The capability of the strain sensor to detect small- to large-scale strains makes it promising for application in robotics and human motion detection for the measurement of facial expressions and movement of limbs and joints. Researchers have also modified the structures of soft materials to enhanc

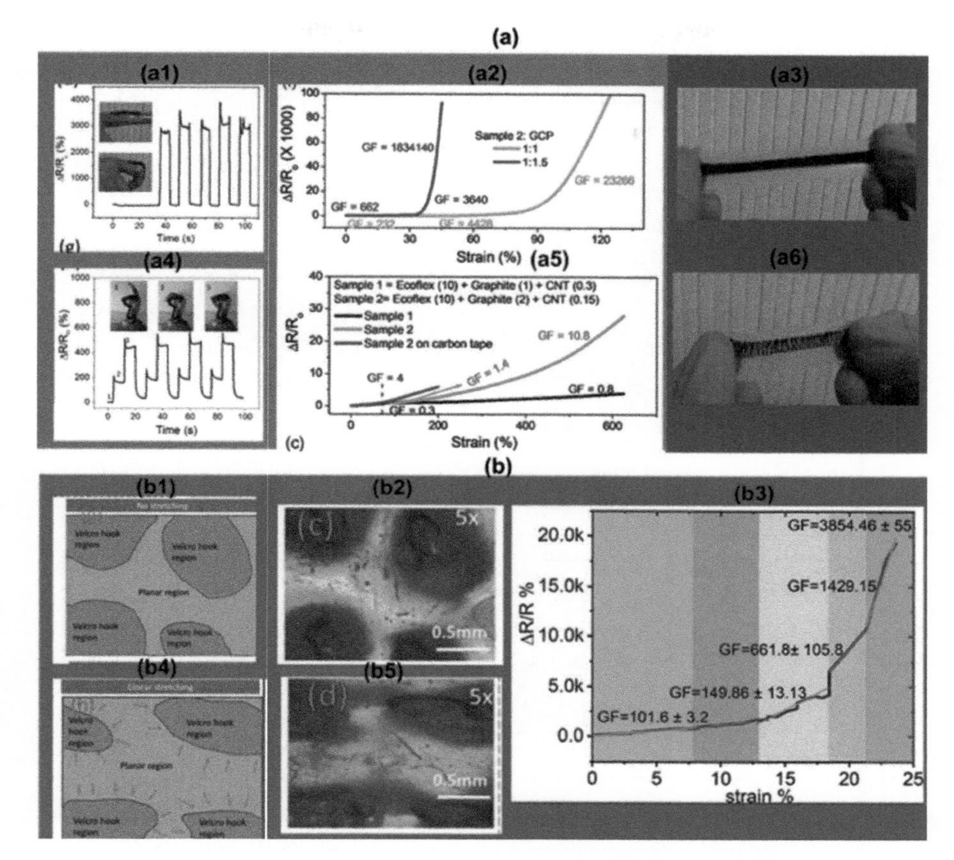

Figure 4.3 (a) Stretchable strain sensor with ultra-high gauge factor. Reproduced with permission under the terms of the ceative commons (CC BY 4.0) [15] Copyright 2022. (b) Graphite-based bioinspired piezoresistive soft strain sensors with performance optimized for low strain value [5].

the performance of the resistive-type strain sensors. For instance, a custom-made graphite-based piezoresistive strain sensor with gecko foot-inspired macroscopic features was realized using a Velcro tape on Ecoflex substrate (Figure 4.3(b)) [5]. 3D-printed strain sensor has also been explored [60, 61]. The use of 3D printing is advantageous as it has the ability for rapid prototyping and creates opportunity for the seamless integration of the strain sensor into robotic structure during printing.

4.3.2 Capacitive-type strain sensors

Capacitive strain sensors respond to changes in the strain field by changing the dielectric properties of sandwiched functional nanocomposite films. Conventionally, the capacitive-type strain sensors could be regarded as a parallel plate capacitor

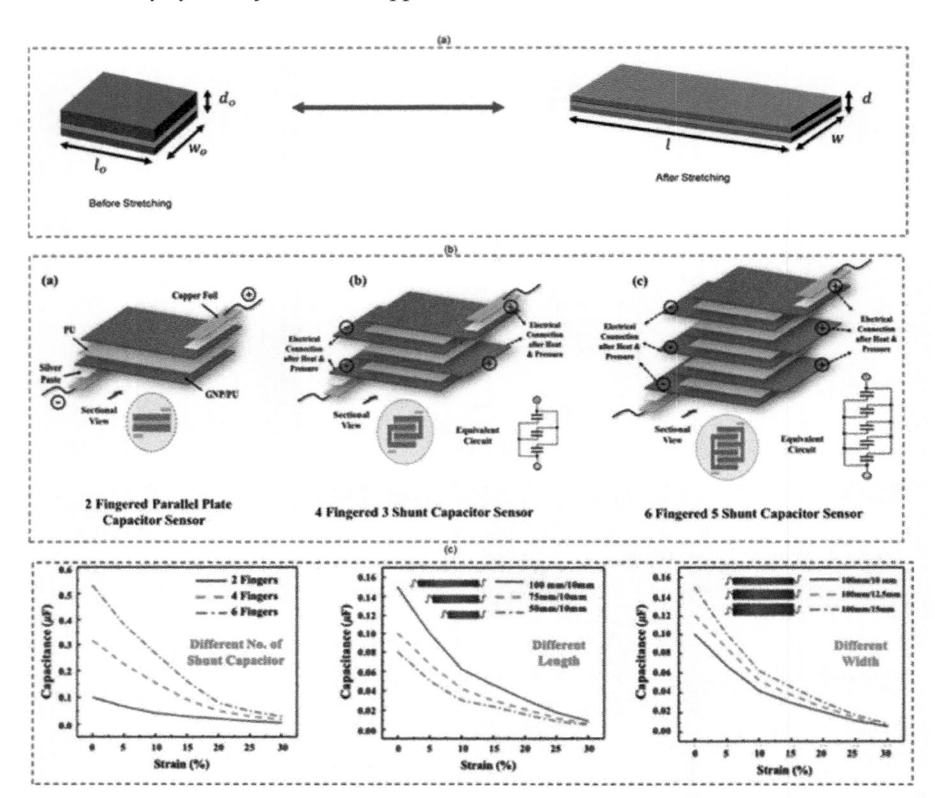

Figure 4.4 *Capacitive-type strain sensor. (a) Structure of a parallel plate capacitive sensor before and after stretching. (b) Schematic diagram of the capacitive sensor with different layout designs. Reproduced with permission [63] Copyright 2020, Elsevier. (c) Capacitance signal change with different layout designs, lengths or widths under different strain rate. Reproduced with permission [63] Copyright 2020, Elsevier.*

whose change in capacitance is based on the geometrical deformation [62]. Flexible strain sensors are currently being used in a multitude of applications in fields from healthcare to sports, soft robotics and smart textiles [33, 62].

Figure 4.4(a) shows the structure and mechanism of a parallel plate capacitive strain sensor [62, 63]. Equation (4.1) shows an expression for the initial capacitance, C_o, of a parallel plate capacitor of length l_o where, w_o, is the width of the plate, d_o is the distance between the electrodes, ε_o is the vacuum permittivity and ε_r is the permittivity of dielectric [62].

$$C_o = \varepsilon_o \varepsilon_r \frac{w_o l_o}{d_o} \tag{4.1}$$

When a uniaxial strain E is applied to the sensor, the length will increase to $(1 + E)\, l_o$, while the width will decrease to $(1 - V_{Electrode}E)\, w_o$, and the thickness of the capacitor will also decrease to $(1 - V_{dielectric}E)\, d_o$.$V_{Electrode}$ and $V_{dielectric}$ are the Poisson ratios of the electrode and dielectric, respectively. Assuming that $V_{Electrode}$ is equal to $V_{dielectric}$, the capacitance, C, and strain can be estimated as a linear relationship [62, 63].

$$C = \frac{\varepsilon_0 \varepsilon_r (1 + E) l_0 (1 - V_{Electrode}E) w_0}{(1 - V_{dielectric}E) d_0} = \varepsilon_0 \varepsilon_r \frac{(1 + E) w_0 l_0}{d_0} = (1 + E)\, C_o \tag{4.2}$$

The GF is given as

$$\mathrm{GF} = \frac{\Delta C}{C_o} \frac{1}{E} = \frac{(1 + E)\, C_o - C_o}{C_o E} = 1 \tag{4.3}$$

Theoretically, it could be seen from (4.3) that the maximum GF for parallel plate capacitive strain sensor is typically equal to one, and the majority of the reported capacitive strain sensors have GF <1 [62, 64, 65]. However, by modifying the sensor structure, capacitive strain sensors with GF >1 have been realized [62, 63]. For instance, in Reference [62], a capacitive-type strain sensor with a GF of 2.07 at 300% was realized. The sensor was fabricated using out-of-plane wrinkles formed by prestretched method. When a uniaxial strain is applied on the sensor, its structure changes from its original compact form to a planar form. The authors also proposed a modified formula for assessing the relationship of capacitance and tensile strain and demonstrated its application for human motion tracking and monitoring of soft robots. Figure 1.4(b) and 1.4(c) shows the layout design and results of a capacitive-type strain sensor fabricated using graphite nanoplatelet. The sensor has a GF of ~3.5. Additionally, the authors compared the performance of the layout design of shunt capacitors and the pseudo-interdigital capacitor sensor and compared it to the parallel plate capacitor [63].

4.3.3 Triboelectric-type strain sensors

Triboelectric-type strain sensors which use triboelectric nanogenerators and operate based on the triboelectric effect [66] are gradually gaining attention. The self-powered capability of this sensor gives it advantage over the resistive-, capacitive- and optical-type strain sensors. Triboelectric strain sensors using both sliding mode and contact mode [67] mechanism have been explored. In Reference [67], a contact-mode triboelectric self-powered strain sensor using an auxetic polyurethane foam, conductive fabric and polytetrafluroethylene (PTFE) was presented. When stretched, the auxetic polyurethane foam would expand into the PTFE, causing contact electrification. The device works as strain sensor because of the larger contact area between the PTFE and the foam when stretched.

Figure 4.5 shows a triboelectric strain sensor developed by introducing helical structure-braided fibers on a stretchable substrate fiber [68]. The sensor was fabricated by twinning the shell fibers (PTFE fiber or nylon fiber) around the core fiber

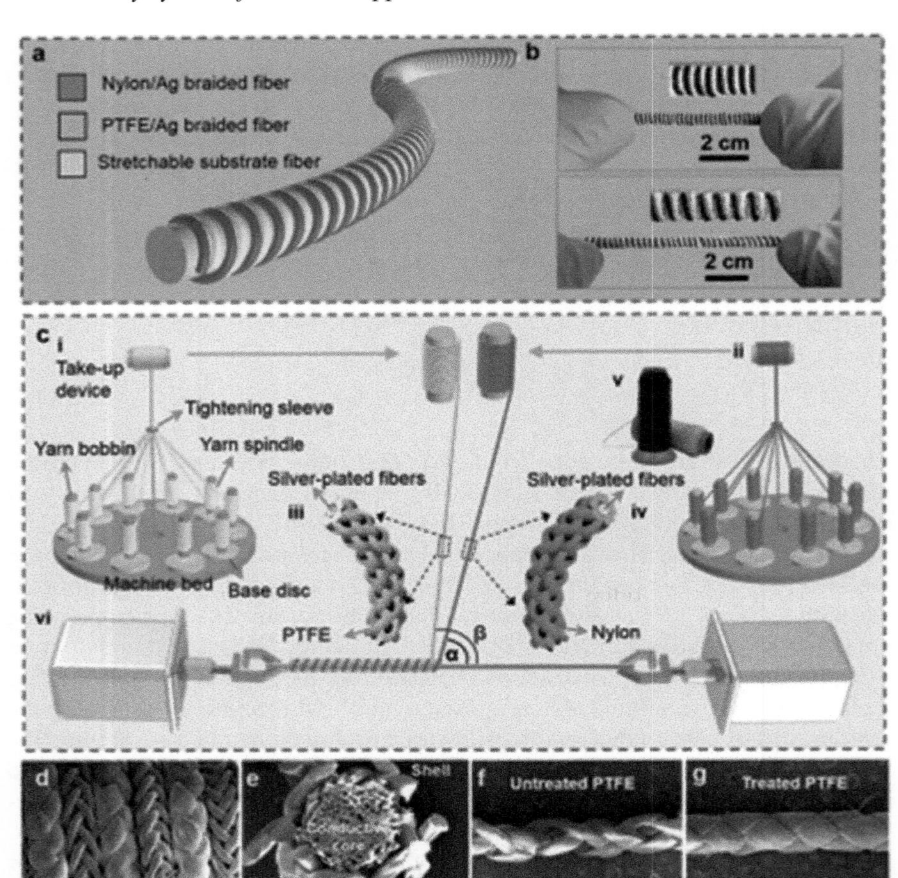

Figure 4.5 Triboelectric-type strain sensor. Reproduced with permission [68] Copyright 2022, American Chemical Society.

(silver-plated fiber) and both interwoven using a multiaxial fiber winding machine. Following this, two similar motors were each attached on the opposite ends of the stretchable substrate. Finally, a pair motors were used to alternately wind the nylon/Ag- and PTFE/Ag-braided fibers on the substrate fiber to form the required fiber.

4.4 Conclusion

This chapter summarizes some recent developments in strain sensors for soft robotic applications including the different strain sensing mechanisms, fabrication approach, and key results. Resistive-type strain sensors are quite common and generally have good sensitivity. However, they often suffer from hysteresis and nonlinear electromechanical response. The performance of resistive-type strain sensors has been improved through the use of advanced nanomaterials, structural engineering, and fabrication approach. On the other hand, capacitive-type sensors offer excellent stretchability, linearity, and negligible hysteresis, but they have poor sensitivity. Further, as the interest in self-powered systems continues to rise, triboelectric-type strain sensors are gradually gaining attention. However, their sensitivity is still low when compared to the resistive-type strain sensors.

For the application of strain sensors for robotic applications, a number of technical challenges which prevent its application for real-life application still exists. First, it is still challenging to realize a stretchable strain sensor that has the ability to measure decoupled multidirectional and multiplane deformations [3]. Solving this challenge will be a breakthrough for applications such as soft robotics, considering the stretchable and conformable nature of soft robots. Rather than the conventional architectures and materials, researches could strive to realize more advanced sensing architectures with 3D structures and metamaterials. Some other key novel features to introduce into the available strain sensors include, high sensitivity, nonlinearity, self-healing, and to adopt a more reliable system integration approach. Most strain sensors are also susceptible to unwanted pressure as well as variations in environmental conditions such as changes in temperature and humidity [3]. The solution to such environment challenges could come from the use of a more advanced packages.

References

[1] Dahiya R. 'E-skin: from humanoids to humans [point of view]'. *Proceedings of the IEEE*. 2020, vol. 107(2), pp. 247–52.

[2] Dahiya R., Akinwande D., Chang J.S. 'Flexible electronic skin: from humanoids to humans [scanning the issue]'. *Proceedings of the IEEE*. 2020, vol. 107(10), pp. 15–2011.

[3] Souri H., Banerjee H., Jusufi A, *et al.* 'Wearable and stretchable strain sensors: materials, sensing mechanisms, and applications [online]'. *Advanced Intelligent Systems*. 2020, vol. 2(8), p. 2000039. Available from https://onlinelibrary.wiley.com/toc/26404567/2/8

[4] Karipoth P., Christou A., Pullanchiyodan A., Dahiya R. 'Bioinspired inchworm- and earthworm-like soft robots with intrinsic strain sensing [online]'. *Advanced Intelligent Systems*. 2022, vol. 4(2), p. 2100092. Available from https://onlinelibrary.wiley.com/toc/26404567/4/2

[5] Karipoth P., Pullanchiyodan A., Christou A., Dahiya R. 'Graphite-based bioinspired piezoresistive soft strain sensors with performance optimized for low strain values'. *ACS Applied Materials & Interfaces*. 2021, vol. 13(51), pp. 61610–19.

[6] Ozioko O., Dahiya R. 'Smart tactile gloves for haptic interaction, communication, and rehabilitation [online]'. *Advanced Intelligent Systems*. 2022, vol. 4(2), p. 2100091. Available from https://onlinelibrary.wiley.com/toc/26404567/4/2

[7] Escobedo P., Bhattacharjee M., Nikbakhtnasrabadi F., Dahiya R. 'Smart bandage with wireless strain and temperature sensors and batteryless NFC tag'. *IEEE Internet of Things Journal*. 2022, vol. 8(6), pp. 100–5093.

[8] Pullanchiyodan A., Manjakkal L., Ntagios M., Dahiya R. 'MnOx-electrodeposited fabric-based stretchable supercapacitors with intrinsic strain sensing'. *ACS Applied Materials & Interfaces*. 2021, vol. 13(40), pp. 47581–92.

[9] Verma R.P., Sahu P.S., Rathod M., Mohapatra S.S., Lee J., Saha B. 'Ultra-Sensitive and highly stretchable strain sensors for monitoring of human physiology'. *Macromolecular Materials and Engineering*. 2022, vol. 307(3).

[10] Tai H., Duan Z., Wang Y., Wang S., Jiang Y. 'Paper-based sensors for gas, humidity, and strain detections: A review'. *ACS Applied Materials & Interfaces*. 2020, vol. 12(28), pp. 31037–53.

[11] Boutry C.M., Kaizawa Y., Schroeder B.C., *et al*. 'A stretchable and biodegradable strain and pressure sensor for orthopaedic application'. *Nature Electronics*. 2018, vol. 1(5), pp. 314–21.

[12] Karipoth P., Pullanchiyodan A., Christou A., Dahiya R. 'Graphite-based bioinspired piezoresistive soft strain sensors with performance optimized for low strain values'. *ACS Applied Materials & Interfaces*. 2021, vol. 13(51), pp. 61610–19.

[13] Bhattacharjee M., Soni M., Escobedo P., Dahiya R. Pedot: 'PSS microchannel-based highly sensitive stretchable strain sensor [online]'. 2020. Available from https://onlinelibrary.wiley.com/toc/2199160x/6/8

[14] Karipoth P., Christou A., Pullanchiyodan A., Dahiya R. 'Bioinspired inchworm and earthworm like soft robots with intrinsic strain sensing'. *Advanced Intelligent Systems*. 2021.

[15] Kumaresan Y., Mishra S., Ozioko O., Chirila R., Dahiya R. 'Ultra-high gauge factor strain sensor with wide-range stretchability'. *Advanced Intelligent Systems*. 2018.

[16] Han F., Li M., Ye H., Zhang G. 'Materials, electrical performance, mechanisms, applications, and manufacturing approaches for flexible strain sensors'. *Nanomaterials (Basel, Switzerland)*. 2021, vol. 11(5), pp. 5–11.

[17] Amjadi M., Kyung K.U., Park I., Sitti M. 'Stretchable, skin-mountable, and wearable strain sensors and their potential applications: a review'. [online] *Advanced Functional Materials*. 2016, vol. 26(11), pp. 1678–98. Available from http://doi.wiley.com/10.1002/adfm.v26.11

[18] Georgopoulou A., Clemens F. 'Piezoresistive elastomer-based composite strain sensors and their applications'. *ACS Applied Electronic Materials*. 2020, vol. 2(7), pp. 1826–42.

[19] Dubey P.K., Yogeswaran N., Liu F., Vilouras A., Kaushik B.K., Dahiya R. 'Monolayer $MoSe_2$-based tunneling field effect transistor for ultrasensitive strain sensing'. *IEEE Transactions on Electron Devices*. 2020, vol. 67(5), pp. 46–2140.

[20] Atalay A., Sanchez V., Atalay O, *et al*. 'Batch fabrication of customizable silicone-textile composite capacitive strain sensors for human motion tracking

[online]'. *Advanced Materials Technologies*. 2017, vol. 2(9), p. 1700136. Available from https://onlinelibrary.wiley.com/toc/2365709x/2/9

[21] Tavassolian M., Cuthbert T.J., Napier C., Peng J., Menon C. 'Textile-based inductive soft strain sensors for fast frequency movement and their application in wearable devices measuring multiaxial hip joint angles during running [online]'. *Advanced Intelligent Systems*. 2020, vol. 2(4), p. 1900165. Available from https://onlinelibrary.wiley.com/toc/26404567/2/4

[22] Kamita G., Frka-Petesic B., Allard A., *et al*. 'Biocompatible and sustainable optical strain sensors for large-area applications'. *Advanced Optical Materials*. 2016, vol. 4(12), pp. 1950–54. Available from http://doi.wiley.com/10.1002/adom.v4.12

[23] Lee S.W., Park J.J., Park B.H. 'Enhanced sensitivity of patterned graphene strain sensors used for monitoring subtle human body motions'. *ACS Applied Materials & Interfaces*. 2017, pp. 83–11176.

[24] Gao Y., Fang X., Tan J., Lu T., Pan L., Xuan F. 'Highly sensitive strain sensors based on fragmentized carbon nanotube/polydimethylsiloxane composites'. *Nanotechnology*. 2018, vol. 29(23), 235501.

[25] Amjadi M., Pichitpajongkit A., Lee S., Ryu S., Park I. 'Highly stretchable and sensitive strain sensor based on silver nanowire-elastomer nanocomposite'. *ACS Nano*. 2014, vol. 8(5), pp. 5154–63.

[26] Chen Q., Sun Y., Wang Y., Cheng H., Wang Q.-M. 'ZnO nanowires–polyimide nanocomposite piezoresistive strain sensor'. *Sensors and Actuators A*. 2013, vol. 190, pp. 67–161.

[27] Zhai W., Xia Q., Zhou K. 'Multifunctional flexible carbon black/polydimethylsiloxane piezoresistive sensor with ultrahigh linear range, excellent durability and oil/water separation capability'. *Chemical Engineering Journal*. 2013, vol. 372, pp. 82–373.

[28] Li W., Guo J., Fan D. '3D graphite–polymer flexible strain sensors with ultrasensitivity and durability for real-time human vital sign monitoring and musical instrument education'. *Advanced Materials Technologies*. 2017, vol. 2(6), p. 1700070. Available from https://onlinelibrary.wiley.com/toc/2365709x/2/6

[29] Dahiya R.S., Valle M. *Robotic Tactile Sensing* [online]. Dordrecht: Springer Science + Business Media; 2013. pp. 1–245. Available from http://link.springer.com/10.1007/978-94-007-0579-1

[30] Fu X., Al-Jumaily A.M., Ramos M., Meshkinzar A., Huang X. 'Stretchable and sensitive sensor based on carbon nanotubes/polymer composite with serpentine shapes via molding technique'. *Journal of Biomaterials Science. Polymer Edition*. 2019, vol. 30(13), pp. 1227–41.

[31] Duan L., D'hooge D.R., Cardon L. 'Recent progress on flexible and stretchable piezoresistive strain sensors: from design to application'. *Progress in Materials Science*. 2020, vol. 114, 100617.

[32] Souri H., Banerjee H., Jusufi A, *et al*. 'Wearable and stretchable strain sensors: materials, sensing mechanisms, and applications [online]'. *Advanced Intelligent Systems*. 2020, vol. 2(8), p. 2000039. Available from https://onlinelibrary.wiley.com/toc/26404567/2/8

[33] Khalid M.A.U., Chang S.H. 'Flexible strain sensors for wearable applications fabricated using novel functional nanocomposites: a review'. *Composite Structures*. 2020, vol. 284, 115214.
[34] Li L., Zheng Y., Liu E. 'Stretchable and ultrasensitive strain sensor based on a bilayer wrinkle-microcracking mechanism'. *Chemical Engineering Journal*. 2020, vol. 437, 135399.
[35] Cheng H.W., Yan S., Shang G., Wang S., Zhong C.J. 'Strain sensors fabricated by surface assembly of nanoparticles'. *Biosensors & Bioelectronics*. 2021, vol. 186, p. 113268.
[36] Yang T., Li X., Jiang X., *et al.* 'Structural engineering of gold thin films with channel cracks for ultrasensitive strain sensing'. *Materials Horizons*. 2016, vol. 3(3), pp. 248–55.
[37] Seyedin S., Zhang P., Naebe M. 'Textile strain sensors: a review of the fabrication technologies, performance evaluation and applications'. *Materials Horizons*. 2016, vol. 6(2), pp. 49–219.
[38] Yi Y., Wang B., Liu X., Li C. 'Flexible piezoresistive strain sensor based on cnts–polymer composites: a brief review'. *Carbon Letters*. 2022, pp. 1–14.
[39] Wang T., Liu Y., Liu H., Liu C. 'Variations of tunnelling resistance between cnts with strain in composites: non-monotonicty and influencing factors'. *Nanotechnology*. 2022, vol. 33(40).
[40] Hu N., Karube Y., Yan C., Masuda Z., Fukunaga H. 'Tunneling effect in a polymer/carbon nanotube nanocomposite strain sensor'. *Acta Materialia*. 2016, vol. 13, pp. 36–2929.
[41] Lv Z., Huang X., Fan D., Zhou P., Luo Y., Zhang X. 'Scalable manufacturing of conductive rubber nanocomposites with ultralow percolation threshold for strain sensing applications'. *Composites Communications*. 2021, vol. 25, p. 100685.
[42] Hu H., Ma Y., Yue J., Zhang F. 'Porous GNP/PDMS composites with significantly reduced percolation threshold of conductive filler for stretchable strain sensors'. *Composites Communications*. 2021, 101033.
[43] Bozyel I., Gokcen D. 'Determining electrical percolation threshold of randomly distributed conductor materials in polymer composites via pathfinding algorithms'. *Composites Science and Technology*. 2021, 109404.
[44] Khan T., Irfan M.S., Ali M., Dong Y., Ramakrishna S., Umer R. 'Insights to low electrical percolation thresholds of carbon-based polypropylene nanocomposites'. *Carbon*. 2021, vol. 176, pp. 602–31.
[45] Lee J., Pyo S., Kwon D.S., Jo E., Kim W., Kim J. 'Ultrasensitive strain sensor based on separation of Overlapped carbon nanotubes [online]'. *Small (Weinheim an Der Bergstrasse, Germany)*. 2019, vol. 15. Available from https://onlinelibrary.wiley.com/toc/16136829/15/12
[46] Cao Y., Lai T., Teng F., Liu C., Li A. 'Highly stretchable and sensitive strain sensor based on silver nanowires/carbon nanotubes on hair band for human motion detection'. *Progress in Natural Science*. 2021, vol. 31(3), pp. 379–86.

[47] Li C., Wang G., Guo S. 'Ag nanowire-based omnidirectional stretchable sensing array for independent pressure–strain detection'. *ACS Applied Nano Materials*. 2022, vol. 5, pp. 88–6980.
[48] Zhang W., Liu Q., Chen P. 'Flexible strain sensor based on carbon black/silver nanoparticles composite for human motion detection'. *Materials (Basel, Switzerland)*. 2018, vol. 11(10), E1836.
[49] Gu J., Kwon D., Ahn J., Park I. 'Wearable strain sensors using light transmittance change of carbon nanotube-embedded elastomers with microcracks'. *ACS Applied Materials Interfaces*. 2019, vol. 9, pp. 10908–17.
[50] Han F., Su R., Teng L., *et al.* 'Brittle-layer-tuned microcrack propagation for high-performance stretchable strain sensors'. *Journal of Materials Chemistry C*. 2021, vol. 9(23), pp. 7319–27.
[51] Miao L., Guo H., Wan J., *et al.* 'Localized modulus-controlled PDMS substrate for 2D and 3D stretchable electronics'. *Journal of Micromechanics and Microengineering*. 2020, vol. 30(4), p. 045001.
[52] Liu Z., Qi D., Guo P., *et al.* 'Thickness-gradient films for high gauge factor stretchable strain sensors'. *Advanced Materials (Deerfield Beach, Fla.)*. 2015, vol. 27(40), pp. 6230–37.
[53] Cao Y., Guo Y., Chen Z. 'Highly sensitive self-powered pressure and strain sensor based on crumpled mxene film for wireless human motion detection'. *Nano Energy*. 2020, 106689.
[54] Liu H., Li Q., Zhang S. 'Electrically conductive polymer composites for smart flexible strain sensors: a critical review'. *Journal of Materials Chemistry C*. 2020, vol. 6, pp. 41–12121.
[55] Chen X., Wang F., Shu L. 'A single-material-printed, low-cost design for a carbon-based fabric strain sensor'. *Materials & Design*. 2020, 110926.
[56] Wang C., Xia K., Wang H., Liang X., Yin Z., Zhang Y. 'Advanced carbon for flexible and wearable electronics'. *Advanced Materials (Deerfield Beach, Fla.)*. 2019, vol. 31(9), e1801072.
[57] Ozioko O., Dahiya R. 'Spray coated piezoresistive bend sensor for controlled movements in soft robots'. *IEEE International Conference on Flexible and Printable Sensors and Systems (FLEPS)*; Vienna, Austria, 2021. pp. 1–4.
[58] Yan C., Wang J., Kang W., *et al.* 'Highly stretchable piezoresistive graphene-nanocellulose nanopaper for strain sensors'. *Advanced Materials (Deerfield Beach, Fla.)*. 2014, vol. 26(13), pp. 2022–27.
[59] He Y., Wu D., Zhou M., *et al.* 'Wearable strain sensors based on a porous polydimethylsiloxane hybrid with carbon nanotubes and graphene'. *ACS Applied Materials & Interfaces*. 2021, vol. 13(13), pp. 15572–83.
[60] Xiao T., Qian C., Yin R., Wang K., Gao Y., Xuan F. '3D printing of flexible strain sensor array based on UV-curable multiwalled carbon nanotube/elastomer composite [online]'. *Advanced Materials Technologies*. 2021, vol. 6(1), p. 2000745. Available from https://onlinelibrary.wiley.com/toc/2365709x/6/1
[61] Nassar H., Dahiya R. '3D printed embedded strain sensor with enhanced performance'. *IEEE International Conference on Flexible and Printable Sensors and Systems (FLEPS)*; Vienna, Austria, 2022. pp. 1–4.

[62] Hu X., Yang F., Wu M., *et al.* 'A super-stretchable and highly sensitive carbon nanotube capacitive strain sensor for wearable applications and soft robotics'. *Advanced Materials Technologies.* 2022, vol. 7(3), p. 2100769. Available from https://onlinelibrary.wiley.com/toc/2365709x/7/3

[63] Xu J., Wang H., Ma T. 'A graphite nanoplatelet-based highly sensitive flexible strain sensor'. *Carbon.* 2022, pp. 27–316.

[64] Deng C., Lan L., He P. 'High-performance capacitive strain sensors with highly stretchable vertical graphene electrodes'. *Journal of Materials Chemistry C.* 2022, vol. 8, pp. 46–5541.

[65] Yao S., Yang J., Poblete F.R., Hu X., Zhu Y. 'Multifunctional electronic textiles using silver nanowire composites'. *ACS Applied Materials & Interfaces.* 2019, vol. 11(34), pp. 31028–37.

[66] Lone S.A., Lim K.C., Kaswan K., *et al.* 'Recent advancements for improving the performance of triboelectric nanogenerator devices'. *Nano Energy.* 2022, vol. 99, p. 107318.

[67] Zhang S.L., Lai Y.C., He X., Liu R., Zi Y., Wang Z.L. 'Auxetic foam-based contact-mode triboelectric nanogenerator with highly sensitive self-powered strain sensing capabilities to monitor human body movement'. *Advanced Functional Materials.* 2017, vol. 25, 1606695.

[68] Ning C., Cheng R., Jiang Y., *et al.* 'Helical fiber strain sensors based on triboelectric nanogenerators for self-powered human respiratory monitoring'. *ACS Nano.* 2022, vol. 16(2), pp. 2811–21.

Chapter 5
Neuromorphic principles for large-scale robot skin

Florian Bergner[1], Emmanuel Dean-Leon[1], and Gordon Cheng[1]

5.1 Classical engineering approaches are reaching their limits

5.1.1 Motivations for robot skin

Providing sensitive skin to robots has been explored since the 1980s [1]. The reasons for artificial robot skin are diverse. Recently, new development and new application of robot skin have seen another boost, because collaborative and interactive robots have been considered as a viable solution (i) to further increase the level of automation in complex industrial scenarios [2]; (ii) for health-care [3]; and (iii) also in household [4] applications. So, why do robots need sensitive skin? Skin is considered to be the key factor to (1) enable robots to recognize textures for contact/object classification and recognition and (2) enable intuitive and safe human–robot interaction and collaboration. Texture recognition allows robots to add feel to objects, which so far are only known visually. For example, the visual knowledge of an object (round and yellow) can be extended with the feel of the object (soft with a smooth surface). The knowledge helps the robot to increase the success rate of interaction and manipulation tasks because the robot can exploit knowledge about the grip properties of the object through a sense of touch. Furthermore, artificial robot skin can guarantee the safety of humans in the robots' workspace. In contrast to visual safeguards, contacts are direct and cannot be occluded. In this way, robot skin already contributes to collaborative robots. Potentially, it makes robots safe enough to remove safety fences and allow humans to touch and interact closely with the robot. In addition to that, robot skin can provide an intuitive interface for manipulating and teaching the robot. With the development of appropriate tactile behaviors, the robot can be guided and taught simply by touching and moving it as desired [5].

[1]Technical University of Munich (TUM), Arcisstraße, Munich, Germany

5.1.2 Robot skin

Different kinds of artificial robot skins have been developed mainly for two specific purposes: (1) skin for texture and contact/object recognition and (2) skin for whole-body control and tactile interaction.

Skin for contact/object recognition For texture and contact/object recognition, robot skin has to be deployed on the most prominent regions (e.g., fingertips, hands) with a high spatiotemporal resolution to capture as many stimulus features as possible (see section 5.4.6.2). The robot skins developed for this task use glabrous skin on human fingertips and hands as reference. The most prominent robot fingertip sensors are the OptoFoce sensor [6–8], the bio-mimetic multi-modal BioTac sensor [9–11], the iCup fingertip sensor [12], and the Shadow Hand fingertip sensor [13]. Commonly, the coverage of robot skin at this level is provided at specific known locations, thus, simplifying their deployment.

Skin for whole-body control and tactile interaction For whole-body control and tactile interaction, robot skin should ideally cover the whole robot. Consequently, robot skin has to be deployed on all body regions. In contrast to skin specifically designed for contact/object recognition, which can profit from concentrated localized solutions such as integrated circuits, concentrated solutions are not applicable in large-scale robot skin covering a whole robot. Such large-scale skin has thousands of sensors distributed on spatially distant locations. The wiring, organization, and acquisition of structural information are challenging (see section 5.4.6.1). Artificial robot skins for whole-body deployment follow different strategies to tackle these challenges. Here, we highlight a few of these recent approaches. A modular approach uses stiff standard PCB modules (skin cells) with standard off-the-shelf electronics. These modules are assembled to skin patches by using flexible and stretchable connections. Modular robot skin is realized in References [14–18]. Large-scale robot skin is also implemented in non-modular approaches with large-area sensor matrices. Reference [19] proposes proximity sensors on a flexible substrate, and References [20, 21] introduce force sensors on a flexible substrate. References [22–25] realize large-scale robot skin by using flexible and stretchable substrates. The large-scale robot skin proposed in References [26, 27], Reference [28] implements deformable robot skin with a soft surface. The robot skin of the APAS Assistant Bosch [29] uses large-area capacitive force sensing, and the robot skin in Reference [30] is implemented on a flexible substrate, is stretchable, and is partially modularized.

5.1.3 Challenges and limits of robot skin

In both kinds of artificial robot skin systems (see section 5.1.2), fingertip skin and large-scale robot skin alike, the acquisition, transmission, and processing of skin information are demanding. This is due to the large amount of skin information that has to be acquired, transmitted, and processed with low latency. But power, transmission bandwidth, and processing power are limited resources such that current robot systems need to find a good tradeoff between the number of sensors and sampling rate. Even though challenges are less demanding for fingertip sensors, of course, fingertip sensors need a high spatial resolution and high sampling frequency to catch all the necessary stimulus features of textures. However, these sensors are deployed in a limited small region. This enables the use of concentrated systems where the bandwidth and wiring issues are less demanding. In contrast to that, large-scale robot skin systems are distributed systems and tactile sensors are spatially distributed. Wiring and information exchange in these distributed systems are complex, and the available transmission bandwidth is limited. Current robot skin systems realized by standard engineering approaches (polling for every sensor) can reach their limits in bandwidth and processing power easily. High-performance large-scale robot skin covering a whole humanoid robot seems currently infeasible. It is estimated that covering a humanoid robot requires around 3 000–5 000 skin cells with 27 000–45 000 sensors [31], which is much more than any robot skin system can currently handle. However, in comparison to nature, this number of sensors seems rather small. Humans employ around 5 million skin receptors innervated by around 1.1 million ascending tactile nerve fibers [32]. Most impressively, humans can handle this large amount of tactile information efficiently with low latency and with unmatched performance in tactile behaviors. The big open question is: how can biological systems be so efficient and outperform every engineered system?

This chapter investigates principles that nature uses to efficiently acquire, transmit, and process sensory information (see section 5.2). Afterward, this chapter surveys how recent developments employ biological principles generally in neuromorphic sensors and more specifically in neuromorphic tactile sensors and robot skin systems (see sections 5.4 and 5.5).

5.2 Biology employs a toolbox full of optimized principles

Biology employs many principles for optimal information acquisition, transmission, and processing [33–35]. This section provides an overview of fundamental biological principles, which can be transferred to engineering to improve the efficiency of their applications.

5.2.1 Skin receptors are tuned to sense specific stimulus features

In biology, skin employs specialized receptors for sensing mechanical, thermal, and noxious (potentially dangerous/destructive) stimuli [32, 35–39]. Different skin receptors are tuned to sense specific stimulus features that focus on distinct pieces of contact/object properties. The most common tactile stimulus features are normal pressure (Merkel cell receptors [36, 37, 39]), horizontal motions and slip (Meissner corpuscle receptors [36, 37, 39]), vibrations (Pacinian corpuscle receptors [37, 39]), stretch (Ruffini endings [37, 39]), and proximity/approach (tylotrich-hair receptors [37, 39]). The receptors' specific stimulus feature selectivity is influenced by the location of skin receptors in different dermal layers, by the deployment pattern, and by mechanical filter mechanisms. Selectivity enables skin receptors to disassemble complex multi-modal stimuli to simple distinct uni-modal stimulus features. Distinct stimulus features allow efficient encoding of complex information and enable selective attention.

5.2.2 Skin receptors transduce stimuli features to binary action potentials

The specialized skin receptors (see section 5.2.1) apply different functional principles to transduce different stimulus features to neural activity. However, the last step of stimulus transduction is similar for all types of skin receptors. Stimuli features either directly or indirectly gate ion channels in the terminals of peripheral nerve fibers. The influx of ions into the peripheral axons of the dorsal root ganglion neurons depolarizes the neurons and eventually creates action potentials/spikes [32, 36]. Action potentials are binary; they are present or not. The action potentials on their own cannot convey sensory information such as intensity, type, etc. [40]. Therefore, biology employs neural codes (see section 5.2.3). In general, neurons transmit information through action potentials and action potentials initiate the processing of information on arrival. The encoding of information in action potentials (events) allows redundancy reduction at the receptor level and efficient novelty-driven (event-driven) processing.

5.2.3 Skin information is encoded by different neural codes

As previously discussed (see section 5.2.2), action potentials are binary and cannot convey sensory information. Sensory information is encoded by different neural coding schemes which are: (1) type code, (2) spatial code, (3) rate code, (4) temporal code [35], and (5) latency code. Different stimulus features are captured by different types of receptors (see section 5.2.1). Nerve fibers of a neuron only innervate receptors of one particular type and consequently apply type code. That is, the type of a stimulus feature is encoded by the nerve fiber itself. Spatially distributed population of receptors of the same type can encode spatial information through activity patterns [35]. Such populations apply spatial code. The spatial coding scheme is extensively employed by Merkel cell and Meissner corpuscle receptors [35, 36, 39, 41]. To preserve the information in type and spatial codes, biological systems have to keep the ordered structure (somatotopic order, see section 5.2.4) in bundles of nerve fibers (nerves). That is, nerve fibers innervating a skin region have to preserve spatial relationships, e.g., nerve fibers innervating neighboring receptors are neighbors, too. Merkel cell receptors and Ruffini endings are innervated by slowly adapting (SA) nerve fibers and encode intensities by proportionally modulating spike rates [35, 36, 39, 41]. SA nerve fibers apply rate codes. Neurons also encode information in the temporal sequence of spikes and apply the temporal coding scheme. The rapidly adapting (RA) nerve fibers of such neurons innervate, for example, Meissner corpuscle receptors and Pacinian corpuscle receptors [32, 37]. The RA nerve fibers are exclusively sensitive to stimulus feature changes and encode the exact time of each change. The relative spike arrival time in a population of neurons can encode moving stimulus features by employing the latency coding scheme. The latency coding scheme combines spatial and temporal information [35]. The different neural codes demonstrate that biology uses structure and time to efficiently encode information through binary activities in parallel nerve fibers.

5.2.4 Skin information ascends somatotopically ordered

Skin afferents (ascending nerve fibers), which ascend from innervated skin receptors via relay centers in the spinal cord, medulla, and thalamus to the primary somatosensory cortex (S1), are somatotopically ordered [37, 39]. Somatotopically ordered means that the structural order of nerve fibers preserves spatial relationships and type information of the innervated skin receptors (section 5.2.3). A mixing of afferents would result in loss of information in spatial and type codes. The information of stimulus feature type and location would be lost. During the ascend, the somatotopically ordered information of different body parts is assembled into a comprehensive sensory representation of the whole body, the homunculus [37, 39, 42].

5.2.5 Skin information is structured and processed hierarchically

The primary somatosensory cortex (S1) processes and represents tactile information in a hierarchical structured way (see Figure 5.1). Different areas in the primary somatosensory cortex abstract information to different kinds of features.

5.2.5.1 Topographical features

The Brodmann areas 3a and 3b in the somatosensory cortex (S1) filter and sharpen the topographical features of proprioceptive and mechanoreceptive (tactile) information, respectively. Topographical features describe the location of stimuli and their spatial relations. The somatosensory cortex (S1) represents topographical

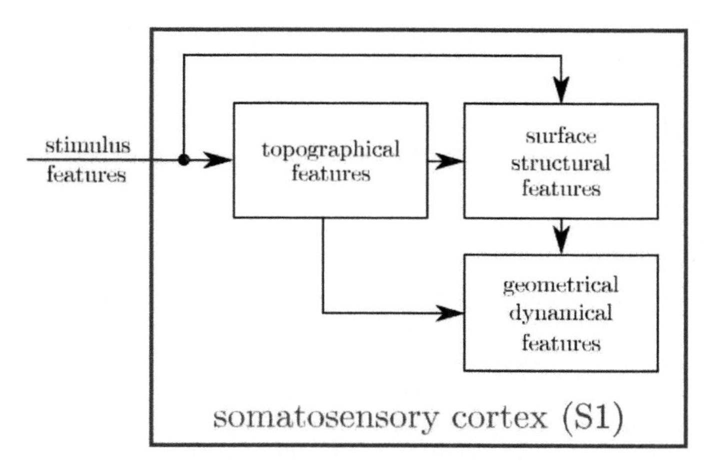

Figure 5.1 Features processed in the primary somatosensory cortex (S1)

features with high spatial resolution in whole-body somatotopic maps [33, 36, 39]. These maps enable higher cortical circuits to precisely locate stimuli and to set them in the spatial context.

5.2.5.2 Surface structural features (texture)

Brodmann area 1 in the primary somatosensory cortex (S1) abstracts tactile information to spatiotemporal surface structural features such as structural contact/object properties (texture) and movement directions of repeated patterns [35, 36, 39]. This area processes static and dynamic information (e.g., force and horizontal movements) and combines structural with temporal information. This includes topographic features, which are processed in a different area of the S1 (see section 5.2.5.1). The neurons in this area process information of all neural codes and abstract it to structural features [33, 35, 36, 39, 43]. The neurons are specialized to encode structural features rather than to encode exact stimuli locations [33, 39].

5.2.5.3 Geometrical and dynamical features

Brodmann area 2 in the primary somatosensory cortex (S1) abstracts proprioceptive information of area 3a and mechanoreceptive information of 3b and 1 to geometrical and dynamical contact/object features [33, 36, 39]. Thus, the processes in this area also consider topographical features (see section 5.2.5.1) and surface structural features (see section 5.2.5.2). Geometrical and dynamical features embrace orientation (e.g., angle of edges), direction of excitation (e.g., stroking direction), surface curvature (e.g., 3D perception of contacts/objects), posture (e.g., position of body parts in space), applied forces, and spatial arrangement of repeated patterns [33, 36, 39]. It is important to note that some geometrical and dynamical features require proprioceptive information because they depend on body posture. Consequently, neurons in Brodmann area 2 fuse proprioceptive information with tactile information. The neurons in this area use population codes (spatial codes, see section 5.2.3) to encode geometrical/dynamical features. The organization of neurons is still somatotopic but has less spatial acuity [33, 36, 39]. Thus, neurons focus on features rather than on location.

5.2.6 The cognitive where

The *where* pathway connects cortical areas that focus on where the contact/object is located [33, 36, 39]. The pathway also considers contact/object dynamics that are tightly coupled with location. Contact dynamics are especially important in object or whole-body interaction/manipulation. These processes need to consider the feel of the contact/object in order to develop reliable interaction/manipulation strategies. The *where* pathway originates in the primary somatosensory cortex (S1). The pathway abstracts location and dynamics of contacts/objects and provides the acquired information to cortical pre-motor and motor areas [33, 36, 39]. Furthermore, parts of the pathway predict the consequences of actions on contact location and dynamics and provide feedback on prediction accuracy [36, 39]. For example, in the action of grasping an object, the anticipation of weight and texture of the object to grasp can increase the success rate of the action.

5.2.7 The cognitive what

The *what* pathway connects cortical areas which try to determine what the contact/object is [33, 36, 39]. The pathway is involved in contact/object recognition and processes tactile features (surface structural features/texture) which are provided by the primary somatosensory cortex (S1) (see section 5.2.5). The *what* pathway tries to classify the contact/object group to which these features belong to. The pathway is connected to higher cortical processes that are involved with the association of contact/object features with memory, with emotion, and with language [33, 36, 39]. In this way, the *what* pathway drives processes that determine whether the perceived contact/object is related to memorized contact/objects, whether the contact/object is related to a certain emotion, or whether the contact/object can be named.

5.3 Biological principles are the key to large-scale robot skin

The neuromorphic approach to efficient biologically inspired robot skin systems targets the limits of state-of-the-art robot skin systems as discussed in section 1.1 by applying optimal biological principles (see section 5.2). The application of two fundamental biological principles induces reductions in power consumption, transmission bandwidth, and processing power of robot skin systems. These two principles are (1) efficient event-driven information acquisition, transmission, and processing (section 5.3.1) and (2) efficient information representation with hierarchical structure and hierarchical abstraction (section 5.3.2). The efficiency gain in neuromorphic robot skin systems will finally enable the up-scaling to large-scale robot skin.

5.3.1 Neuromorphic event-driven sensors

Neuromorphic event-driven sensors take advantage mainly of three biological principles: (1) all biological sensing systems disassemble complex stimulus information into simple stimulus features by using specialized transducers in specialized receptors (see section 5.2.1), (2) biological sensors (receptors) transduce stimulus features directly and continuously into events (binary spike trains) (see section 5.2.2), and (3) biological information acquisition, transmission, and processing are event-driven, i.e., the novelty of information drives all processes (see section 5.2.2). Neuromorphic vision sensors have been engineered first (see section 5.4.1). They successfully implement all these three principles and demonstrate superior performance in comparison to conventionally engineered synchronous systems. This indicates the potential of biological principles for the design of neuromorphic event-driven robot skin systems. Section 5.4.1 provides an overview of how different neuromorphic vision sensors realize biological principles. Sections 5.4.2 and 5.4.3 introduce two different methods for generating, representing, and transmitting information in event-driven systems as events. Finally, section 5.4.4 addresses the progress in the development of neuromorphic event-driven robot skin.

5.3.2 Neuromorphic information representation in hierarchical structures

Neuroscientific research results, anatomy, and physiology (see section 5.2) strongly indicate that biology efficiently processes proprioceptive and tactile (cutaneous) information (somatosensory information) in an extensively structured in a hierarchical manner. The abstraction of information grows within the hierarchy. The biological somatosensory system organizes and structures information through all levels. This includes information acquisition, transmission, and processing and also considers information representation. Structuring, organizing, and building hierarchies increase system efficiency in three ways: (1) information that is represented in a well-structured way increases efficiency through reducing the complexity of accessing information; (2) the stepwise parallel condensation of essential information through abstraction enables cognitive processes to efficiently access and combine relevant information of different abstraction levels; and (3) the concentrated concurrent processing of information in specialized centers increases efficiency and reduces latency. Section 5.4.5 discusses how tactile information is represented in current robot skin systems. Afterward, section 5.4.6 outlines the differences between the tactile information abstraction and processing structures in the cortex and in robot skin systems.

5.4 Neuromorphic systems realize biological principles

5.4.1 Neuromorphic event-driven vision has been engineered first

Emerging neuromorphic systems first successfully mimicked the visual sense [44]. The development of neuromorphic vision systems has been mainly motivated by the big gap between the properties of state-of-the-art vision systems and the requirements of robot systems. Standard vision systems sample image frames synchronously in the form of a constant rate of image frames. The information generated by these vision systems is enormous. For example, an HD frame has 1,920 × 1,080 RGB pixels. Each pixel represents colors with a depth of 24 bits. In this case, one single HD frame would have 6.2 MB, and a video stream with a frame rate of 30 Hz would approximately require a bandwidth of 1.5 Gbit/s. The video stream contains a large amount of redundant information. Meaningful information such as optical flow has to be extracted. This requires powerful processing systems in order to deal with all this information in real time. Such systems are expensive and hard to embed into power- and space-constrained autonomous robot platforms. Consequently, visual information is often processed in low frame rates. Essential information is lost, which results in poor and slow visual responses. Neuromorphic vision sensors reduce redundant information at the pixel sensor level. They use biological principles and produce, instead of a synchronous stream of image frames, asynchronous streams of pixel events (see sections 5.4.1.1–5.4.1.3). Neuromorphic vision sensors convince through rather low transmission bandwidths and super fast response time. Different successful applications demonstrate the efficiency of event-driven sensing, transmission, and processing of visual information (see section 5.4.1.4).

5.4.1.1 The dynamic vision sensor

The first neuromorphic vision sensor developed has been the dynamic vision sensor (DVS) [44]. This vision sensor transduces in each pixel logarithmic intensity changes directly to trains of ON/OFF events. The transduction takes place asynchronously in continuous analog circuits. Comparator circuits generate ON/OFF events whenever the accumulated intensity change exceeds a specified threshold. ON events encode increasing light intensity and OFF events decreasing light intensity (see Figure 5.2). The neuromorphic vision sensor uses the address event representation (AER) (see section 5.4.2) for representing and transmitting events. The asynchronous pixel events are arbitrated and time-multiplexed onto a common high-speed asynchronous

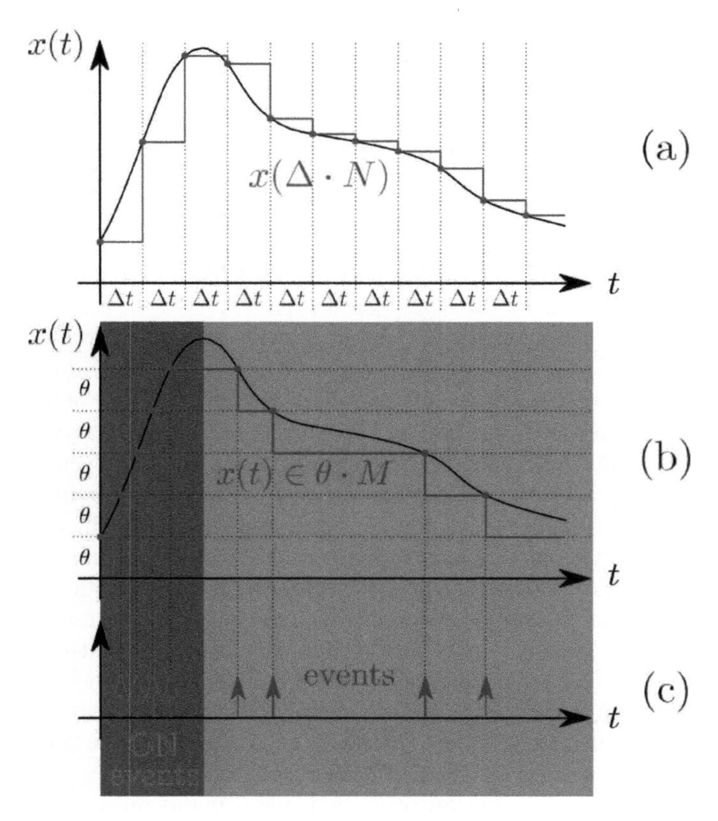

Figure 5.2 The differences between time synchronous sampling (a) and novelty-driven sampling (b) and (c); adjusted from Reference [44]

event bus that also uses AER. The presented neuromorphic vision sensor implements all the three basic biological principles for efficient event-driven sensor systems that have been discussed in section 5.3.1. The DVS is sensitive to a specific stimulus feature that in this case is the logarithmic light intensity change, it directly transduces stimuli features into trains of events, and it uses intensity change as an indicator for novel information and for driving sampling and transmission. The pixel event generators of the DVS behave like the RA neurons innervating skin receptors (see section 5.2.3) such as Meissner corpuscles. These skin receptors are particularly sensitive to changes and insensitive to static stimulus features. In contrast to their biological counterparts, DVS pixels do not use a binary representation for events (see section 5.2.2). They code event type (ON/OFF) and location (the pixel row and column address) into the event itself using the AER. Biology encodes type and location by using separate nerve fibers, i.e., one nerve fiber for each type and location. The AER exploits the high bandwidth capabilities of digital buses (in comparison to nerve fibers) and reduces the number of wires through time-multiplexing events onto a common bus (see section 5.4.2). As a drawback, events in a common bus have to be tagged to keep information of type and location and have been sorted (demultiplexed) afterward for further processing. The ON/OFF events of DVS only carry spatial and temporal information along with the polarity of the intensity change but no absolute intensity values or color information. To encode absolute intensity and color into events, engineers pursue two different approaches: the asynchronous time-based vision sensor (ATIS) (see section 5.4.1.2) and the (color) dynamic and active vision pixel sensor (DAVIS, C-DAVIS) (see section 5.4.1.3).

5.4.1.2 The asynchronous time-based vision sensor (ATIS)

References [45, 46] extend the DVS (see section 5.4.1.1) with the ability to sense absolute intensity at the pixel level. The proposed sensor, the asynchronous time-based vision sensor (ATIS), transduces light intensities directly into two types of events: the absolute intensity events and the temporal contrast events. All events use the AER (see section 5.4.2) and are arbitrated onto a common asynchronous bus that also uses AER. The temporal contrast events (ON/OFF events) encode intensity changes just like the DVS. The absolute intensity events, however, encode absolute continuous intensity values into continuous inter-event times and thus apply the biological principle of neural temporal codes (see section 5.2.3). Pixels of the ATIS do not generate absolute intensity events continuously (such as in neural rate codes, see section 5.2.3). Their generation is coupled with the generation of temporal contrast events. In this way, the novelty of information can also be guaranteed for absolute intensity events. Thus, redundancy suppression at pixel level applies for both event types, for absolute intensity events and for temporal contrast events. The absolute intensity event generators of the ATIS pixels behave like SA neurons innervating skin receptors (see section 5.2.3), such as Merkel cells. These receptors are particularly sensitive to static and insensitive to dynamic stimulus features. However, SA neurons encode different intensities into different event rates such that activity correlates to intensity. For biological systems, rate codes are efficient and robust because events are transmitted in many parallel nerve fibers and neurons can directly use rate codes for their analog processing schemes. For neuromorphic systems, which use the AER protocol and where events share a common bus, rate codes are highly inefficient. Rate codes encode redundant information and negatively influence the efficiency of event-driven systems. Event-driven systems derive their efficiency from the fact that only novel information is transmitted and processed. This does not apply for rate codes. The ATIS pixels use relative time codes to encode absolute information. They efficiently combine the encoding of absolute continuous intensities with the biological principle of novelty/event-driven sensing, transmission, and processing. So far, the ATIS only encodes gray levels but no colors.

5.4.1.3 The dynamic and active vision pixel sensor

The (color) dynamic and active vision pixel sensor (DAVIS/C-DAVIS) proposed in References [47, 48] combines a standard frame-synchronous vision sensor with integrated DVS pixels (see section 5.4.1.1). This vision sensor generates temporal contrast (ON/OFF) events at the pixel level and concurrently samples (colored) image frames with a constant update rate. The (C)-DAVIS incorporates two digital communication buses: one synchronous bus for transmitting image frames and one asynchronous bus implementing AER for transmitting temporal contrast events. The nice feature of the (C-)DAVIS is that it allows to tag events with absolute intensity information or even color information. However, the highly redundant stream of image frames needs to be transmitted from the periphery to the event-driven processes.

5.4.1.4 Applications prove the efficiency of neuromorphic vision sensors

Neuromorphic vision sensors are already used in many applications where they successfully demonstrate the superior computational efficiency and low latency of event-driven vision systems. Neuromorphic vision sensors have been used in humanoid robot platforms, e.g., iCub [49] and for efficient calculation of optic flow [50, 51]. Reference [52] demonstrates superior control performance in the application of a pencil balancer that uses low-latency visual feedback from two DVSs. Finally, Ghaderi *et al.* [53] recently presented an embedded vision-to-audio guidance system for the blind. This system fully takes advantage of efficient event-driven processing of visual information, resulting in real-time feedback and rather low power consumption.

5.4.2 The neuromorphic AER is a standard for transmitting events

5.4.2.1 The AER

The AER (see Figure 5.3) [54, 55] evolved as a standard for representing and transmitting events in neuromorphic systems. Many neuromorphic event-driven sensing systems [44, 56, 57] (see sections 5.4.1 and 5.4.4) use AER to encode sensory information into events and to transmit information to event-driven neuromorphic processing systems, e.g., SpiNNaker [58]. Originally, the AER has been developed for the

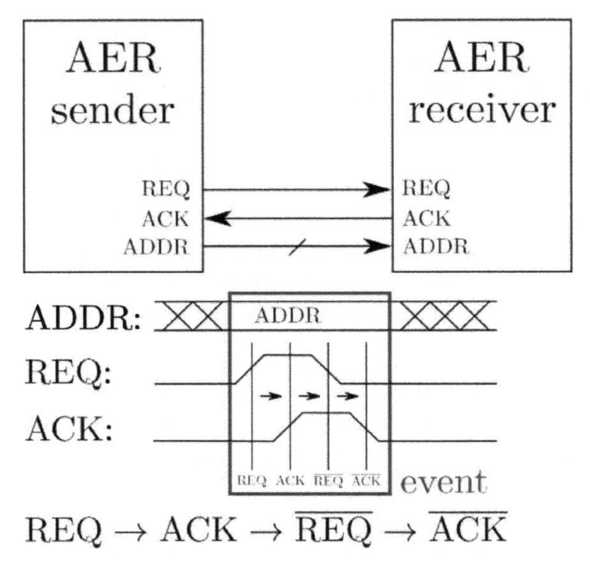

Figure 5.3 The AER; adjusted from Reference [54]

communication between artificial neurons in VLSI ICs. AER rigorously takes advantage of integrated high-speed digital asynchronous parallel bus systems. The transmission of one digital pulse on a standard digital interface, e.g., a PCIe 3.0 lane with a bit rate of up to 10 Gbit/s is much faster than the transmission of a spike in a neural fiber. The fastest neurons can fire up to 1 000 times per second [39, 59], which roughly corresponds to 1 kbit/s. The AER makes use of the large gap between event rates and bus speed. Instead of one-by-one modeling of nerve fibers to digital connections, the AER time-multiplexes the activity of a whole nerve, with several thousand nerve fibers, to a common self-timed asynchronous parallel bus. Time-multiplexing exploits the gap between event rates and bus speed and reduces the number of wires for N sending neurons from N to $1 + \log N/\log 2$ without losing temporal precision. The asynchronous self-timed parallel bus guarantees the high temporal precision of events because clock synchronization is not necessary. The transmission of an event in AER is solely driven by the occurrence of the event. An asynchronous digital bus using AER is a parallel address bus accompanied by a request (REQ) and an acknowledge (ACK) line. The address identifies the neuron that generated the event (location and type). The occurrence of an event on the bus encodes its creation time in the neuron. Like biological events (spikes, action potentials), events in AER only encode where the event originated (location and type) and when it occurred. However, events in biological systems are binary and encode type and location with the nerve fiber itself (see section 5.2.3). All other information, including intensity, intensity changes, frequency, phase, spatial relations, etc., have to be encoded in neural codes (see section 5.2.3). Neural codes employ different principles. Neural codes can encode information in distributed activities in a population of neurons (population code), in the precise timing of events (time code), and in the number of events per time frame (rate codes) [35, 40]. Time codes require rather high temporal precision but can encode continuous values with analog-like precision into the continuous domain of time. That is, a continuous value x can be transcoded into a continuous Δt between two consecutive events. Because of its high temporal precision, AER implementations can use time codes. References [45, 46, 60] (see sections 5.4.1.2 and 5.4.4.2) demonstrate the feasibility of transducing sensory values directly, without prior quantization, to time codes in AER. It is important to note that neurons, which continuously generate time-coded AER events, actually apply rate codes. In rate codes, the event rate f directly correlates to the mean absolute value $\bar{x}$ of the encoded value x. Rate codes are highly inefficient within neuromorphic event-driven systems as they do not apply the principle of redundancy reduction at the sensor level; they do not consider the novelty of information.

5.4.2.2 AER for distributed sensors and processing

Utilizing AER in event-driven systems offers these systems asynchronous event-driven acquisition, transmission, and processing of information, high temporal precision, and continuous, analog-like representation of information through time codes. However, using AER currently has two major drawbacks. First, systems using AER require special hardware that is still under development, not standardized, and not readily available. Second, AER implementations rely on high-speed asynchronous parallel bus systems, which are readily available in concentrated and dense systems such as ICs and circuit boards, but which are hard to realize in distributed systems with many spatially distant nodes and many connections, e.g., robot skin. Wiring and connecting these nodes via parallel high-speed buses is infeasible. Recent developments try to tackle these drawbacks. Special AER-to-synchronous system bridges, as proposed in References [44, 45], quantize time and convert events in AER to time-stamped event packets. These bridges enable standard synchronous processing systems to profit from event-driven sensors. Furthermore, different network architectures for AER, such as flat AER, boardcast AER, prestructured AER, hierarchical AER, routed-mesh AER, and multicast-mesh AER, have been proposed in Reference [61]. These networks enable the communication between distributed event-driven nodes using AER. Serializing the parallel AER buses has been considered in several works [62–66]. Serial AER reduces wire count and facilitates connecting distributed nodes. The development of serial AER started with using standard state-of-the-art synchronous serial signaling technologies with 8B/10B coding schemes and two-wire LVDS (low-voltage differential signaling) [62]. However, such a system relies on precise clocks. References [63, 64] propose Manchester coding with clock recovery for serial AER. Their solution is more reliable against clock jitter, but the serial transmission is still synchronous. The recent works [65, 66] demonstrate that asynchronous serial AER systems are feasible. While the serial AER proposed in Reference [65] requires a three-wire bidirectional connection, the system proposed Reference [66] requires only one differential LVDS wire pair. Serial AER based on asynchronous LVDS is highly appealing, because it is self-driven and thus robust against clock jitter. The asynchronous self-driven signaling preserves the high temporal precision property of AER and differential signaling allows high-speed transmission over long distances.

5.4.3 The send-on-delta principle allows event-driven transmission and processing in synchronous systems

5.4.3.1 The send-on-delta principle

Efficient and power-saving sampling, processing, and transmission of information are of first importance in battery-powered and widely distributed large wireless sensor networks. Efficiency and power-saving mechanisms increase the life expectancy of these distributed sensors. Transmitting information, wire-bound or wireless, often consumes more power than information acquisition and processing—simply due to the fact that signal energy has to overpower noise energy at distant receivers. The distributed sensors save power most efficiently when they only become active and acquire, process and transmit information when it is strictly necessary. In the rest of the time, power can be saved in sleep modes. References [67, 68] present the send-on-delta principle (SoDP) that allows distributed networks of sensors to do this. The SoDP for distributed sensors works quite similar to the previously discussed neuromorphic AER event generator of the DVS (see section 5.4.1.1). The SoDP is able to realize two of the three biological principles for neuromorphic event-driven sensors (see section 5.3.1). The SoDP links the necessity to acquire, process, and transmit information to the novelty of information. The novelty of information is evaluated by monitoring the change of sensor signals (stimulus features). Whenever the change exceeds a specified limit (i.e., the threshold θ), then the sensor becomes active, acquires the monitored sensor information, processes it, and finally sends it. The SoDP changes the representation of information. Instead of sampling information $x(N \cdot \Delta T)$ at periodic time intervals $t \in N \cdot \Delta T$, $N \in \mathrm{N}$, a sensor using the SoDP now presents $x(t) \in M \cdot \theta$ continuous in time but only at certain levels $x \in M \cdot \theta, M \in \mathrm{Z}$ (see Figure 5.2). However, the SoDP does not strictly implement the second biological principle introduced in section 5.3.1. The SoDP does not directly transduce stimulus features into event trains, and generated events are not binary but contain the absolute sensor value at event occurrence time. The SoDP still allows the development of highly efficient event-driven systems. Nevertheless, sensors using SoDP require more transmission bandwidth and might have a slightly lower temporal precision in comparison to neuromorphic event-driven sensors that fully implement the second biological principle; the direct and continuous transduction of stimulus features into events (see section 1.3.1).

5.4.3.2 Comparison of SoDP with AER

The SoDP represents events differently than the AER (see section 5.4.2). An event produced by applying the SoDP is just an event packet. That is completely different from events in AER where events are represented by activity on an asynchronous parallel bus. Event generators applying the SoDP generate event packets when the novelty of information is guaranteed and thus ensure redundancy reduction at the sensor level. Similar to events in AER, an event packet contains an id to identify its type and origin. However, event packets also contain the absolute sensor value at event generation time. Thus, event packets contain information, and there is no need to encode information into the precise timing of events. This has advantages and drawbacks. Event packets are self-contained and can be transmitted in standard state-of-the-art networks. The transmission not necessarily needs to be asynchronous but asynchronous communication increases temporal precision. Furthermore, a timing breach (e.g., non-deterministic delay) does not induce information loss. This results in a lot of flexibility and robust systems. However, event packets can only transport discrete information, induce more transmission overhead, and have lower temporal precision. Events in AER can transport continuous information with analog-like precision. The big advantage of using the SoDP is that it can be easily implemented in existing sensor systems by applying compound architectures [68]. In compound architectures, standard sensors sample as fast as necessary and event generators apply the SoDP to generate and transmit event packages. Compound architectures further improve the flexibility of the SoDP at the cost of power and computational efficiency at the sensor level. The sensor system is less efficient because sampling and processing have to run at sample rate. Event-driven systems, whether they use SoDP or AER, reduce redundancy at the sensor level, and transmission and processing are driven by novel information. This increases efficiency in transmission bandwidth usage and computational costs. It is also worth noting that the AER-to-synchronous bridge [44, 45] (see section 5.4.2) basically converts AER events to event packets. These packets are similar to packets generated by applying the SoDP. So, if one uses standard synchronous systems for event-driven processing of event-driven information, then using SoDP or AER at the sensor level will have no influence on the efficiency gain and the precision of these event-driven processes.

5.4.4 Neuromorphic event-driven skin is under development

The development of neuromorphic skin started recently with the goal to breach the limits of state-of-the-art robot skin systems. As discussed in section 5.1, state-of-the-art robot skin systems lack the ability to efficiently acquire, transmit, and process information. Limits in transmission bandwidths, caused by constrained wiring in networks of distributed sensor nodes, set a distinctive limit to the spatiotemporal resolution of robot skin. Furthermore, the flood of information in large-scale robot skin covering a whole robot definitely requires efficient tactile information processing schemes. Currently, the development of efficient neuromorphic skin focuses on event-driven tactile sensors. These sensors have the potential to extend the advantages of event-driven systems from the sense of vision (see section 5.4.1.1) to the sense of touch, i.e., they have the potential to contribute to large-scale robot skin.

5.4.4.1 Transduction of tactile stimuli to events in quasi-digital AER

References [69, 70] propose a concentrated event-driven tactile sensing system that uses discrete touch sensors in a robot hand. Their system directly transduces sensor values without quantization to quasi-digital AER (QD-AER) events. The system transduces measured capacitance or resistance to periodic digital signals. The frequencies of these signals correspond to absolute tactile stimulus intensities. Positive signal edges represent the quasi-digital event. The system tags these events with addresses that encode the origin and type of the event. The authors name this event representation QD-AER. In contrast to the standard AER (see section 5.4.2), QD-AER represents an event on the asynchronous bus with a positive edge, while AER uses request/acknowledge (REQ/ACK) signals. Furthermore, QD-AER events are periodic and not driven by information novelty. The proposed system serializes QD-AER events to one wire and transmits the serialized events to a receiver system. The serial event receiver deserializes the events and quantifies time differences between events. Then, the receiver system thresholds the change of time differences and creates novelty-driven ON/OFF events in standard AER. A nice feature of QD-AER is that it encodes absolutes intensity values directly to event rate code. Absolute intensities are encoded with analog-like precision. However, rate codes are inefficient as they contain a large amount of redundant information and require more transmission bandwidth than necessary. The proposed system serializes events and transmission can take place via one wire or even wireless. This eases information transmission. The receiver of the proposed system produces ON/OFF events in AER that allows its direct connection to efficient event-driven processing systems. Nevertheless, the proposed system cannot be easily applied to large-scale robot skin systems. The proposed system requires many connections from distributed sensors to a centralized event converter, which results in a centralized architecture with a tight wiring problem.

5.4.4.2 Neuromorpic event-driven tactile sensors generate events in AER

A neuromorphic integrated array of event-driven tactile sensors has been proposed in References [57, 60, 71]. The authors apply the same principles used in neuromorphic vision sensors (see section 5.4.1) and transduce tactile force stimuli to events in AER at the sensor unit level. These sensor units are arranged in a matrix structure, and the authors name them taxels (tactile pixels). Just like in neuromorphic vision sensors, the proposed neuromorphic tactile sensor arbitrates taxel events to a common asynchronous parallel bus that also implements AER. The authors started with taxels that are sensitive to sustained pressure [57, 71]. These taxels act like the SA Merkel cell receptors in human skin (see section 5.2.3). The taxels directly transduce pressure/release to trains of ON/OFF events. The taxels apply rate codes and absolute pressure/release corresponds to the ON/OFF event rates. The authors have also introduced taxels that are sensitive to pressure transients [60]. The biological counterparts of these taxels are the RA Meissner corpuscle receptor in glabrous human skin (see section 5.2.3). The neuromorphic pressure transient taxels resemble the pixels of DVS (see section 5.4.1.1). These taxels monitor accumulated force changes. When the change exceeds a predefined threshold, then the taxels create an ON/OFF event. In contrast to the sustained force taxels, the transient force taxels generate only events when the novelty of information is guaranteed. The transient force taxels apply redundancy reduction at the taxel level. A nice feature of the proposed sensor is that it directly transduces forces into events in AER. This enables high temporal resolution and analog-like precision for absolute force values. However, rate codes are inefficient in event-driven systems because they contain a large amount of redundant information and they are not driven by novelty.

5.4.4.3 Neuromorphic event-driven tactile sensors with AER for large-scale robot skin

The neuromorphic event-driven tactile sensor, which has been explained in section 5.4.4.2, is also discussed in the surveys of References [72, 73]. At first glance, using neuromorphic event-driven sensors in robot skin seems to be appealing. AER has shown good efficiency in vision systems (see section 5.4.1.4), and if this efficiency improvement could be exposed to robot skin, then large-scale robot skin would finally become feasible. Improvements, such as decreasing transmission bandwidth and increasing computation efficiency, would target the missing links of large-scale robot skin (see section 5.1). However, the surveys do not discuss how AER could be implemented in robot skin. Robot skin is basically a large collection of distributed sensors. The problems of AER in distributed systems and possible solutions have been discussed in section 5.4.2.2. The solutions indicate that self-driven serial AER with differential signaling in combination with meshed and routed AER networks seems feasible for modularized self-organizing robot skin systems [14, 15]. However, these systems would need a complete revision of their hardware designs.

5.4.4.4 Neuromorphic force sensors with high spatiotemporal resolution using SoDP

The authors of References [74, 75] propose a new event-driven neuromorphic force sensor with high spatiotemporal resolution. They developed a flexible resistive sensor matrix with 4 096 sensing elements (taxels). The sensor matrix has an active area of 1 610 cm^2, a spatial resolution of 0.4 cm, and a sensing bandwidth of 5 kHz. The taxels in the matrix transduce forces to voltages, and the voltages of all taxels are sampled at high speed. The sensor time-multiplexes columns and samples voltages in parallel. The sampling mechanism converts voltages into time differences. These time differences are directly proportional to the sampled voltages and correspond to taxel forces. Thus, the force applied to a taxel is represented by a time difference. A fast programmable gate array (FPGA) monitors the force changes of each taxel and when the change exceeds a specified threshold, then the FPGA creates an event packet. These event packets contain a time stamp, the location of the event (matrix row and column), and the absolute force value. The event packets are transmitted from the FPGA to the PC via a standard Ethernet connection. The proposed neuromorphic event-driven force sensor implements the previously discussed SoDP (see section 5.4.3). The sensor demonstrates the feasibility of tactile sensors with high temporal resolution and the efficiency of on-site redundancy reduction at the sensor level. However, the sensor's event generation mechanism is centralized in an FPGA and not implemented in the taxels. This prevents the sensor's application in modularized distributed sensor systems such as large-scale robot skin. The wiring complexity from the distributed sensors to the centralized event-generation system would be infeasible. Furthermore, sensor matrices are susceptible to propagate wiring damages, which make them fragile and lack robustness.

5.4.5 Neuromorphic information representations mimic the primary somatosensory cortex

5.4.5.1 Robot skin systems structure tactile information somatotopically

Structuring at sensor level Biological tactile systems (part of the somatosensory system) disassemble contact/object stimuli at the receptor level to multi-modal stimulus features (see section 5.2.1). This fundamental biological principle has been proposed in section 1.3.1 as one of the three key principles of efficient neuromorphic event-driven sensors. The disassembling of stimulus features describes how information can be efficiently structured and represented in a modular way already at the sensor level. This principle is applied by all robot skin systems—simply due to the fact that electrical sensors rely on specialized transduction mechanisms that are optimized to transduce one particular stimulus feature. Robot skin systems transduce stimulus features such as absolute force, force changes, vibrations, stretch, proximity, temperature, etc. (see section 5.1).

Structuring in peripheral communication paths Biological skin receptors are innervated individually by separate nerve fibers. These nerve fibers themselves encode the type and the location of receptors. Nerve fibers are bundled to peripheral nerves and finally enter the spinal cord. To preserve the type and location information, biological systems order and structure nerve fibers throughout the whole pathway from skin receptors to the somatosensory cortex (see section 5.2.4). Nerve fibers are somatotopically structured. This means that the ordering and structuring of nerve fibers preserve spatial relationships between skin receptors. The nerve fibers of neighboring receptors are kept close together. Neuromorphic event-based sensor systems (see sections 5.4.1 and 5.4.4), which use AER (see section 5.4.2) or SoDP (see section 5.4.3), exploit the superior transmission speeds of technical systems in comparison to biology to reduce wire count (see section 5.4.2.1). They time-multiplex events of different locations and types onto a common high-speed communication bus. The events are tagged with type and location information to preserve the same. However, events on a common bus are no longer somatotopically ordered, and the following sorting of individually tagged information to topographical maps is cumbersome and computationally expensive. Neuromorphic systems need to consider strategies to embed more structure in peripheral communication paths in order to ease the construction of topographical maps.

Structuring in central processing system The primary somatosensory cortex (S1) represents whole-body tactile information in complete somatotopic maps (see sections 5.2.4 and 5.2.5.1). These maps represent spatial relationships between tactile information and form the internal representation of the body from the tactile point of view. Topographical maps are important for the efficiency of processes that depend on neighbor information. Such processes abstract topographical features (represented in somatotopic maps, see section 5.2.5.1) to structural, geometrical, or dynamical features (e.g., tactile flow, motion, and shape, see section 5.2.5.3). References [76–80] demonstrate how robots can automatically (re-)acquire the structural and spatial information of their robot skin. The structural and spatial information is used to eventually map the unordered stream of location- and type-tagged tactile information to a structured somatotopic map. Currently, different kinds of somatotopic representation are used: (1) a map that associates skin patches (a connected functional group of skin cells) with rigid robot links (e.g., lower arm, upper arm, shoulder), (2) a map that represents neighborhood relations of skin cells, and (3) a map of kinematic chains from every skin cell to the robot base frame. The works demonstrate that robot skin systems can automatically acquire structural information and structure stimuli features to topographical maps. However, neuromorphic mechanisms, which allow more structural organization in the periphery, will improve the efficiency of mapping and representing information in such somatotopic maps.

5.4.5.2 Real-time robot skin systems with generalized information representations

Reference [81] introduces a synchronous real-time framework for generic tactile information acquisition, representation, and processing. The authors call this framework Skinware. The framework implements a hardware abstraction layer (HAL) that abstracts different robot skin systems. HAL abstracts different information acquisition mechanisms and information representations to general, abstract descriptions. The framework also provides generic definitions for tactile information, for structural information of robot skin, and for exchanging and sharing synchronized information between concurrent processes in real time. The synchronous acquisition of skin information drives the framework's processing of information. The framework allows information consumers become active on the arrival of new information, which avoids the inefficient polling of information. A nice feature of the proposed frameworks is that it abstracts how tactile information is acquired and consequently relaxes the interdependency between information acquisition and processing. In this way, the framework contributes to the development of a general tool set of tactile information processing software that can be used for different robot skin systems. The framework could be used for the development of a tactile information processing architecture that hierarchically condenses and abstracts information just like the primary somatosensory cortex (S1). However, the framework only provides tools to develop such a system. It does not provide a generalized hierarchical description for assembling tactile stimuli features into intermediate abstract representations for cognitive processes (see sections 5.3.2 and 5.4.6). Unfortunately, the proposed framework is not yet ready for asynchronous event-driven information processing and would need major adjustments in its generalized information acquisition and exchange mechanisms. Nevertheless, the framework's mechanism to drive processing on the update of information rather than polling information can be used to drive processing through the arrival of events (see section 5.5).

5.4.6 Neuromorphic parallel information streams of the cognitive where and what

Neuroscientific studies indicate that tactile information leaves the primary somatosensory cortex (S1) in two separate pathways (see sections 5.2.6 and 5.2.7). In these pathways, higher cortical processes process and abstract different aspects of information. These two pathways are (1) the *where* pathway that focuses on the *where* property of the contact/object (see section 5.2.6) and (2) the *what* pathway that focuses on the *what* property of the contact/object (see section 5.2.7).

5.4.6.1 The where pathway

In robotics, the research area of tactile reactive control [5, 15, 27, 29, 82, 83] resembles the cortical where pathway (see sections 5.1.2 and 5.2.6). Tactile reactive control uses contact location and dynamics to implement desired robot behaviors enabling tactile interaction. Tactile interaction requires that large areas of the robot are covered with robot skin. For this reason, research in tactile reactive control focuses on self-organizing large-scale robot skin systems. Spatial self-calibration, which results in somatotopic maps, is strictly required because the manual calibration of thousands of sensors is infeasible (see section 5.4.5).

5.4.6.2 The what pathway

In robotics, the research area of tactile texture and object recognition [74, 84–87] correlates to the what pathway (see sections 5.1.2 and 5.2.7). Tactile recognition algorithms use structural and temporal tactile information to extract structural features and texture properties that enable contact/object classification with high success rates. The reliable extraction of structural features (texture) requires tactile sensors with a high spatiotemporal resolution in small areas placed on the most important locations, e.g., fingertips. In contrast to large-scale robot skin, tactile sensors for object recognition are not distributed on spatially distant location but concentrated on well-known locations (see section 5.1.2). Self-calibration is not needed because sensor locations (i.e., the somatosensory map, see section 5.4.5) are already known. However, tactile sensors need a high spatial resolution and high sampling frequencies to capture all the stimuli features necessary for reliable tactile recognition [74].

5.4.6.3 The connection of the where and the what pathway

As discussed in sections 5.2.6 and 5.2.7, the cortex divides the processing of different contact/object features into two dominant parallel processing pathways. However, these two principal pathways are not strictly separated; on contrary, they are heavily interconnected to each other [33]. Thus, the cortical processing architecture demonstrates that information processed for a specific goal follows a principal pathway. This allows the cortex to efficiently abstract information for specific goals in parallel with low latency. The communication between pathways allows information exchange. In this way, inefficient duplicate processing of features is avoided and additional information is available. For example, surface structural features processed in the what pathway can support tactile control in the where pathway, or contact/object dynamic features processed in the where pathway can support contact/object recognition in the what pathway (see Figure 5.4). So far, the tactile research community completely separates the two pathways (see sections 5.2.6 and 5.2.7). Robot skin systems used for tactile control (where pathway) lack the spatiotemporal resolution for fine recognition tasks and robot skin systems used in object recognition (what pathway) lack the properties for scaling to large-scale robot skin. The integration of both systems would be desirable and would result in increased representation and processing efficiency. It would also improve tactile control and recognition performance. The advantages of combining both processing pathways have been recently shown in Reference [88]. The authors combine dynamic contact/object information (e.g., weight, moving center of mass) with textural information and improve dexterous manipulation tasks with robot hands. In the near future, the realization of dextrous whole-body control will depend on the efficient combination of information about robot body dynamics, tactile contact/object dynamics, and contact/object recognition. The efficient integration of the where and the what pathway will be essential to fully exploit the robots' maneuverability and manipulation capabilities in whole-body robot control.

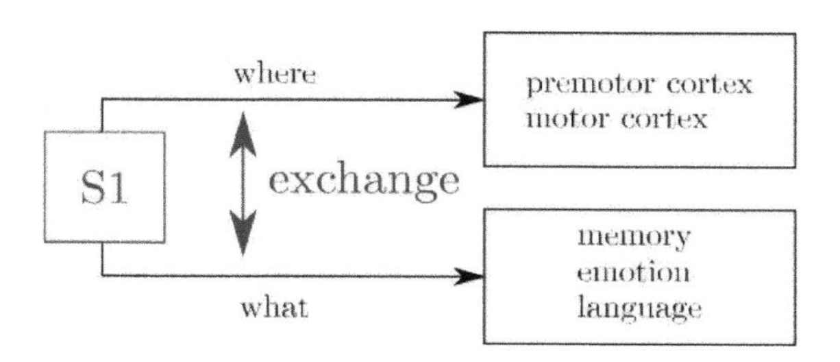

Figure 5.4 The where and what pathways

5.5 The realization of an event-driven large-scale robot skin system

References [31, 89, 90] present an event-driven modular multi-modal robot skin system and an event-driven tactile reactive controller. The authors demonstrate how large-scale robot skin systems profit from redundancy reduction at the skin cell level. They also demonstrate how event-driven processing improves the processing efficiency in tactile reactive control.

5.5.1 Robot skin system

5.5.1.1 Robot skin

The introduced event-driven robot skin system is based on a state-of-the-art modular multi-modal robot skin system [14, 15] (see Figure 5.5). The skin is modular and uses identical, hexagonally shaped skin cells. The skin cells can be assembled to groups (skin patches) that can cover arbitrarily shaped surfaces. Each skin cell employs the same group of multi-modal sensors and a microcontroller on its back (see Figure 5.5). A skin cell can provide tactile information in the form of nine different stimuli features. These stimulus features are transduced by a 3D accelerometer (three vibration stimuli), three capacitive force sensors (three normal force stimuli), two temperature sensors (two thermal stimuli), and one proximity sensor (one distance stimulus). The microcontrollers on the back of the skin cells manage information acquisition, noise filtering, data packet generation, and network routing. Skin cells in a skin patch are interconnect via asynchronous universal asynchronous

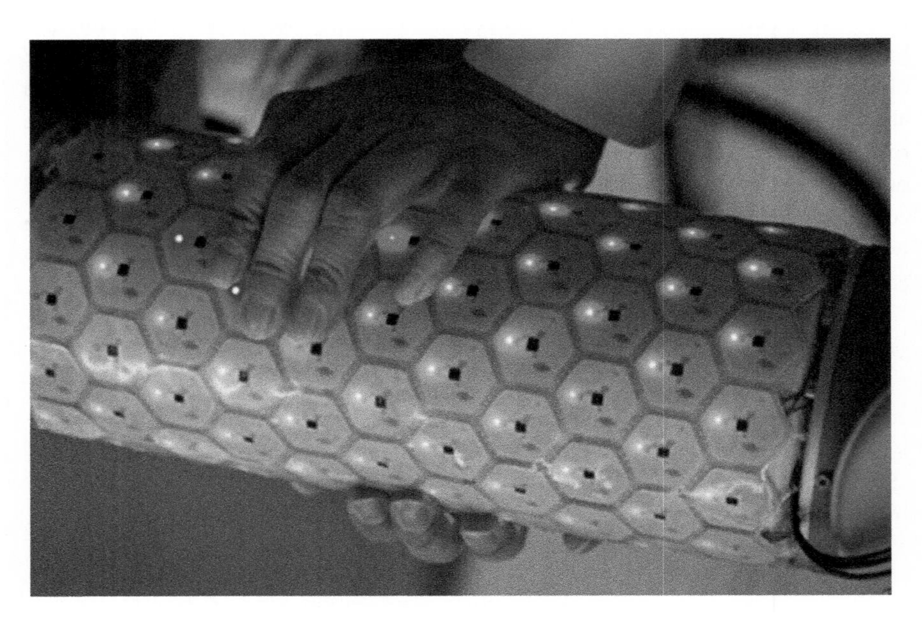

Figure 5.5 Robot skin; Copyright Astrid Eckert/TU München

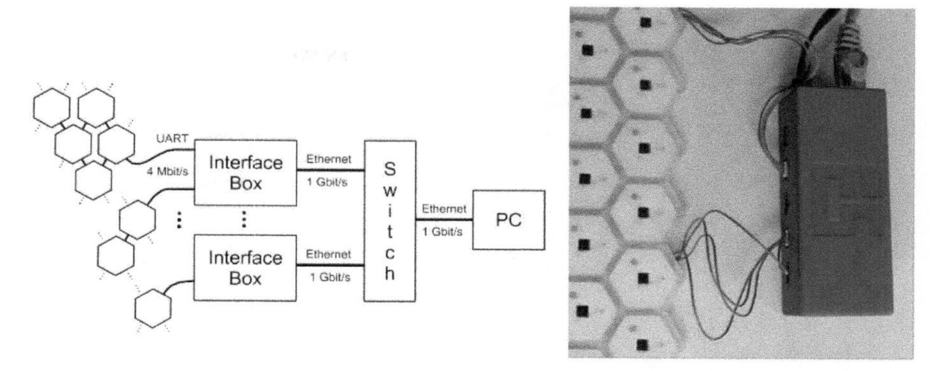

Figure 5.6 A skin patch connected to a TSU; copyright ICS/TU München

receiver transmitter (UART) connections and form a routedmeshed skin cell network. The network is self-organizing and automatically findsbidirectional communication paths between each skin cell and the tactile sectionunit (TSU, see Figure 5.6). The TSU translates skin network packets to standarduser datagram protocol (UDP) Ethernet packets and vice versa. UDP-based communication links ease the connection of TSUs to standard PCs. No additional specialhardware or low-level kernel drivers are needed.

5.5.1.2 Structural self-calibration

The modular robot skin system is fully capable of self-calibrating [76–78]. The skin system can automatically determine static homogenous transformations ${}^{i}T_{k} \in R^{4\times4}$ of skin cell i on joint k. In combination with the robot kinematics, the system can build up the kinematic chain of every skin cell i with respect to a common reference frame, e.g., the world reference frame 0 (see Figure 5.7). The feature to self-calibrate is very important for large-scale robot skin, because the manual calibration of thousands of skin cells is totally infeasible. The kinematic chain of every skin cell, thus ${}^{i}T_{0}$, is needed in tactile reactive control in order to calculate and represent skin cell wrenches with respect to a common reference frame [5].

5.5.1.3 Event-driven robot skin

To increase the transmission and processing efficiency and to reduce the latency, the authors transformed their standard robot skin system into an event-driven robot skin system [31, 89]. The skin cells support two different modes: (1) the synchronous data sampling mode and (2) the event mode. The data mode implements the standard skin information acquisition process. Skin cells in data mode sample and transmit tactile information with a constant sample rate. In event mode, however, the skin cells let information novelty drive the generation of event packets. The skin cells realize a compound architecture (see section 5.4.3.2) and their event generators implement the SoDP (see section 5.4.3). The two different modes allow

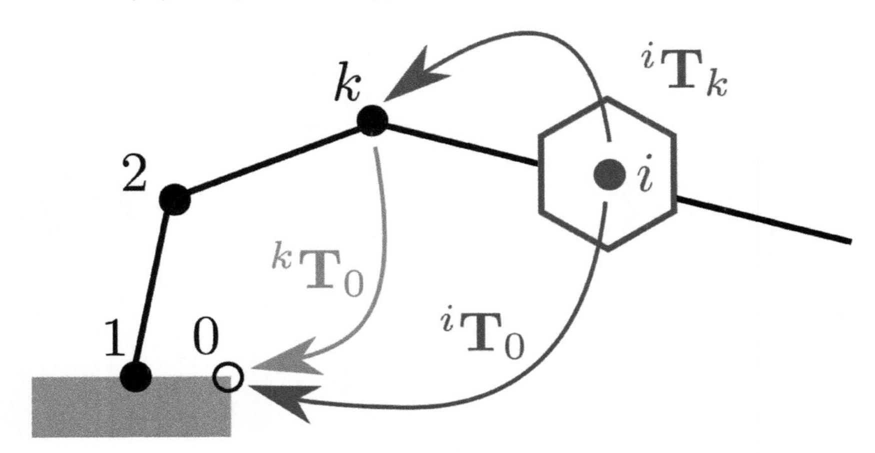

Figure 5.7 Kinematic chain from skin cell i *on joint* k *to the world reference frame 0*

the authors to fully investigate and compare the efficiency of event-driven information generation, transmission, and processing in one and the same robot skin system. The skin cells in event mode increase their efficiency by applying two of three possible biological principles for neuromorphic event-driven sensors (see section 5.3.1). They disassemble complex stimulus information into simple stimulus features, and information acquisition, transmission, and processing of these features are event-driven. The robot skin is the first neuromorphic large-scale skin that successfully applies these two biological principles. Effectively, in this way, skin cells increase their efficiency through redundancy reduction at the skin cell level (see section 5.5.3).

5.5.1.4 Neuromorphic skin system

The authors revised their skin system for efficiently supplying host applications with skin information. As the skin cells (see section 5.5.1.3), the system supports the data and event mode. The system provides a modularized skin driver library for host applications (see Figure 5.8). The low-level processes, which acquire and process raw skin cell packets, are implemented in the skin HAL. The skin HAL generalizes the interface between connected skin cells and host applications. The HAL provides a set of interfaces for different hardware interfaces like Ethernet or USB. The interfaces convert the raw skin cell packets to general skin cell packets. They also handle the initialization process of the skin cell network and the different skin cell operation modes. The high-level driver (HLD) of the skin driver library provides a generalized interface to skin cells for host applications. The new skin system enables neuromorphic event-driven processing of tactile information on standard PCs. The neuromorphic event-driven processing of tactile information is more efficient because processes become only active when novel information is available (see sections 5.2.2 and 5.3.1). Furthermore, the representation

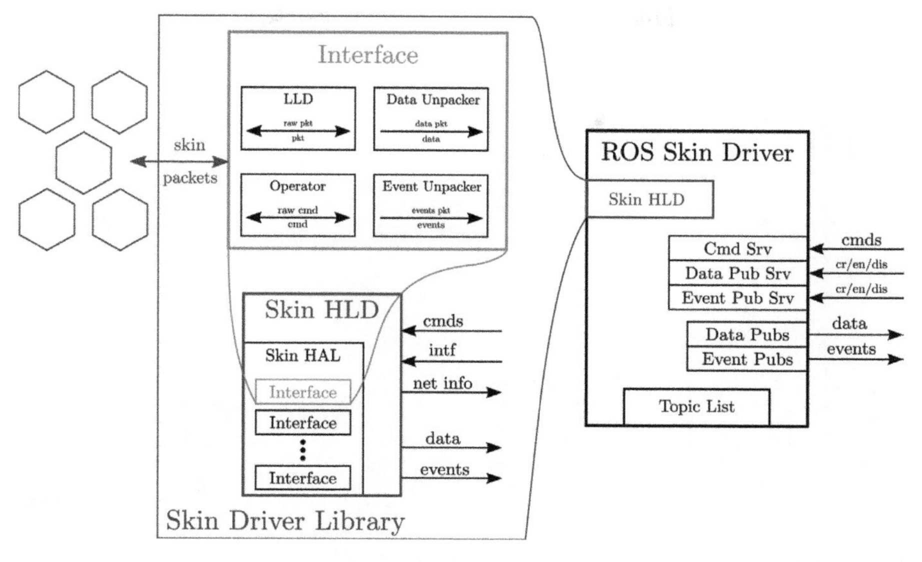

Figure 5.8 *Skin driver and ROS skin driver; the low-level driver converts raw skin cell packets to general skin cell packets; the HLD provides a generalized interface to host applications, e.g., ROS skin driver*

of information in events avoids the inefficient searching of activity in large data sets because the event itself is salient and signifies activity. The novel system skin system applies biological principles for processing and representing information in a hierarchical structured way (see section 5.3.2). The system employs the three biological principles discussed in section 5.3.2, namely, structure, hierarchical abstraction, concurrent specialized processing, and benefits from increased computational efficiency (see section 5.5.3). The neuromorphic skin system hierarchically represents skin information in somatotopic order and provides topographical features (see section 5.2.5.1) and geometric/dynamical features (see section 5.2.5.3). These features are generally available in the system and can be accessed concurrently. Specialized neuromorphic event-driven control processes can access these features and implement efficient tactile behaviors with large-scale robot skin (see sections 5.5.2 and 5.5.3).

5.5.2 Event-driven reactive skin control

Recently, the authors presented an event-driven tactile reaction controller [90] (see Figure 5.9) that can take full advantage of their novel neuromorphic skin system (see section 5.5.1.4). In contrast to the reactive skin control system presented in their previous work [31], the new system now directly uses events in the control loop. The previous system converted events back to the standard synchronous skin data representation for a standard synchronous reactive controller. The novel reactive skin controller itself is event-driven—the arrival of

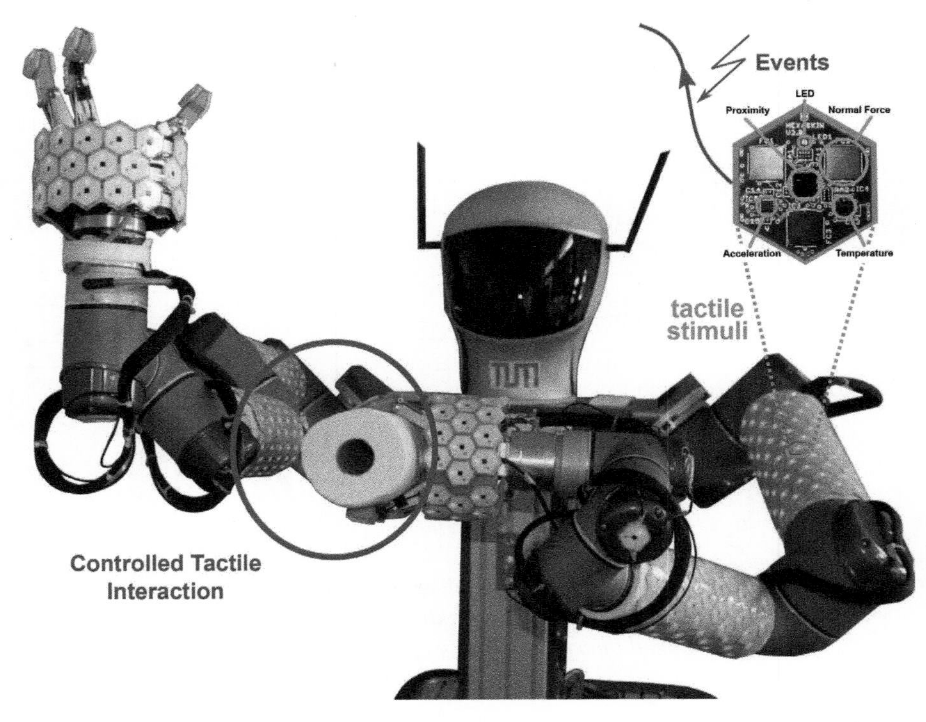

Figure 5.9 *The robot TOMM [91] with two arms and two grippers covered with skin; the robot holds a paper towel roll in its left gripper which it uses in the experiment to push the right arm; the right arm is controlled by a novel event-driven tactile reaction controller [90]; it tries to avoid contacts and moves to the right*

new events drives its processing (see sections 5.2.2 and 5.3.1). This increases the computational efficiency of the controller. Tactile information can now be processed inside the control loop without negative influence on controller stability. For the design of the reactive skin controller, the authors use a new separated formulation of the Jacobian torque calculation. For the validation of their novel system, the authors constructed a repeatable experimental setup, which ensures the comparability of tactile interaction experiments. In this way, experiments can be conducted separately in data or event mode with different controllers and compared afterward. The experimental setup is as follows (see Figure 5.9): the authors use the left arm of TOMM to excite its right arm. The left gripper holds a paper towel roll that is moved against the lower right arm. The authors execute the reactive skin controllers on the right arm and a Cartesian position controller on the left arm. The left arm moves along a predefined trajectory. When the paper towel roll touches the right arm, then the right arm moves away and avoids the contact. The authors provide a video as an attachment in Reference [90].

5.5.3 The benefits

A comprehensive analysis of the proposed robot skin system shows that the event-driven robot skin reduces the network usage by at least 21.2% and the CPU usage in a standard tactile control system by at least 17% [31]. The novel event-driven controller introduced in Reference [90] (see section 5.5.2) reduces the overall system CPU usage by at least 66%. The proposed event-driven skin system has the previously discussed drawbacks of compound architectures (see section 1.4.3.2). The system cannot take full advantage from the power-saving and high temporal precision properties of event-drivensensing. However, the proposed robot skin system is the first distributed event-driven multi-modal robot skin system developed. The neuromorphic system demonstrates the flexibility in its ability to easily switch between modes for evaluation purposes and proves the efficiency of neuromorphic event-driven transmission, processing, and control.

References

[1] Harmon L.D. 'Aautomated tactile sensing'. *International Journal of Robotics Research*. 1982, vol. 1(2), pp. 3–32.

[2] Bogue R. 'Europe continues to lead the way in the Collaborative robot business'. *Industrial Robot*. 2016, vol. 43(1), pp. 6–11.

[3] Nejat, G.. 'Assistive robots in health care settings'. *Home Health Care Management & Practice*. 2009, vol. 3, pp. 177–87.

[4] Dautenhahn K., Woods S., Kaouri C., Walters M.L., Kheng L.K., Werry I. 'Presented at 2005 IEEE/RSJ international conference on intelligent robots and systems;' edmonton, AB, 2005'.

[5] Dean-Leon E., Ramirez-Amaro K., Bergner F., Dianov I., Lanillos P., Cheng G. 'Robotic technologies for fast deployment of industrial robot systems'. *IECON 2016 - 42nd Annual Conference of the IEEE Industrial Electronics Society*; Florence, Italy, Manhattan, New York, U.S: IEEE, the Institute of Electrical and Electronics Engineers, 2016.

[6] Tar Á., Cserey G. 'Development of a low cost 3D optical compliant tactile force sensor'. *EEE/ASME International Conference on Advanced Intelligent Mechatronics (AIM)*; 2011. pp. 40–236.

[7] OptoForce. 'Optoforce manual for general DAQs with USB, UART or can interface'. *Budapest, Hungary: Optoforce Kft*. 2016, vol. 16, p. H–1043.

[8] OptoForce. 'Optoforce 3 axis frorce sensor OMD-20-SE-40N datasheet.' Budapest, Hungary: Optoforce Kft; 2016.

[9] SynTouch *Biotac product manual*. 222 S. Los Angeles CA: SynTouch LLC; 2015.

[10] Fishel J.A., Santos V.J., Loeb G.E. 'Presented at EMBS international conference on bomedical robotics and biomechatronics (biorob); scottsdale, AZ, 2008'. Germany,

[11] Fishel J.A., Gerald E. L. 'Sensing tactile microvibrations with the bio-tac—comparison with human sensitivity'. *4th IEEE RAS & EMBS International*

Conference on Biomedical Robotics and Biomechatronics (BioRob); Germany, 2012.

[12] Schmitz A., Maggiali M., Natale L., Bonino B., Metta G.. 'IEEE/RSJ International Conference on Intelligent Robots and Systems (Iros 2010); Taipei'. 2010.

[13] Risto K., Matthias Z., Carsten S., Robert H., Helge J. 'A highly sensitive 3D-shaped tactile sensor'. *IEEE/ASME International Conference on Advanced Intelligent Mechatronics (AIM)*; Manhattan, New York, U.S: IEEE, the Institute of Electrical and Electronics Engineers, 2013. pp. 89–1084.

[14] Mittendorfer P., Cheng G. 'Humanoid multimodal tactile-sensing modules'. *IEEE Transactions on Robotics*. 2011, vol. 27(3), pp. 401–10.

[15] Mittendorfer P., Yoshida E., Cheng G. 'Realizing whole-body tactile interactions with a self-organizing, multi-modal artificial skin on a humanoid robot'. *Advanced Robotics*. 2015, vol. 29(1), pp. 51–67.

[16] Maggiali, M., Giorgio C., Perla M. 'Embedded distributed capacitive tactile sensor'. *11th Mechatronics Forum Biennial International Conference*; Germany, 2008.

[17] Baglini E., Youssefi S., Mastrogiovanni F., Cannata G. 'A real-time distributed architecture for large-scale tactile sensing'. *IEEE/RSJ International Conference on Intelligent Robots and Systems (IROS 2014)*; Chicago, IL, 2014.

[18] Maiolino P., Maggiali M., Cannata G., Metta G., Natale L. 'A flexible and robust large scale capacitive tactile system for robots'. *IEEE Sensors Journal*. 2013, vol. 13(10), pp. 3910–17.

[19] Lumelsky V.J., Shur M.S., Wagner S. 'sensitive skin'. *IEEE Sensors Journal*. 2001, vol. 1(1), pp. 41–51.

[20] Strohmayr M.W., Schneider D. 'IEEE/RSJ international conference on intelligent robots and systems (IROS 2013)'. 2013. tokyo,

[21] Strohmayr M.W., Worn H., Hirzinger G. 'IEEE International Conference on Robotics and Automation (ICRA)'; Karlsruhe, Germany, IEEE. 2013.

[22] Alirezaei H., Nagakubo A., Kuniyoshi Y. 'IEEE symposium on 3D user interfaces; lafayette, LA'. Manhattan, New York, U.S: IEEE, the Institute of Electrical and Electronics Engineers, 2009. pp. 87.

[23] Yogeswaran N., Dang W., Navaraj W.T., *et al*. 'New materials and advances in making electronic skin for interactive robots'. *Advanced Robotics*. 2015, vol. 29(21), pp. 1359–73.

[24] Buscher G., Meier M., Walck G., Haschke R., Ritter H.J. ' Aaugmenting curved robot surfaces with soft tactile skin '. Presented at IEEE/RSJ International Conference on Intelligent Robots and Systems (IROS); Hamburg, Germany, 2015.

[25] Büscher G.H., Kõiva R., Schürmann C., Haschke R., Ritter H.J. 'Flexible and stretchable fabric-based tactile sensor'. *Robotics and Autonomous Systems*. 2015, vol. 63, pp. 52–244.

[26] Mukai T., Onishi M., Odashima T., Hirano S., Luo Z. 'Development of the tactile sensor system of a human-interactive robot " ri-man"'. *IEEE Transactions on Robotics*. 2008, vol. 24(2), pp. 505–12.

[27] Fritzsche M., Elkmann N., Schulenburg E. 'Tactile sensing: a key technology for safe physical human robot interaction'. Proceedings of the 6th International Conference on Human-robot Interaction; 2011.

[28] Schenk M. Taktile sensorsysteme. Vol. 22. Sandtorstrasse, Magdeburg: Fraunhofer IFF; 2016.

[29] Bosch R. 'APAS intelligent systems for man-machine collaboration'. *Robert Bosch GmbH, Postfach*. 2016, vol. 30.

[30] Ohmura Y., Kuniyoshi Y., Nagakubo A. 'Conformable and scalable tactile sensor skin for curved surfaces'. *IEEE International Conference on Robotics and Automation (ICRA)*; Orlando,FL: The publisher is IEEE, the Institute of Electrical and Electronics Engineers, 2006. pp. 53–1348.

[31] Bergner F., Dean-Leon E., Cheng G. 'Event-based signaling for large-scale artificial robotic skin - realization and performance evaluation'. *2016 IEEE/RSJ International Conference on Intelligent Robots and Systems (IROS)*; Daejeon, South Korea, Manhattan, New York, U.S: The publisher is IEEE, the Institute of Electrical and Electronics Engineers, 2016. pp. 24–4918.

[32] Gardner E.P., Kenneth O. J. 'The somatosensory system: receptors and central pathways' in Principles of Neural Science. Fifth Edition. New York, NY, 2013: The MacGraw-Hill Companies; pp. 97–475.

[33] Benarroch E.E. 'Processing information in the cerebral cortex' in *Basic neurosciences with clinical applications*. Philadelphia, PA, USA: Elsevier Health Sciences; 2006. pp. 67–427.

[34] Kandel E.R., JamesH.S., Thomas M.J., Steven A.S., Hudspeth A.J. *Principles of Neural Science*. Fifth Edition. New York, NY: The MacGraw-Hill Companies; 2013.

[35] Saal H.P., Bensmaia S.J. 'Touch is a team effort: interplay of submodalities in cutaneous sensibility'. *Trends in Neurosciences*. 2014, vol. 37(12), pp. 689–97.

[36] Gardner E.P. Touch in encyclopedia of life sciences. New York City, NY: John Wiley & Sons; 2010.

[37] Abraira, V.E., David D. G. 'The sensory neurons of touch'. *Cell*. 2013, vol. 79, pp. 618–39.

[38] Roudaut Y., Lonigro A., Coste B., Hao J., Delmas P., Crest M. 'Touch sense: functional organization and molecular determinants of mechanosensitive receptors'. *Channels (Austin, Tex.)*. 2012, vol. 6(4), pp. 234–45.

[39] Gardner E.P., Kenneth O. J. *Principles of Neural Science*. Fifth Edition. New York, NY: The MacGraw-Hill Companies; 2013. pp. 428–529.

[40] Kandel E.R., Barres B.A., Hudspeth A.J. 'Nerve cell, neural circuitry and behavior' in *Principles of Neural Science*. Fifth Edition. New York, NY: The MacGraw-Hill Companies; 2013. pp. 21–38.

[41] Gardner E.P., Kenneth O J. ' principles of neural science ' in *Sensory Coding*. Fifth Edition. New York, NY: The MacGraw-Hill Companies; 2013. pp. 49–74.
[42] Amaral D.G. 'The functional organization of perception and movement' in *Principles of Neural Science*. Fifth Edition. New York, NY: The MacGraw-Hill Companies; 2013. pp. 69–356.
[43] Kandel E.R. 'The internal representations of space and action' in *Principles of Nneural Science*. Fifth Edition. New York, NY: The MacGraw-Hill Companies; 2013. pp. 91–370.
[44] Lichtsteiner P., Posch C., Delbruck T. 'A 128 × 128 120 dB 15 μs latency asynchronous temporal contrast vision sensor'. *IEEE Journal of Solid-State Circuits*. 2008, vol. 43(2), pp. 566–76.
[45] Posch C., Daniel M., Rainer W. 'An asynchronous time-based image sensor'. *ISCAS, IEEE International Symposium on Circuits and Systems*; Seattle,WA, 2008.
[46] Posch C., Teresa S.G. 'Retinomorphic event-based vision sensors: bioinspired cameras with spiking output'. *Proceedings of the IEEE*. 2014, vol. 10, pp. 1470–84.
[47] Brandli C., Berner R., Minhao Y., Shih C.L., Delbruck T. 'A 240 × 180 130 dB 3 μs latency global shutter spatiotemporal vision sensor'. *IEEE Journal of Solid-State Circuits*. 2014, vol. 49(10), pp. 2333–41.
[48] Li C., Brandli C., Berner R, *et al*. *Design of an RGBW color VGA rolling and global shutter dynamic and active-pixel vision sensory*. Lisbon, Portugal: 2015 IEEE International Symposium on Circuits and Systems (ISCAS); 2015.
[49] Bartolozzi C., Rea F., Clercq C, *et al*. Presented at EEE computer society conference on computer vision and pattern recognition workshops (CVPR workshops); colorado springs, CO, USA, 2011.
[50] Benosman, R. 'Asynchronous frameless event-based optical flow'. *Neural Networks*. 2012, vol. 27, pp. 32–37.
[51] Benosman R. 'Event-based visual flow'. *IEEE Transactions on Neural Networks and Learning Systems*. 2014, vol. 25, pp. 17–407.
[52] Conradt J., Cook M., Berner R., Lichtsteiner P., Douglas R.J., Delbruck T. 'Presented at 2009 IEEE International Symposium on circuits and systems– ISCAS 2009; Taipei, Taiwan, 2009'.
[53] Ghaderi V.S., Mulas M., Pereira V.F.S., Everding L., Weikersdorfer D., Conradt J. 'A wearable mobility device for the blind using retina-inspired dynamic vision sensors'. *Annual International Conference of the IEEE Engineering in Medicine and Biology Society. IEEE Engineering in Medicine and Biology Society. Annual International Conference*. 2015, vol. 2015, pp. 3371–74.
[54] Boahen K.A. 'Point-to-point connectivity between neuromorphic chips using address events'. *IEEE Transactions on Circuits and Systems II*. 2000, vol. 47(5), pp. 416–34.

[55] Mahowald M. 'VLSI analogs of neuronal visual processing: a synthesis of form and function'. [PhD thesis]. Pasadena, CA, California Institute of Technology, 1992.
[56] Schaik A., Shih Chii L. 'AER ear: a matched silicon cochlea pair with address event representation interface'. *IEEE International Symposium on Circuits and Systems (ISCAS)*; Kobe, Japan, 2005. pp. 16–4213.
[57] Caviglia S., Valle M., Bartolozzi C. 'Asynchronous, event-driven readout of posfet devices for tactile sensing'. *IEEE International Symposium on Circuits and Systems (ISCAS)*; Melbourne VIC, IEEE, 2014. pp. 51–2648.
[58] Furber S.B., Lester D.R., Plana L.A, *et al.* 'Overview of the spinnaker system architecture'. *IEEE Transactions on Computers*. 2013, vol. 62(12), pp. 2454–67.
[59] Siegelbaum S.A., Koester J. 'Ion channels in principles of *Neural Science*' in Fifth edition. New York, NY: The MacGraw-Hill Companies; 2013. pp. 25–100.
[60] Caviglia S., Pinna L., Valle M., Bartolozzi C. 'An event-driven POSFET taxel for sustained and transient sensing'. *2016 IEEE International Symposium on Circuits and Systems (ISCAS)*; Montréal, QC, Canada, Manhattan, New York, U.S: The publisher is IEEE, the Institute of Electrical and Electronics Engineers, 2016. pp. 52–349.
[61] Zamarreno-Ramos C., Linares-Barranco A., Serrano-Gotarredona T., Linares-Barranco B. 'Multicasting mesh AER: a scalable assembly approach for reconfigurable neuromorphic structured aer systems. Application to convnets'. *IEEE Transactions on Biomedical Circuits and Systems*. 2013, vol. 7(1), pp. 82–102.
[62] Berge H., Philipp H. 'High-speed serial AER on FPGA'. IEEE International Symposium on Circuits and Systems (ISCAS); 2007.
[63] Zamarreno-Ramos C., Serrano-Gotarredona R., Serrano-Gotarredona T., Linares-Barranco B. 'IEEE International Symposium on Circuits and Systems-ISCAS 2008; Seattle, WA'. 2008.
[64] Zamarreno-Ramos C., Serrano-Gotarredona T., Linares-Barranco B. 'An instant-startup jitter-tolerant manchester-encoding serializer/deserializer scheme for event-driven bit-serial IVDS interchip AER links'. *IEEE Transactions on Circuits and Systems I*; 2011. pp. 60–2647.
[65] Rovere G., Bartolozzi C., Imam N., Manohar R. 'Design of a QDI asynchronous AER serializer/deserializer link in 180 nm for event-based sensors for robotic applications'. *IEEE International Symposium on Circuits and Systems (ISCAS)*; Lisbon, Portugal, IEEE, 2015. pp. 15–2712.
[66] Ros P.M., Crepaldi M., Bartolozzi C., Demarchi D. 'Asynchronous DC-free serial protocol for event-based AER systems'. *IEEE International Conference on Electronics, Circuits, and Systems (ICECS)*; Cairo, Egypt, 2015. pp. 51–248.
[67] Mario N., Klaus K. 'A new protocol for a low power sensor network'. *IEEE International Conference on Performance, Computing, and Communications*; Phoenix, United States, 2004. pp. 99–393.

[68] Miskowicz M. 'Send-on-delta concept: an event-based data reporting strategy'. *Sensors*. 2006, vol. 6(1), pp. 49–63.

[69] Paolo Motto R., Crepaldi M., Bonanno A., Demarchi D. 'Wireless multichannel quasi-digital tactile sensing glove-based system'. Presented at Euromicro Conference on Digital System Design (DSD); Los Alamitos, CA, 2013.

[70] Ros P.M., Crepaldi M., Demarchi D. 'A hybrid quasi-digital/neuromorphic architecture for tactile sensing in humanoid robots'. 6th EEE International Workshop on Advances in Sensors and Interfaces (IWASI); Gallipoli, 2015.

[71] Caviglia S., Pinna L., Valle M., Bartolozzi C. 'Spike-based readout of posfet tactile sensors'. *IEEE Transactions on Circuits and Systems I*. 2016, vol. 64(6), pp. 1421–31.

[72] Seminara L., Pinna L., Ibrahim A, *et al*. 'Electronic skin: achievements, issues and trends'. *Procedia Technology*. 2014, vol. 15, pp. 549–58.

[73] Seminara L., Luigi P., Ali I, *et al*. 'Towards integrating intelligence in electronic skin'. *Mechatronics*. 2015, vol. 34, pp. 84–94.

[74] Lee W.W., Kukreja S.L., Thakor N.V. 'A kilohertz kilotaxel tactile sensor array for investigating spatiotemporal features in neuromorphic touch'. *2015 IEEE Biomedical Circuits and Systems Conference (BioCAS)*; Atlanta, GA, USA, Manhattan, New York, U.S: The publisher is IEEE, the Institute of Electrical and Electronics Engineers, 2015. pp. 1–4.

[75] Lee W.W. '*A Neuromorphic Approach To Tactile Perception*'. [PhD thesis]. National University of Singapore, 2016

[76] Philipp M., Cheng G. '3D surfacere construction forroboticbody parts with artificial skins'. IEEE/RSJ International Conference on Intelligent Robots and Systems (IROS 2012); Vilamoura-Algarve, Portugal, IEEE, 2012.

[77] Philipp M., Cheng G. 'Open-loop self-calibration of articulated robots with artificial skins'. Presented at 2012 IEEE International Conference on Robotics and Automation (ICRA); St Paul, MN, IEEE, 2012.

[78] Philipp M., Dean E., Cheng C. 'Automatic robot kinematic modeling with a modular artificial skin'. *IEEE-RAS 14th International Conference on Humanoid Robots (Humanoids 2014)*; Madrid, Spain, Manhattan, New York, U.S: The publisher is IEEE, the Institute of Electrical and Electronics Engineers, 2014. pp. 54–749.

[79] Giorgio C., Denei S., Mastrogiovanni F. 'Tactile identification of objects using bayesian exploration''. Presented at IEEE International Symposium on Robot and Human Interactive Communication; Lisbon, Portugal, 2010.

[80] Denei S., Mastrogiovanni F., Cannata G. 'Towards the creation of tactile maps for robots and their use in robot contact motion control'. *Robotics and Autonomous Systems*. 2015, vol. 63, pp. 293–308.

[81] Youssefi S., Denei S., Mastrogiovanni F., Cannata G. 'A real-time data acquisition and processing framework for large-scale robot skin'. *Robotics and Autonomous Systems*. 2015, vol. 68, pp. 86–103.

[82] Wosch T., Wendelin F. 'Reactive motion control for human-robot tactile interaction'. *IEEE International Conference on Robotics and Automation (ICRA)*; Lisbon, Portugal, 2002.

[83] Nori F., Traversaro S., Eljaik J., Romano F., Del Prete A., Pucci D. 'iCub whole-body control through force regulation on rigid non-coplanar contacts'. *Frontiers in Robotics and AI*. 2015, vol. 2(6), pp. 1–18.

[84] Sinapov J., Vladimir S. 'Vibro-tactile recognition and categorization of surfaces by a humanoid robot'. *IEEE Trans- Actions on Robotics*. 2011, vol. 27(3), pp. 488–97.

[85] Fishel J.A., Loeb G.E. 'Bayesian exploration for intelligent identification of textures'. *Frontiers in Neurorobotics*. 2012, vol. 6(4), pp. 15–34.

[86] Xu D., Loeb G.E., Fishel J.A. 'IEEE International Conference on Robotics and Automation (ICRA); Karlsruhe, Germany, 2013'.

[87] Kaboli M., Walker R., Cheng G. 'Ieee international conference on robotics and automation (icra)'. Presented at Re-using prior tactile experience by robotic hands to discriminate in-hand objects via texture properties; Manhattan, New York, U.S: The publisher is IEEE, the Institute of Electrical and Electronics Engineers,

[88] Kaboli M., Yao K., Cheng G. 'Tactile-based manipulation of deformable objects with dynamic center of mass'. Presented at 2016 IEEE-RAS 16th International Conference on Humanoid Robots (Humanoids); Cancun, Mexico, 2016. Manhattan, New York, U.S: The publisher is IEEE, the Institute of Electrical and Electronics Engineers,

[89] Bergner F., Mittendorfer P., Dean-Leon E., Cheng G. 'Event-based signaling for reducing required data rates and processing power in a large-scale artificial robotic skin'. *2015 IEEE/RSJ International Conference on Intelligent Robots and Systems (IROS)*; Hamburg, Germany, Manhattan, New York, U.S: The publisher is IEEE, the Institute of Electrical and Electronics Engineers, 2015. pp. 29–2124.

[90] Bergner F., Dean-Leon E., Cheng G. 'Efficient event-driven reactive control for large scale robot skin'. *2017 IEEE International Conference on Robotics and Automation (ICRA)*; Singapore, Singapore, Manhattan, New York, U.S: The publisher is IEEE, the Institute of Electrical and Electronics Engineers, 2017. pp. 394–400.

[91] Dean-Leon E., Pierce B., Bergner F. 'TOMM: tactile omnidirectional mobile manipulator'. IEEE International Conference on Robotics and Automation (ICRA); Singapore, 2017.

Chapter 6
Soft three-axial tactile sensors with integrated electronics for robot skin

Alexander Schmitz[1], Sophon Somlor[2], Tito Pradhono Tomo[2], Lorenzo Jamone[3], Richard Sahala Hartanto[2], Harris Kristanto[2], Wai Keat Wong[2], Jinsun Hwang[2], Alexandre Sarazin[2], Shuji Hashimoto[2], and Shigeki Sugano[2]

This chapter introduces recent work of the Sugano lab on distributed, soft, 3-axial tactile sensors for robot skin, based on capacitive [1–3] and Hall-Effect [4–8] sensors, with distributed electronics.

6.1 Introduction

In the future, robots should share their workspace with humans and work in unstructured environments, and for safety and to perform adaptive tasks, a rich set of sensors is needed in such scenarios. Tactile sensors on the robot's surface can provide the most detailed and direct information about the contacts with the environment and are therefore a crucial component. Yet, the fact that they should be spread over the surface makes the implementation difficult.

When designing tactile skin sensors, we consider that they should have the following characteristics: (i) soft, (ii) distributed, (iii) sensing the three-axis force vectors, and (iv) integrated electronics. (i) Softness of the robot's skin adds to the safety of the robot, because it adds a level of security that cannot be achieved with other methods. Active control schemes alone are typically too slow for impacts. Compliant joints help and are a crucial component as they can typically absorb more energy than soft skin, but they can only decouple the various links of the robot, while the inertia of the robot segments in contact with

[1]Graduate School of Creative Science and Engineering, Waseda University, Shinjuku City, Tokyo, Japan
[2]Waseda University, Shinjuku City, Tokyo, Japan
[3]Queen Mary University of London, Bethnal Green, London, United Kingdom

the human could still lead to harmful forces. Only soft skin can absorb impact energy directly at the collision site. Furthermore, soft skin in robot hands can enhance the grasp stability, and our group has found that soft skin also aids during manipulation [9–11]. (ii) Distributed skin sensors provide detailed information about the contacts with the environment, which can be used, e.g., for tasks like tactile servoing or tactile object recognition [12]. Alternatives exist, like force sensors in the robot's joints, but they provide less information, as each sensor can only measure the sum of all the forces acting on it. (iii) While most distributed skin sensors only measure single-axis force, measuring the force vector is required for tasks such as tactile servoing, and our lab has used such information, e.g., for robust in-hand manipulation [10, 13–15]. Distributed force vector measurements are useful in scenarios with more than one contact, a situation that is typically challenging for existing solutions that cannot measure distributed force vectors. (iv) When distributing many sensors on the robot's surface, without integrated electronics, a lot of wires would be required, which would require a lot of space and which would make the integration in the robot and their maintenance challenging. Distributed electronics can help, and several sensors should share a data line. Sending digital instead of analog information also makes the sensor signals less prone to interference from the environment.

Table 6.1 General comparison between different transducing technologies

Sensing principle	Advantage	Disadvantage
Piezoelectric	• High sensitivity • Dynamic sensing • Slip detection	• Thermal sensitive
Optical	• Immune to electromagnetic interference • Flexible • Sensitive	• Often large size • Need of light source and light receiver
Hall effect	• Small 3D chip available • Simple construction	• Significantly interfered by another magnetic source
Piezoresistive	• Linear • Robust	• Amplified noise • Thermal sensitive
Capacitive	• Small CDC chip available • Good sensitivity • Wide dynamic range • Robust	• Stray capacity • Thermal sensitive
MEMS	• Small in size • Excellent sensitivity	• Fragile • Small measurement range • Complicated production • Production inconsistency

Our lab has produced distributed soft tactile sensors for robot skin, based on capacitive and Hall-effect sensors. For both sensors, we use small available chips that have a digital output, which enable the integration of the sensors with fewer wires and less effort. For the capacitive sensor, we use an innovative arrangement of the transducers for three-axis sensing. The Hall-effect sensor chip already provides three-axis measurements, which enables a smaller and easier production; yet it has more hysteresis and is more prone to magnetic interference, which is difficult to shield (unlike stray capacitance).

Section 6.2 will provide an overview of existing soft three-axis sensors. Section 6.3 shows an overview of our work on capacitive sensing and section 6.4 on Hall-effect-based sensors. Section 6.5 compares them and shows the benefits and downsides for each, draws conclusions, and provides possible directions for future work.

6.2 Related work

Tactile sensing can be implemented with various sensing principles [16, 17]. Table 6.1 shows the summarized advantages and disadvantages of different sensing principles. While most distributed tactile sensors that have been introduced in the literature can measure normal force, the review here will focus on sensors that can measure a multi-axis force vector. We will also discuss related capacitive and Hall-effect sensors. A more detailed review of tactile skin sensors can be found in References 18–20 and a more recent one in Reference 21. A discussion on the importance of tactile sensing for robot manipulators can be found in Reference 22. More recently, an overview of tactile sensing in both human and humanoid was provided in Reference 16. Reviews on tactile sensing for human–robot interaction were provided in References 23 and 24.

6.2.1 Piezoelectric-based sensors

Piezoelectric sensors are sensitive to changes in force, including the vibrations caused by sliding them across different surfaces. As such, they have been used to detect the onset of slip and recognize surface textures. In many implementations, they are sensitive to shear forces as well as normal forces but cannot differentiate them. Piezoelectric sensors based on polyvinylidene fluoride (PVDF) have been used for robot hands [25–28] [an example is shown in Figure 6.1(a)], and also the body of the humanoid robot CB2 is covered with 197 sensors based on PVDF films, which are put between a layer of a soft silicone outer skin and inner urethane foam [29] [see Figure 6.1(b)]. The sensor detects the rate of change of its bending; the readout of all sensors can be done at 100 Hz. Nevertheless, only few information about the performance of the sensor is available. While in the aforementioned skin sensors, the force vector cannot be measured, a company named Touchence sells a thin, small-sized three-axis tactile sensor based on MEMS piezoelectric elements [30] [see Figure 6.1(c)]. However, using this sensor for distributed skin sensing is challenging, as the sensor is rigid and the required additional readout electronics are rather bulky.

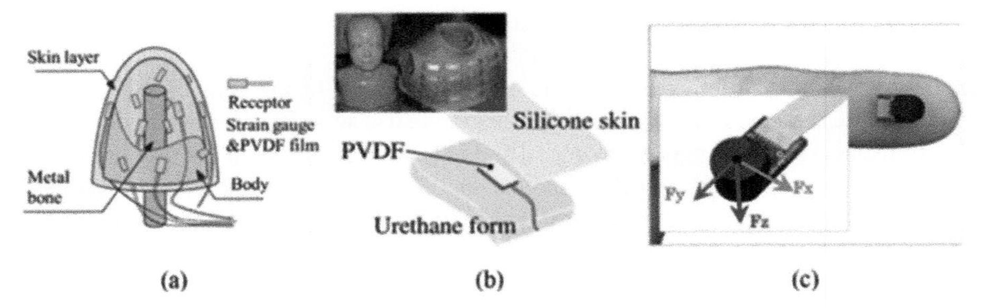

Figure 6.1 *Examples of existing piezoelectric-based sensors. (a) Randomly placed PVDF tactile sensors. (b) PVDF tactile sensors installed on the whole body of CB2. (c) Three-axis piezoelectric MEMS sensors by Touchence Company. The pictures are taken from References 28, 29, and 30, respectively*

6.2.2 *Optical-based sensors*

Two different types of optical-based sensors are described in the literature, and both of them have the potential to measure the force vector. The first type uses a camera to detect deformations in the skin membrane, e.g., References 31, 32. This type can achieve distributed force vector sensing with a high spatial density. It is suitable for the fingertip of robot grippers but is often too thick for other robot parts where the skin should be only several millimeters thick, because it requires a minimum thickness for the camera to be able to focus. The highly sensitive three-axis distributed sensors in Reference 33 use optical fibers but are still 42.6 mm high and have 27 mm diameter. Moreover, the non-uniform thickness of the cover silicone could lead to challenging contact force control. The second type of optical-based sensors uses separate photoreceptors (not in the form of a camera) to measure light intensity; this type can measure the force vector in a slimmer package; however, it does not achieve the high spatial density sensing of the first type. The concept of a small-sized three-axis sensor for distributed sensing was introduced in Reference 34. Cubic soft optical three-axis sensors have also been integrated into a $20 \times 20 \times 20$ mm^3 lightweight polyurethane foam in Reference 35 [see Figure 6.2(a)]. Similar sensors are sold by Touchence. These sensors, however, are still rather thick that could drastically increase the overall size of the robot when they are used as a soft skin. A 10-mm-wide and 8-mm-high optical sensor that can measure the force vector is currently available from OptoForce [36] [see Figure 6.2(b)]. However, an additional analog-to-digital (ADC) unit that has a bigger size compared to the sensor itself is required for each sensor.

6.2.3 *Hall-effect-based sensors*

A Hall-effect sensor and magnet can also be used together for tactile sensing. A one-axis soft sensor was recently integrated on a robot's hand fingertips and phalanges

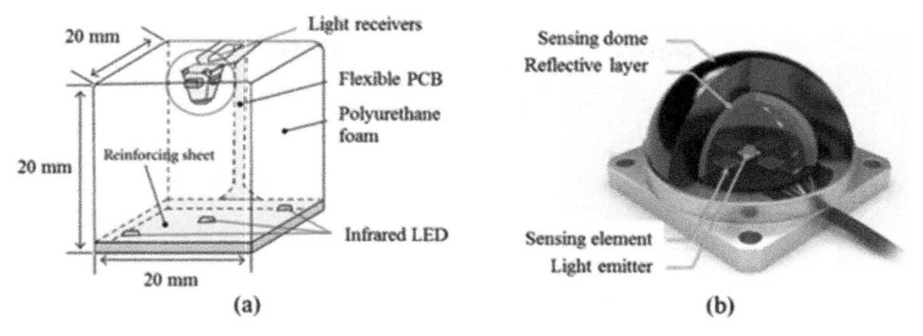

Figure 6.2 Examples of existing optical-based tactile sensors. (a) Three-axis soft sensor flesh. (b) Three-axis dome-shaped sensor from Optoforce Company. The pictures are taken from References 35 and 36, respectively

and used for classifying several kinds of objects [37]. An extended study on the characteristic of the sensor was published in Reference 38; the sensor uses an air gap for increased deformability and thereby higher sensitivity [see Figure 6.3(a)]. Furthermore, the studies in References 39, 41 proposed a three-axis sensor made of dome-shaped silicone rubber; it consists of a magnet in the top of the dome and four Hall-effect sensors located at the base of the dome [see Fig. Figure 6.3(b)]. The sensor was installed on various parts of several robots such as Obrero. The sensor prototype in Reference 42 instead introduced the use of a single chip that can measure a magnetic field in three dimensions [see Figure 6.3(c)]. However, only preliminary experiments on one sensor were performed and no integration on a robot is reported.

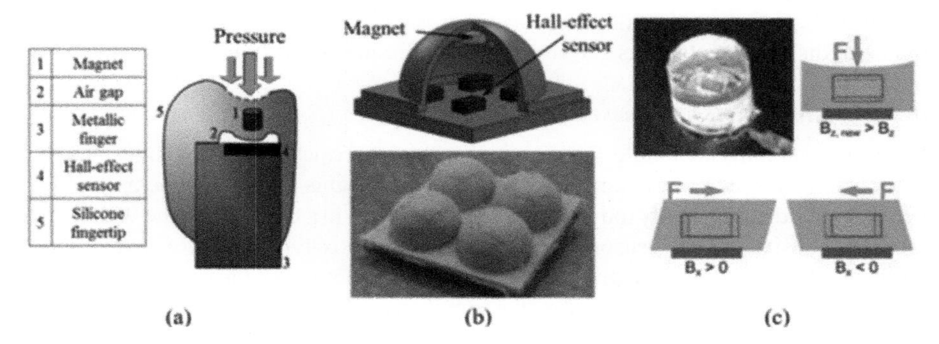

Figure 6.3 Examples of existing Hall-effect-based sensors. (a) One-axis sensor with an air gap. (b) Three-axis dome-shaped sensor. (c) Three-axis sensor prototype with a flat surface. The pictures are taken from References 38, 39, and 40, respectively

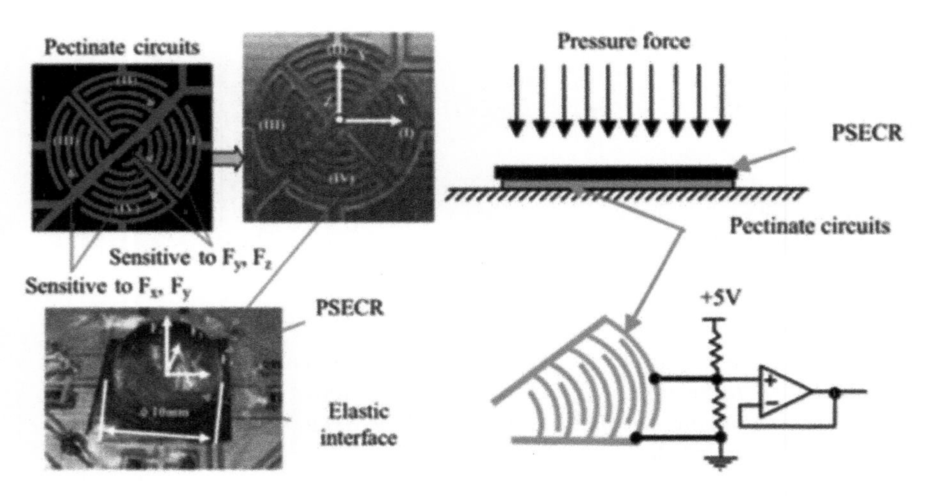

Figure 6.4 *An example of existing sensors based on PSECR. Its dome shape enables three-axis force measurement. The pictures are taken from Reference 43*

6.2.4 PSECR-based sensors

Furthermore, in Reference 43, a dome-shaped sensor based on pressure-sensitive electric conductive rubber (PSECR) and a pectinate circuit design is described (see Figure 6.4). The resistivity of the pectinate circuit changes when pressure is applied to the sensor's top. The sensor prototype is 10 mm in diameter. In general, a dome-like structure enables flat pressure sensors to sense shear forces; however, the size of the dome is the determinant of the smallest size of its detectable object. Additionally, a skin with a flat surface is preferable in many applications. The dome shape discussed previously in Reference 39 probably enhances the shear force sensitivity; however, the sensor could still sense shear forces without it, as the sensor is sensitive to lateral displacements of its surface, unlike Reference 43 without a dome.

6.2.5 Piezoresistive-based sensors

Most commercially available force sensors use piezoresistive strain gauges. They need rather complex arrangements to measure three-axis or even three-axis F/T, which makes them typically too bulky for the integration in a thin skin. In the WENDY robot [44], a six-axis F/T sensor was integrated into a box-type structure enveloping a robot link such as its forearm and acting as the arm's cover [see Figure 6.5(a)]. Additionally, force-sensitive resistor pads were put on the cover. Similarly, the MAC hand used a $10 \times 7.5 \times 5.5$ mm^3 three-axis joystick that is connected to a rigid hollow cover to detect contact forces. Additional electronics are integrated inside each finger link as well for the digitization of the sensor measurements [50, 51]. Each soft and curved fingertip of TWENDY-ONE [42] had a six-axis force/torque sensor installed, in addition to 241 distributed sheet-type tactile sensing elements that cover

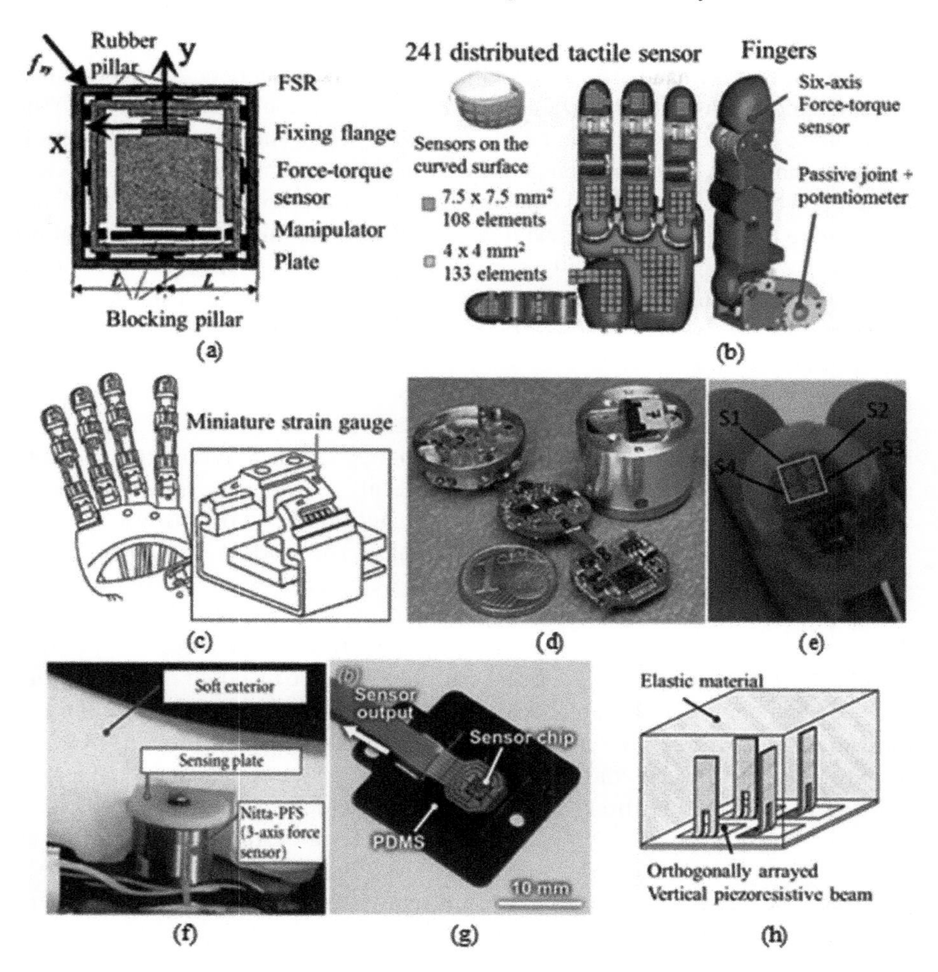

Figure 6.5 Examples of existing piezoresistive-based sensors. (a) Six-axis force/torque sensor and the hard shell. (b) TWENDY's fingertip with six-axis force/torque sensor. (c) Integrated six-axis force/torque sensor used in Robonaut 2. (d) Small digital six-axis force/torque sensor used in DLR-HIT hand. (e) A fingertip with four tri-axial sensors. (f) Three-axis force sensor with soft skin installed on Macra. (g) Small and thin three-axis sensor. (h) A shear force sensor using vertical piezoresistive beams. The pictures are taken from References 35, 42, 44–48, and 49, respectively

the hand [see Figure 6.5(b)]. The sensors allow the robot to pick up and manipulate small objects. However, additional electronics are required for digitizing the sensor signals. The Robonaut 2 hand [45, 52] used a specially-designed six-axis force/torque sensor with miniature strain gauges itself as a phalange resulting in a compact robotic hand [see Figure 6.5(c)]. The ADC unit of the sensors, however, was located in the palm of each hand. The aforementioned sensors are not compliant and most require remotely installed digitization electronics, which could lead to a susceptibility to noise and increases the amount of required wires. The DLR-HIT hand [46], however, has a 16 ×20 mm^2 compact cylindrical six-axis sensor installed in each fingertip, which provides digital output directly, but it is not commercially available [see Figure 6.5(d)]. Importantly, with these F/T sensors, only the summation of all the forces acting on the link can be sensed.

However, a fingertip with four tri-axial sensors is presented in Reference 47 [see Figure 6.5(e)]. Moreover, three-axis force sensors were integrated into the soft skin of the robot Macra; the sensors are covered by a relatively thick layer of polyurethane foam [35] [see Figure 6.5(f)]. Recently, a small-sized sensor based on piezoresistive beams is described in Reference 48. The 2.0 × 2.0 × 0.3 mm^3 sensor was fabricated and embedded in polydimethylsiloxane (PDMS) sheet [see Figure 6.5(g)]. The test result showed the crossover of sensor readout when subjected to normal stress, but the thermal drift was reduced significantly compared to a typical piezoresistor. The work in Reference 49 and its extended work in Reference 53 proposed a concept to sense two-directional shear forces using vertical beams suspended in an elastic material with an additional liquid layer in the later work [Figure 6.5(h)]. Only a shear force perpendicular to the bending axis is sent by the piezoresistor installed at the bending part of each beam. However, the sensor cannot measure normal force and the maximum shear force the sensor can measure was only approximately 0.02 N. Due to its vertical beam configuration, it could be prone to damage when a pure normal force is acting directly on the beam. The sensor fabrication is rather complicated; a magnetic field was used to bend each beam up, and PDMS was used to cover the beams. Moreover, the ADC unit is not mentioned in these works [48, 49, 53]. Also, small-sized ADC converter chips for strain gauges are currently not available; hence, additional space would be necessary for containing the digitization electronics.

6.2.6 Capacitive-based sensors

Capacitive sensors in general have good sensor characteristics such as high sensitivity [16] but are also susceptible to stray capacitance if not properly shielded. Furthermore, they are often used with viscoelastic materials, which introduces hysteresis (this is further discussed in section 6.3.2). In References 54, 55, a bump was added on top of an array of four capacitive sensors to allow the shear forces measurement, but both sensors can sense only millinewtons force that is too little for an application in a humanoid robot [see Figure 6.6(a)]. Flexible capacitive sensors that can measure the shear forces in a flat form factor were suggested in References 56, 60; there were four capacitors that share one common electrode on the top, as shown in Figure 6.6(b). However, the measurements electronics were not included; therefore, the use of distributed sensing is challenging.

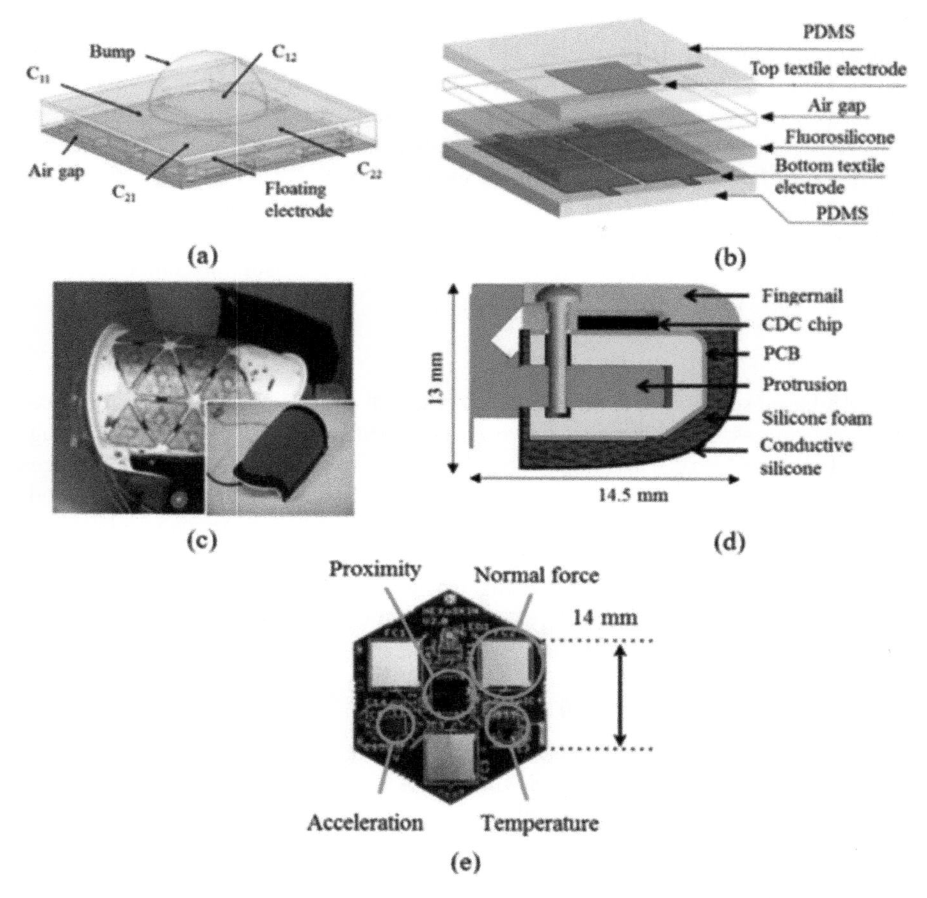

Figure 6.6 *Examples of existing capacitive-based sensors. (a) Three-axis small-sized sensors utilizing bump to detect force vector. (b) Three-axis small-sized sensors using no bump to detect force vector. (c and d) One-axis flexible tactile sensors installed on iCub robot. (e) Multimodal Hex-O-Skin including one-axis force sensor using $CuBe_2$. The pictures are taken from References 54, 56–58, and 59, respectively*

CDC chips, such as the AD7147 from Analog Devices [61], make it possible to provide a digital signal in a small form factor close to the capacitive force transducer. The chip has been used for large-scale robot skin with few wires and robot fingertips [57, 58] [see Figure 6.6 Figure 6.6(c) and 6.(d)]. Hysteresis was present in both sensors due to the soft dielectric used. A pressure-insensitive capacitor was used in Reference 57 to countermeasure the thermal drift. The time-dependent behavior of the silicone used in Reference 58 was compensated by modeling the relaxation of the material. A capacitive-type force sensor that uses copper beryllium ($CuBe_2$)

plates was described in Reference 59 as part of a multi-modal skin sensor; it can also sense proximity, temperature, and three-axis acceleration [see Figure 6.6(e)]. There were four connection ports to reduce the total amount of necessary wires. Nevertheless, the sensors in References 57–59 can measure only normal force.

The work described in section 6.3 uses the AD7147 chip, like References 57, 58 but uses tilted transducers to measure shear forces. We also use $CuBe_2$ plates, but compared to Reference 59, the shape of the plates was modified to foster a parallel deformation for enhanced sensitivity and/or range [1]. Furthermore, our sensors are embedded in soft silicone, while in Reference 59, the sensors are embedded in a rather hard 3D printed material.

6.2.7 MEMS-based sensors

MEMS-based sensors are another ongoing research area. They can detect tiny forces of 1 mN or less. However, the drawback of MEMS sensors is that they are fragile and easy to break if overload force is applied to them [59, 62], which makes them not suitable for integration into a robotic fingertip or hand. While the previous sections covered different sensing principles, MEMS sensors can be based on various physical principles (Reference 30 is based on piezoelectric transducers and Reference 62 is based on capacitive sensing); however, MEMS sensors are produced in a particular way and have different characteristics to other sensors and are therefore mentioned in their own section.

6.2.8 Proximity detection

Instead of force sensors, some research such as in References 63, 64 focuses instead on proximity detection of obstacles with 500 pairs of infrared light-emitting diodes (IRED) and light-receiving PIN diodes, to cover an industrial robotic arm to allow it to work in an unknown environment. The multi-modal skin sensor in Reference 59 also included proximity detection in each sensor module. Proximity sensing has the benefit that it can detect collisions before they occur, but self-sensing and sensing of nearby but not colliding objects have to be taken into account.

6.2.9 Summary of related work

Overall, currently existing sensors have the problem that it is difficult to incorporate all the electronics to provide digital output in a small sensor package. However, if the sensors provide only analog output, many wires are required, which makes the integration of a large number of sensors in a robot difficult.

6.3 Three-axis capacitive soft skin sensor

6.3.1 Concept

In order to measure a force vector while still being compliant and having a smooth and flat outer surface, the proposed skin sensor has four capacitive-type force sensing units tilted in different directions and is covered with a soft material as can be

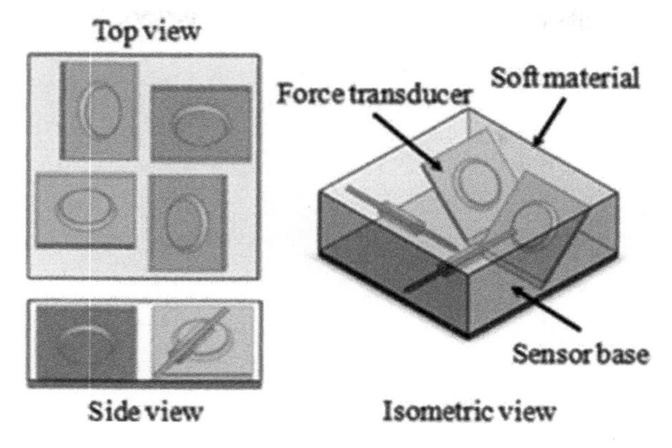

Figure 6.7 *The conceptual design of the proposed three-axis distributed skin sensor*

seen in Figure 6.7; each unit senses the force acting perpendicular to it. In particular, each tilted unit uses two copper beryllium plates for the force transducer. The copper beryllium plates have a bump to foster the parallel deformation (see Figure 6.8). Thereby, the benefit of the bump (the cylindrical extrusion of the copper beryllium seen in Figure 6.9) is to enhance the sensitivity and/or range of the sensor.

6.3.2 Implementation

Many capacitive skin sensors have severe hysteresis in their measurements, as they use viscoelastic materials such as silicone as a deformable dielectric to be able to

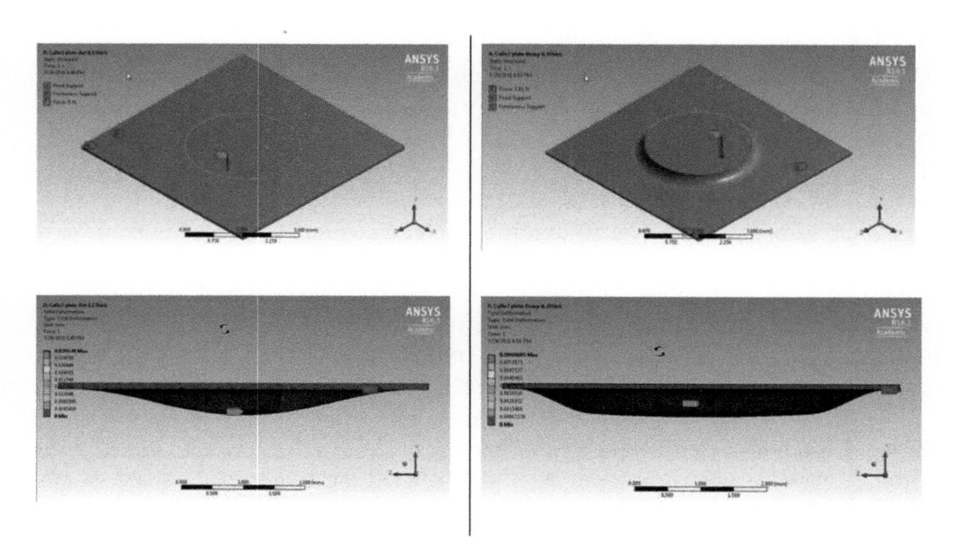

Figure 6.8 *The deformation of a plate without (left) and with (right) a bump. The bump fosters a parallel deformation*

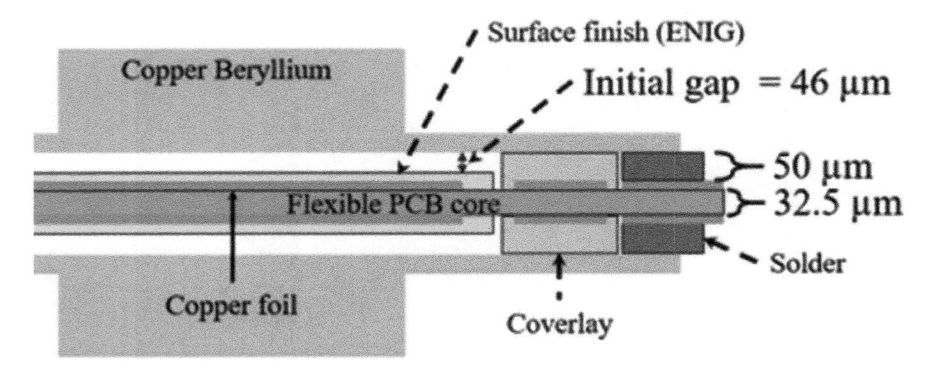

Figure 6.9 *Section view of the transducer. The thickness of the coverlay and the resulting initial gap between the $CuBe_2$ and the PCB at the sensing area can be seen*

measure forces. In general, many skin sensors incorporate soft materials, as a soft skin enhances the safety of the robot and its capability to work in unstructured environments. A certain level of hysteresis is therefore hard to avoid; however, by making the transducer itself free of viscoelastic materials, the hysteresis in the sensor measurements can probably be significantly reduced. Therefore, in this work, we use deformable metal as one of the capacitive plates, separated by the other plate only by air, hoping to reduce the overall hysteresis of the system. However, the transducers are embedded in silicone.

6.3.2.1 Copper beryllium plate with bump

Copper beryllium ($CuBe_2$) is used as the top deformable capacitive plate due to its high strength, spring properties, high electric conductivity, and its solderability. It has been used as a flat capacitive plate for a force sensing module in Reference 59. The variant of copper beryllium used here is CW101C R580 due to its availability in the form of thin sheets. The initial distance between the $CuBe_2$ and the sensing electrode is implemented by a 50-μm-thick coverlay on the PCB (see Figure 6.9).

Regarding the shape of the plate, when a fully-flat plate is loaded at its center with a circular shape object, the deformation of the plate will be in a bell shape where the maximum deformation is at the center (where the load is applied). The finite-element simulation (ANSYS R16.1) of a $6 \times 6 \times 0.1$ mm^3 $CuBe_2$ plate that is subject to a load of 5 N is shown in Figure 6.8 (left).

Furthermore, this bell-shaped deformation will result in a non-uniform change in capacitance of a parallel plate capacitor where the overlap area and the relative distance between two capacitive plates vary toward the center of the plate. If the initial gap between the two capacitive plates is small, the center of the plate will eventually be in contact with the other capacitive plate, and therefore the plate cannot deform more, thereby limiting the range of the sensor. This means that with just a flat plate, the two plates need to be distant in order to have a broader range of measurement. However, the large gap can result in a less sensitive sensor since there is a only minor change in the sensor's capacitance that could be difficult to distinguish by the capacitance readout circuit.

Therefore, in order to attenuate the effect of the bell-shaped deformation, a cylindrical bump was added at the center of the flat plate. This bump strengthens the center portion of the plate, which results in a more parallel and more homogeneous deformation of the whole plate when the bump is pushed by an external force. The simulation result of merging a 0.31 mm high, Ø3 mm bump on the $6 \times 6 \times 0.1$ mm^3 copper-beryllium plate is shown in Figure 6.8 (right). It can be seen that more uniform deformation is achieved.

With the same magnitude of the normal force, the maximum displacement of the plate with a bump toward the other electrode will be less, so it can be placed closer to the signal pad of the PCB in order to gain more sensitivity. Counterintuitively, a smaller initial distance of the capacitive plates is beneficial for the sensitivity/range. For example, with half the initial distance, double the range with the same sensitivity can be achieved. It should also be considered that in this case, a thicker $CuBe_2$ plate has to be used (in our example with double the stiffness), in order to achieve the desired sensitivity and range.

6.3.2.2 Tilt double-sided transducers

The ability of the sensor to detect the force vector can be explained as follows (see also Figure 6.10). When there is only a normal force acting on the surface of the sensor, all four transducers will sense the same magnitude of force equally (if the contact area covers all four transducers completely and/or equally). When a shear force additionally acts on the sensor, the transducer that points in the same direction as the shear force will sense more force, while the other transducer that points in the

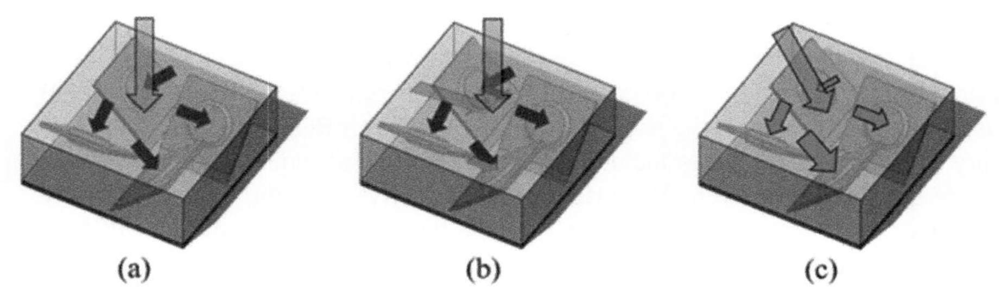

Figure 6.10 *The force detection principle of the proposed skin sensor. (a) When a normal load (blue) is applied to the top of the sensor, all four sensing units sense the force. (b) When a shear force (red) is added, the force sensed by the pair of two perpendicular sensing units becomes increased and decreased, respectively. (c) The resultant forces of input force (green) and the sensed forces of all four units are shown*

opposite direction will sense less force. The other two transducers that are perpendicular to the shear force will not sense the shear force.

When the sensor is being exerted by a force, not only the copper beryllium plates are elastically deformed, but also the bending angle of the transducer is expected to temporarily change (this is due to the fact that the transducers are surrounded by a compliant material such as soft silicone rubber). This is beneficial because it allows the transducers to be closer to the surface, which increases their sensitivity and response time and at the same time maintains the overall compliance of the sensor. As a further clarification, our sensor does not sense the bending angle of the transducers, unlike, e.g., Reference 49.

Furthermore, a double-sided capacitor was used because it basically doubles the size of the capacitive plate, thereby enhancing the sensitivity of the sensor, without increasing the overall size of the sensor. A double-sided capacitor was previously also used in Reference 65.

6.3.2.3 Temperature compensation pad

Capacitive sensors are susceptible to thermal change. Therefore, in order to sense the influence of temperature on the sensor, a pair of copper foils is put in the third and the bottom layer of an area of the PCB not used for the force transducers to create a pressure-insensitive capacitor called temperature compensation pad (TCP) as can be seen in Figure 6.11 (a). Its capacitance changes only according to temperature. With the measurements from the TCP, the effect of temperature on the force measurement can be compensated, as has been done in Reference 20: temperature compensation can be achieved by adding to the force measurement the change in TCP multiplied by a certain gain. The validity of this approach will be evaluated in section 6.3.3.4.

6.3.2.4 Manufacturing

The manufacturing process of the sensor is shown in Figure 6.11. The flexible printed circuit board (FPCB) was order-made. Solder paste was applied to the FPCB using a stencil to control the amount of solder and the $CuBe_2$ plates, and the necessary electronic components including the AD7147 were soldered to the FPCB in house using a reflow oven. “Electric Paint” conductive ink from Bare Conductive was used to cover any possible gaps between the $CuBe_2$ plates and the FPCB to ensure sealing for the subsequent molding and good electronic shielding against stray capacitance; the ink was left to completely dry for 30 minutes. The FPBC was then fixed temporary in the mold with a double-sided tape. The transducers were slightly bent and a silicone-made triangular support was inserted under each transducer to maintain the tilting angle. The angle of tilting was confirmed visually using a protractor, and the mold was closed. The Ecoflex 30 was mixed thoroughly and degassed, and the silicone was poured into the mold slowly. The mold filled with liquid silicone was degassed again to minimize the air bubbles. The mold was left for the silicone to completely cure for 4 hours. The molded sensor was taken out of the mold afterward.

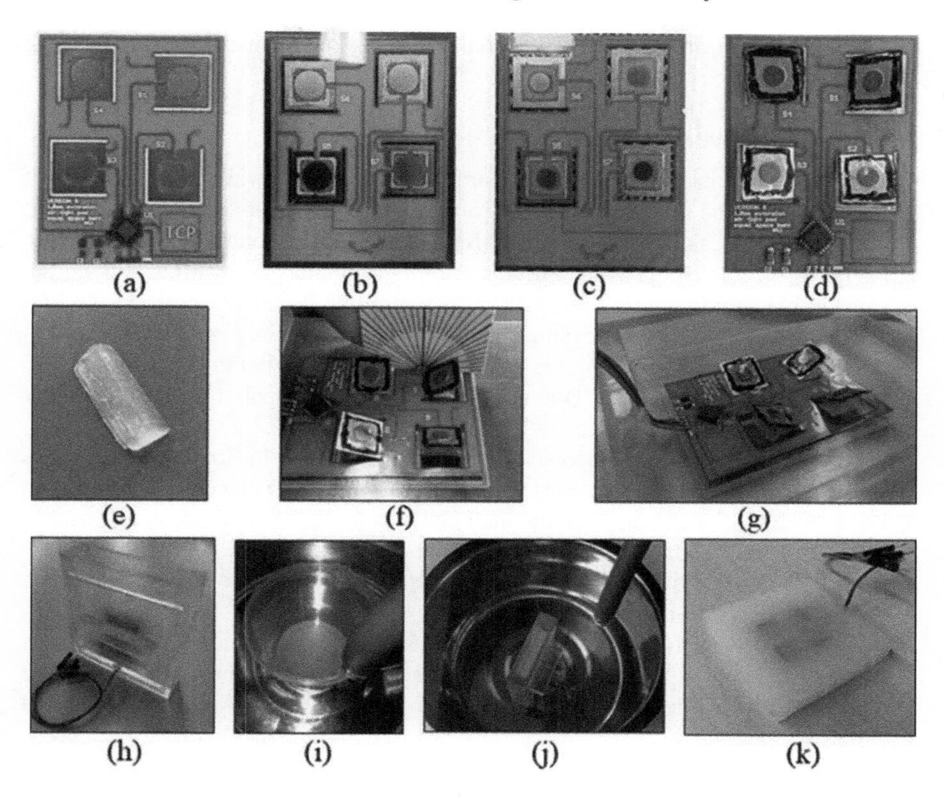

Figure 6.11 *The manufacturing process of the sensor. (a) The FPBC received from the manufacturer. (b) Solder paste was applied to the FPBC using a stencil. (c) The FPBC with the* $CuBe_2$ *plates soldered on it. (d) The FPBC with the electronic components mounted on. The transducers are also sealed with the conductive ink. (e) One of the silicone triangular supports. (f) The supports were inserted under the transducer; the tilting angles were confirmed with a protractor. (g) The FPBC with all the transducers tilted and ready for molding. (h) The mixed silicone is being degassed before pouring it into the mold. (i) The mold with the liquid silicone and the sensor inside. (j) The degassing right after pouring the liquid silicone into the mold. (k) The sensor after the molding*

The tilting angle of the transducers was chosen to be 45° to allow for a good balance of sensitivity for shear and normal forces. Therefore, 45° triangular supports also made from Ecoflex were produced first. They were molded with a 3D printed plastic mold. However, due to the springback of the flexible PCB, a 30° bending angle instead of a 45° angle was achieved (determined visually with the help of a protractor). Hence, in the current version of the sensor, we used this 30° bending angle.

6.3.3 Experiments

A set of experiments was performed in order to test the following characteristics of the sensor (after the molding of the silicone rubber): a linearity and hysteresis test, tri-axial force test, signal-to-noise ratio (SNR) test, and temperature drift test.

The AD7147 chips provide digital sensor measurements via an I2C bus. In the current implementation, an Arduino Due was used as the microcontroller and all transducer measurements were collected with a 40 Hz sampling rate. A higher sampling rate can be set by reducing the digital filter in the chip, but this would also lead to higher noise. The Arduino Due recorded the measurements on an SD card for later offline analysis.

For the linearity and hysteresis test, we applied force by loading the sensors with different weights. The weights were placed on a weight placement plate, and a Ø6-mm pushing shaft transferred the load onto the sensor. A linear bushing ensured that the pushing shaft is vertical to the sensor's surface and applies only normal load. The weights of the weight placement plate and the shaft were also considered for the force acting on a transducer.

For the tri-axial force experiment, a test setup shown in Figure 6.12 (a) was used; it consists of a VM5050-190 linear voice coil motor from Geeplus (called VCM hereafter) applied to varying amounts of forces, which were also measured

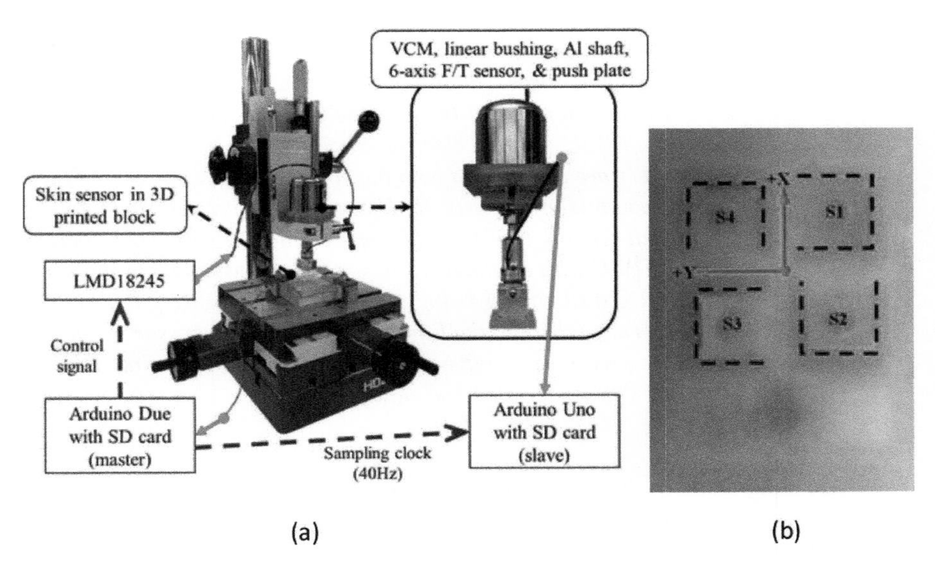

Figure 6.12 *(a) The experimental setup of the tri-axial force test. (b) The location of the four force sensing units and the axes and origin of the skin sensor are shown*

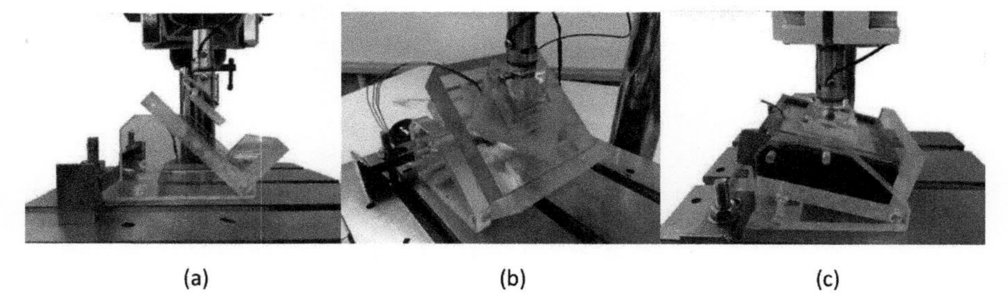

(a) (b) (c)

Figure 6.13 (a) The adjustable angle stage fixed at 45° with respect to the x–y stage surface and the corresponding 45° push plate. (b) A 3D printed plastic block that allows the sensor to be aligned at 45° with respect to the outer edge of the block. This allows the shear force to act in both x and y directions at the same time. (c) The adjustable angle stage is set at 15° with respect to the x–y stage surface and the corresponding 15° push plate

by a six-axis F/T sensor (Nano1.5/1.5 from BL Autotech) as the reference measurements. A 30 × 30 mm^2 acrylic plate transferred the force onto the skin sensor. The force measurements of the F/T sensor were divided by the contact area and used as the reference pressure measurements. In our previous publications, we reported force, but it can be assumed that our sensor measures pressure rather than force if the contact area is bigger than the sensor (this will be further investigated in future work). Therefore, pressure is more suitable to be able to compare the response of the capacitive and Hall-effect sensors, as they have different sizes and were used with different pushing plate sizes. In any case, as we state the size of the contact area, force can be easily converted to pressure and vice versa.

A motor driver (LMD18245 from Texas Instrument) was used for regulating the force generated by the VCM, and the motor driver was controlled by the same Arduino Due that was recording the skin sensor measurements. The six-axis F/T sensor requires a different voltage (5 V) than the skin sensor (3.3 V) and was therefore connected to an Arduino Uno, which recorded the six-axis F/T measurements onto an SD card. The Arduino Due and Uno were synchronized: the Arduino Due was set as the master and provided the clock signal to the Uno. To apply shear force, the skin sensor was tilted (as shown in Figure 6.13), and also the pusher had a corresponding slope. The axes of the sensor and its origin are defined as shown in Figure 6.12 (b).

6.3.3.1 Linearity and hysteresis

In this experiment, the sensor was stepwise loaded and then unloaded; the loading steps can be seen in Table 6.2. The weight was increased and subsequently decreased every 5 seconds. The weight of the weight placement plate and the shaft are 45 g, corresponding to the first step. The load was applied with a 6-mm shaft on the center

Table 6.2 *The weights applied to the sensor for different steps. The loading sequence goes from 0 to 17 and the unloading sequence goes from 17 to 0*

Step	**0**	**1**	**2**	**3**	**4**	**5**	**6**	**7**	**8**	**9**	**10**	**11**	**12**	**13**	**14**	**15**	**16**	**17**
Weight (g)	0	45	46	48	53	63	73	93	113	133	153	203	253	353	453	553	753	1253
Pressure (kPa)	0.000	15.599	15.945	16.639	18.372	21.838	25.305	32.237	39.170	46.103	53.036	70.368	87.700	122.364	157.028	191.692	261.020	434.339

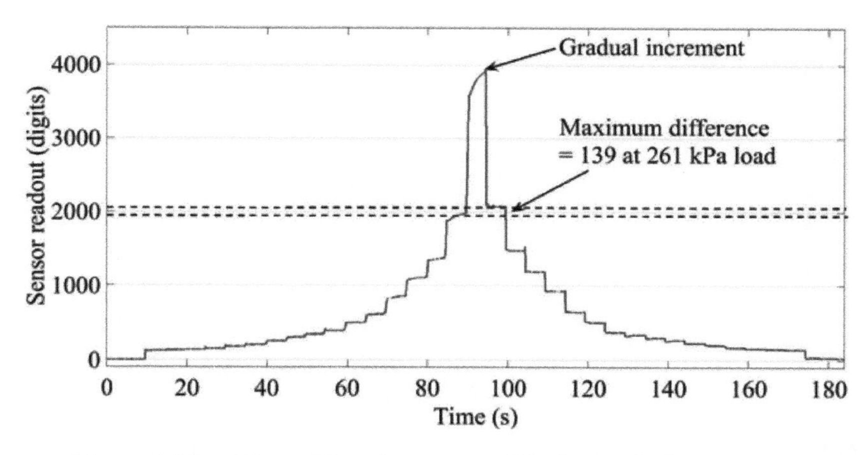

Figure 6.14 The unfiltered response of S3 during the linearity test

of one of the tilted plates (S3), and the response of S3 will be analyzed in the following. This was done because in our original analyses, the response of each sensing unit before and after the silicone molding was compared [3]. However, the four double-sided transducers or sensing units (S1–S4) react roughly similar to normal force, and therefore the hysteresis and linearity for normal force of S3 can be seen as demonstrative for the whole sensor.

Figure 6.14 shows the sensor readout of S_3. At each step, the sensor readout slowly and gradually changes, most obviously at step 17, where the highest weight change between steps occurs (173.32 kPa). Some hysteresis can be seen especially at the 261-kPa load of around 14.2 kPa difference when comparing loading and unloading.

The response of S_3 to both the loading and unloading sequences (average for each step) is shown in Figure 6.14. The responses are almost linear, which is confirmed by the small constants for the quadratic term. The hysteresis was calculated as 3.4% using 1 (the minimum pressure P_{min} was 0 kPa and the maximum pressure P_{max} was 434 kPa):

$$\text{Hysteresis\%} = \left| \frac{(P_{mu} - P_{ml})}{(P_{max} - P_{min})} \right| \times 100\% \tag{6.1}$$

where P_{ml} and P_{mu} were the pressure values calculated from the sensor readings during the loading and unloading cycles taken at the midpoint pressure of (434 – 0 kPa)/2 = 217 kPa. Both P_{ml} and P_{mu} were calculated using the same model, i.e., the inverse of the equation for the loading cycle in Figure 6.15. This corresponds to the calibration method later used for the Hall-effect sensor: the calibration model is calculated from training data with stepwise loading. The hysteresis was wrongly reported as 1.27 % in Reference 3.

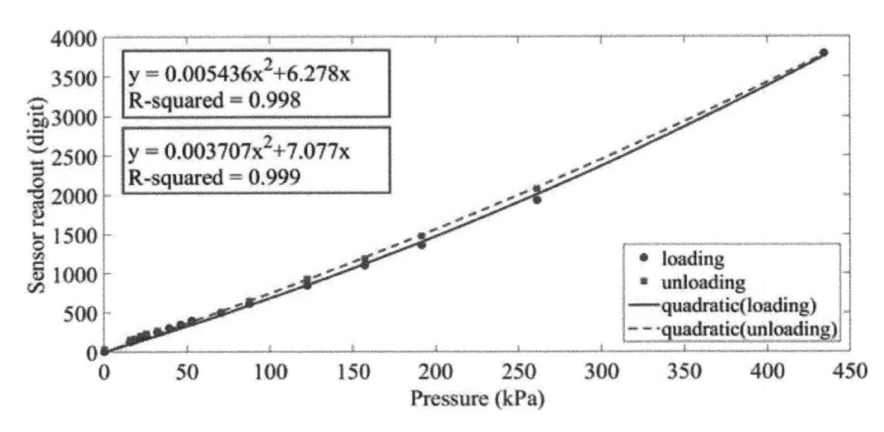

Figure 6.15 *The relationship between the load and sensor readout of S3 and its quadratic approximation line of both loading and unloading sequences. The values shown in the blue box are for loading and those in the green box are for unloading*

6.3.3.2 Tri-axial force

6.3.3.2.1 Normal force calibration

First, the sensor was calibrated for normal pressure. Normal pressure was applied with the VCM and the 30 × 30 acrylic plate stepwise, i.e., six equally-spaced steps from around 3.7 kPa (340 g) to 15.7 KPa (1 440 g), and each step lasted for 10 seconds. All collected data were filtered with a Savitzky–Golay filter. Figure 6.16(a),(b) show the response of the sensor and the F/T sensor (divided by the contact area to calculate pressure), respectively. S_1–S_4 clearly respond differently. Each sensor unit was calibrated to measure a quarter of the received pressure, so that the sum would correspond to the overall normal pressure $P_{z,skin}$. A quadratic model was used to convert the raw sensor measurements S_i of sensing unit i to P_i, the pressure measured by sensor unit i according to 2. The converted P_i (1–4) were all then combined into $P_{z,skin}$ according to (6.3)

$$P_i(t) = a_i S_i(t)^2 + b_i S_i(t) \tag{6.2}$$

$$P_{z,skin}(t) = P_1(t) + P_2(t) + P_3(t) + P_4(t) \tag{6.3}$$

The calibration parameters a and b were calculated with the curve fitting toolbox of Matlab, using all data except the first second of each push to remove the transient response. Four separate equations were used, one for each sensor unit, with an averaged root-mean-square error (RMSE) of 0.0632 kPa and an averaged R-squared value of 0.9776. The sensor units' calibrated sensor measurements can be seen in Figure 6.16 (c), and the sum of them closely resembles the reference sensor's measurements as can be seen in Figure 6.16 (d) with an RMSE value of 0.2362.

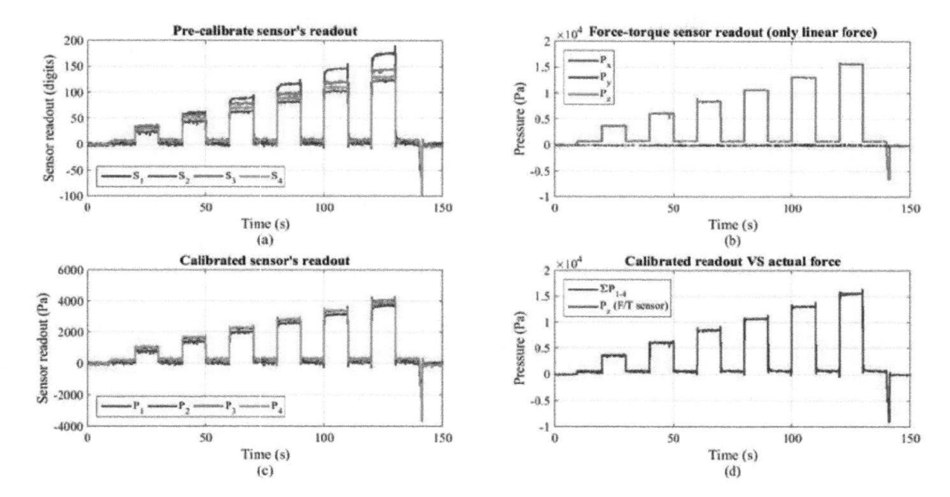

Figure 6.16 *The result of the sensor calibration experiment. (a) Pre-calibrate sensor's readout. (b) Force-torque sensor readout (only linear force). (c) Calibrated sensor's readout. (d) The comparison between converted sensor's readout and actual pressure readout*

6.3.3.2.2 Shear force calibration

Data were collected in the same way as for the normal force test (i.e., six steps of force), but the sensor and the pusher were tilted 45° in either +x, −x, +y, or −y

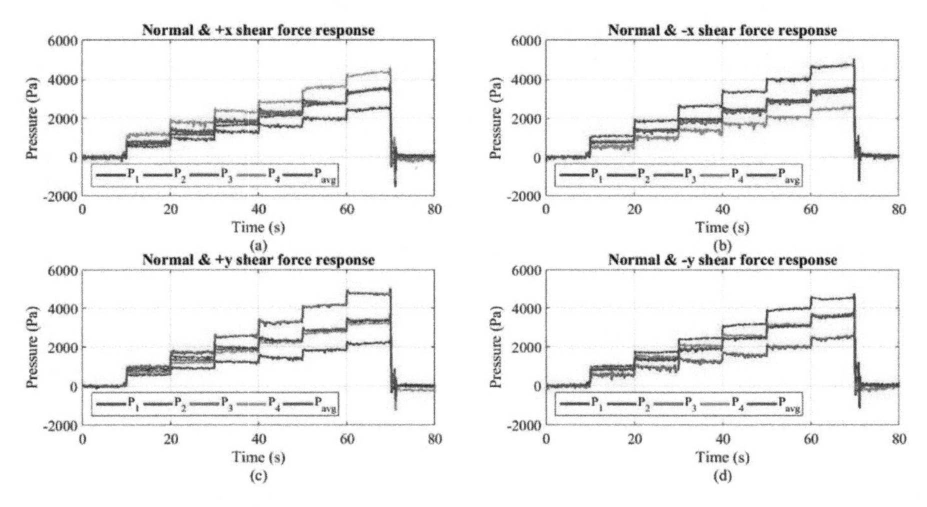

Figure 6.17 *The resultant force sensed by all our units when being pushed with a normal and a shear force acting in different directions. (a) With +x shear force. (b) With −x shear force. (c) With +y shear force. (d) With −y shear force*

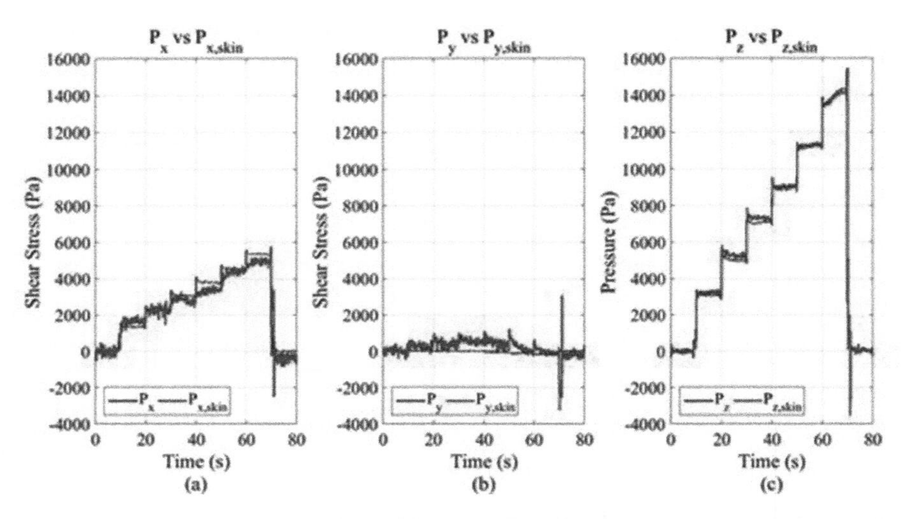

Figure 6.18 The comparison between the calculated force vector and the actual force vector when normal force and +x shear force are applied. The adjustable angle stage is set to 45°

direction. Thereby, four datasets were collected. All data were filtered again with a Savitzky–Golay filter, and the skin unit measurements were then converted into pressure values with the corresponding quadratic equations obtained from the previous experiment. Figure 6.17 shows the results. As expected, the unit that faces the force vector responds stronger, while the unit that points in the opposite direction responds about the same amount weaker than the average response, while the other two units respond like the average.

Therefore, the sensor can be calibrated as(6.4) and (6.5) in to measure the force vector. We used the data Figure 6.17 (a) and (b)fromto determine the calibration coefficient ax and Figure 6.17(c) and (d) for ay. The first and last second of each step was removed to eliminate the transient response. The reference pressure was calculated from the reference sensor measurements (including a conversion to the same reference frame).

$$P_{x,skin}(t) = a_x(P_4(t) - P_2(t)) \tag{6.4}$$

$$P_{y,skin}(t) = a_y(P_3(t) - P_1(t)) \tag{6.5}$$

The results of the calibration for the data previously shown in Figure 6.17 (a) and the pressure calculated from the reference sensor are shown in Figure 6.18, with RMSE values of 0.3593, 0.4306, and 0.1986 kPa for $P_{x,skin}$, $P_{y,skin}$, and $P_{z,skin}$, respectively. It should be noted that these data were part of the data used for finding the calibration parameters for $P_{x,skin}$ and $P_{y,skin}$. Interestingly, the z-axis measurements still closely corresponded. Therefore, it can be concluded that there is little crosstalk between normal pressure and shear stress.

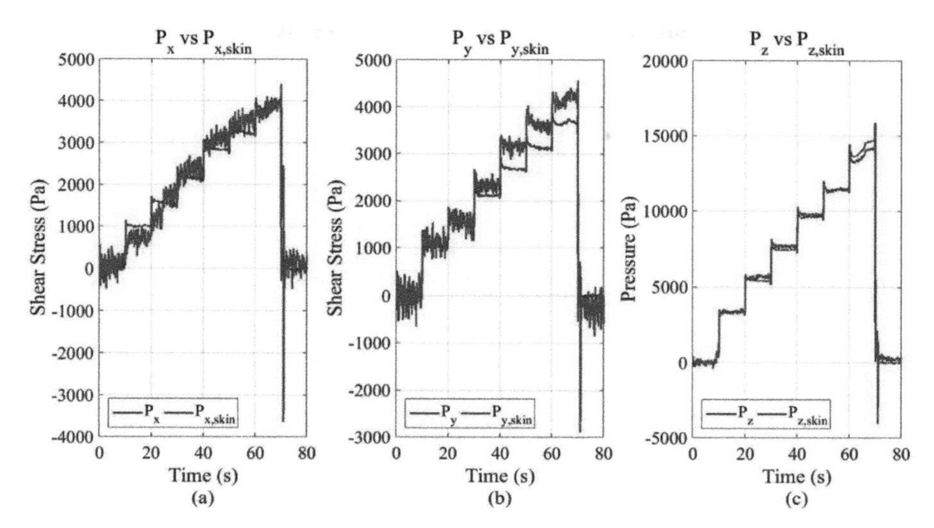

Figure 6.19 *The comparison between the calculated force vector and the actual force vector when the normal force and +x+y shear force are applied. The adjustable angle stage is set to 45°*

X–Y shear force push

To evaluate the calibration performance for new kind of data, x and y force was applied simultaneously with the configuration shown in Figure 6.13 (b). Again, force was applied stepwise with the VCM. Corresponding to the orientation of the skin sensor, four datasets with shear force acting in +x+y, +x−y, −x+y, and −x−y directions were collected. The result of pushing the sensor with shear force in the direction of +x+y can be seen in Figure 6.19 with RMSE values of 0.2658, 0.3483, and 0.2259 kPa, while the averaged RMSEs of all four-direction shear tests are 0.3921,0.4213, and 0.2170 kPa for Px,skin, Py,skin and Pz,skin, respectively.

30° and 15° pushes

The skin sensor was tilted to 30° and 15°, and push plates with the same angle were used [refer to Figure 6.13 (c)]. For both 30° and 15°, four different shear forces were applied in addition to the normal force, i.e., +x, −x, +y, and −y. Again, the VCM produced the same steps of force as in the previous experiments.

The comparison of the force vector calculation accuracy in various experiments is shown in Table 6.3 as the values of averaged RMSE and R-squared for each axis at each test. When the pushing angle is decreasing, the ratio of shear force to normal force decreases too. According to the result when the sensor was pushed at 15° in +y direction, the maximum Py,skin was only 1.307 kPa, while the maximum Py,skin was 15.246 kPa. Figure 6.20 shows the result of pushing the sensor at 15° with shear force in the direction of +y.

Table 6.3 The comparison of the force vector calculation's accuracy at all experiments in the tri-axial force test section

Experiment		**RMSE (kPa)**			**R-squared**		
		$P_{x,skin}$	$P_{y,skin}$	$P_{z,skin}$	$P_{x,skin}$	$P_{y,skin}$	$P_{z,skin}$
Calibration at 45°		0.4549	0.3741	0.2718	0.93878	0.93882	0.99618
With both x and y shear		0.3921	0.4213	0.2171	0.88953	0.86793	0.99721
Varied normal and shear ratio	30 degree	0.5406	0.2770	0.4216	0.73647	0.90931	0.9905
	15 degree	0.4161	0.3202	0.5118	0.43489	0.64024	0.98067

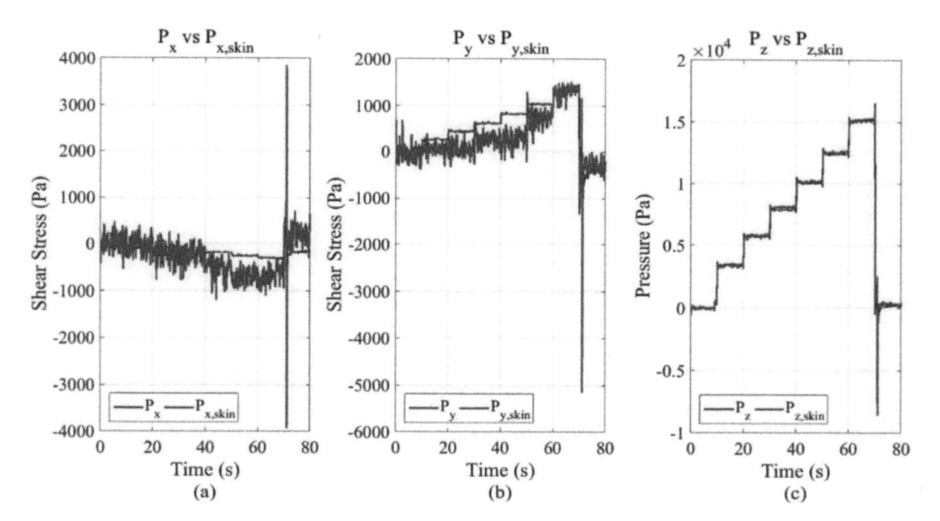

Figure 6.20 The comparison between the calculated force vector and the actual force vector when normal force and +y shear force are applied. The adjustable angle stage is set to 15°

6.3.3.3 SNR

Furthermore, we calculated the SNR using the data from Figure 6.16 (d), but no filter was used for this calculation to take into consideration the actual noise of the sensor. Furthermore, the sensor measurements were converted to z-axis pressure using the calibration parameters found in section 3.3.2.1. The transient was removed from the data by removing the first and last second for each pressure. The SNR was calculated according (6.6)to 6 [66]:

$$SNR_{dB} = 20\log_{10}\left(\frac{\mu_U - \mu_p}{\sigma_u}\right) \text{dB} \tag{6.6}$$

Here, μ_U is the mean z-axis pressure calculated from skin sensor measurements when not loaded, μ_p is the mean z-axis pressure when being loaded with a certain pressure, and σ_u is the standard deviation of the z-axis measurement when not being loaded.

Table 6.4 shows the results for different pressures.

Regarding the SNR of the sensor, the values are rather low. However, when looking at the raw values of the transducers when subjecting to the loads, it can be seen from Figure 6.21 that the transducers' responses were at approximately 170 digits at most while the maximum measurable range of the AD7147 is at 65 535 digits (16 bits).

The $CuBe_2$ plates could be designed to be more sensitive to external pressure in order to exploit more the measurement ability of the CDC chip. This will be investigated in future work.

Table 6.4 The SNRs of the calibrated sensor when subjected to different normal pressures

Pressure (Pa)	SNR (dB)
686	6.40
3 691	23.86
6 126	28.61
8 419	31.50
10 629	33.48
13 091	35.21
15 661	36.63

6.3.3.4 Temperature influences compensation and minimum detectable pressure

The complete sensor was exposed to varying temperatures, starting from 15°C and gradually increased to 40°C, and then brought back to 15°C. During this process, the values of S_1, S_2, S_3, S_4, and TCP were recorded.

Figure 6.22 shows the plot of the thermal drift found in all sensing units, the TCP, and the actual temperature near the sensor as measured by a TMP102 temperature sensor. It can be seen that the measurements of the force sensing units change with temperature and follow the profile of the TCP. Indeed, the TCP measurements are more related to the sensor measurements S_1, S_2, S_3 and S_4 than the TMP102 temperature sensor just a few millimeters away from the sensor. Therefore, the thermal effect on the force sensing transducers can be reduced by (6.7) using 7:

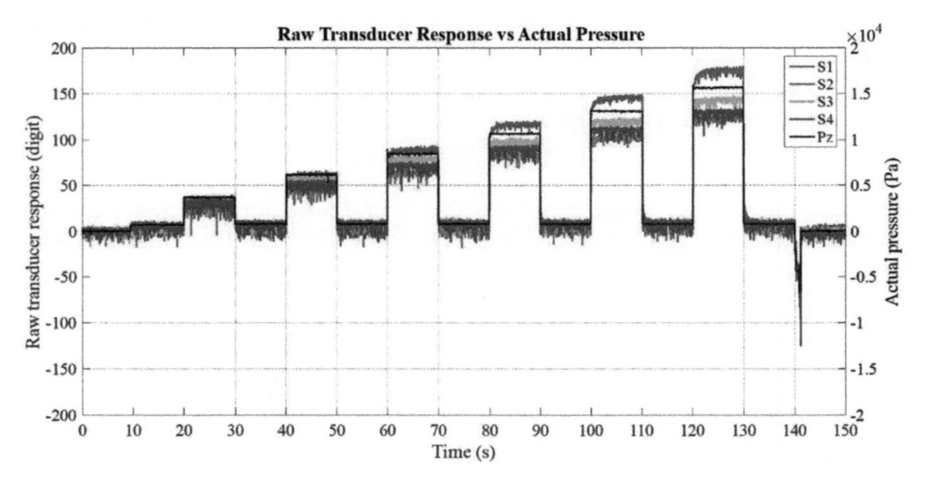

Figure 6.21 The comparison between the raw transducer responses and the actual pressure exerted on the sensor. The figure shows the unfiltered data of Figure 6.16 (a).

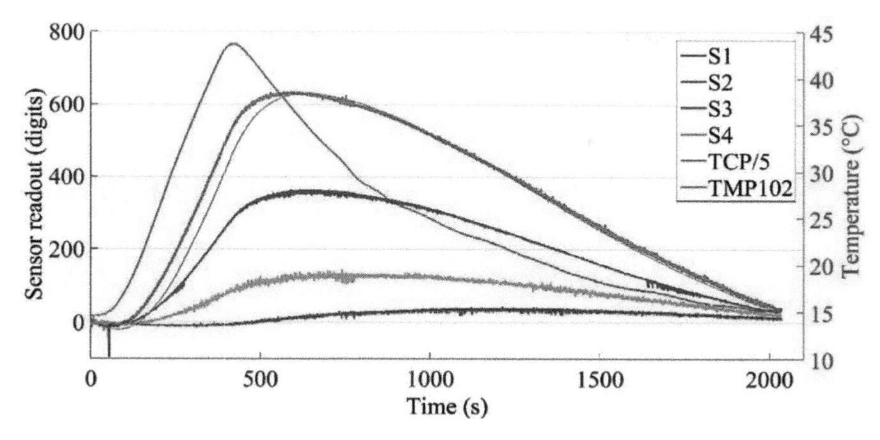

Figure 6.22 *The thermal drift of all sensing units after the molding including TCP (gray, scaled down by 5) are shown together with the actual temperature (purple) sensed by the TMP102 sensor*

$$\widehat{S}_i(t) = S_i(t) - K_i \times TCP(t) \qquad (6.7)$$

where $S_i(t)$ is the raw value of the sensor i at time t, and $\hat{S}_i(t)$ is the temperature compensated value. TCP(t) is the value taken from the TCP at time t. Lastly, K_i is the gain factor, set differently for each sensing unit. This method has been used previously in References 1, 57.

The result of the compensation can be found in Figure 6.23, showing that the thermal drift can be successfully suppressed. The drift can be reduced from around +600 digits to ±60 digits. For calibrated sensor values, this corresponds to a reduction from 150 to 20 kPa Figure 6.4(a) and (b) see.

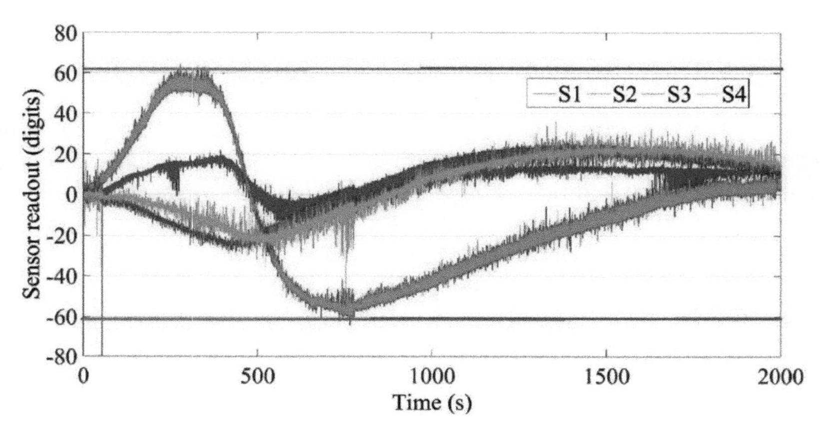

Figure 6.23 *The sensor readout of all four transducers after the temperature compensation*

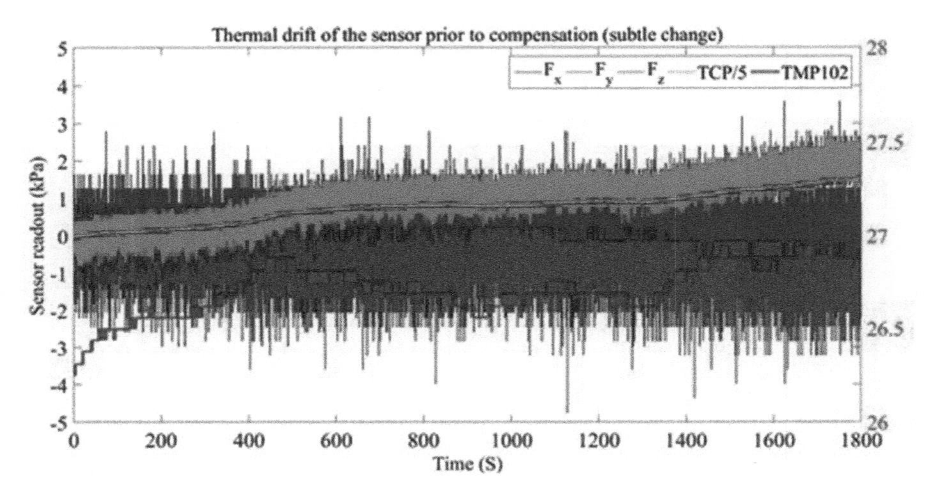

Figure 6.24 The thermal drift of the measurement of the calibrated sensor when subjected to subtle temperature change including TCP (cyan, scaled down by 5) is shown together with the actual temperature (purple) sensed by the TMP102 sensor

In a second test, the sensor was left at room temperature (26–27°C) for 30 minutes and the measurement of the sensor was monitored with the actual temperature and humidity. This test aimed to verify if the use of the TCP for compensating thermal drift is also valid when the temperature only changes slightly. The gain parameters from the previous temperature compensation experiment were also used here. The result Figures 6.24 and 6.25) showed that also in this case, the drift could be slightly reduced, especially for the z-axis, even though the drift was already low before the compensation. In any case, the gain parameters found for a much more extreme situation did not influence the sensor negatively.

Preliminary tests with a humidity sensor (HIH6130 from Sparkfun), which was placed next to the TMP102, showed that the humidity in our tests was closely related (inversely proportional) to the temperature.

Regarding the minimum detectable pressure, looking at Figure 6.25, which shows unfiltered data recorded for 30 minutes, it can be concluded that after applying a filter that eliminates low amplitude outliers, a pressure increase of more than 2 kPa can be detected (in room temperature). By using more advanced detection algorithms or more filters (such as the Savitzky–Golay filter used in section 6.3.3.2), we could achieve minimum detectable pressures of about 0.4 kPa for the x and y axes and 0.1 kPa for the z-axis. However, filters increase the response time of the sensor.

6.3.4 Summary

The skin sensor proposed a novel arrangement of the capacitive transducers that enable the tri-axial force measurement while providing a soft and flat-surface layer

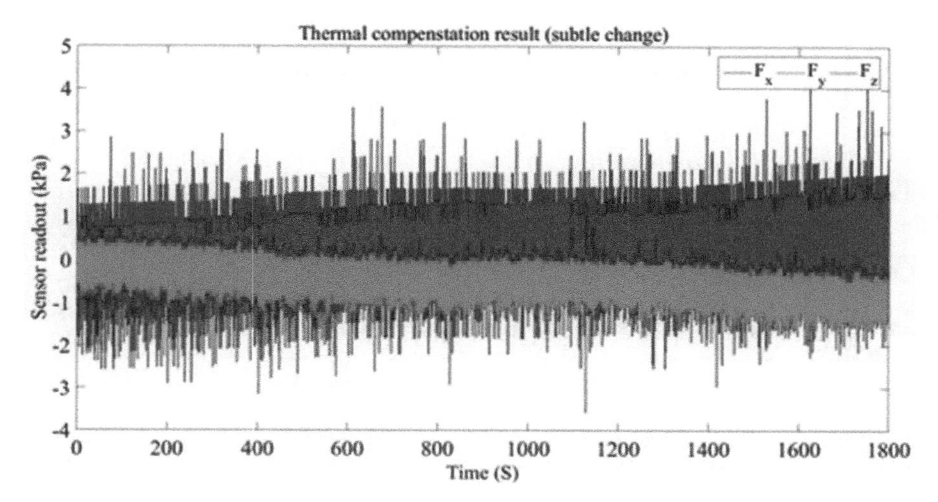

Figure 6.25 *The readout of the calibrated sensor after the temperature compensation when subjected to subtle temperature change (Figure 6.24)*

and digital output. The sensor showed that the measurement of the force vector can be successfully done after the calibration. The pressure measurement provided satisfactory results when compared to the reference pressure sensed by a commercially available 6-axis force/torque sensor.

The sensor has only approximately 3.4% hysteresis even though the transducers were suspended inside 7-mm-thick silicone (Ecoflex SuperSoft 30). The silicone rubber introduces crosstalk (when we push on top of one transducer with a 6-mm-diameter pusher, the other transducers also respond, data shown in Figure 6.13 of Reference 3) and decreases the sensitivity of the sensor (a comparative study before and after molding was also shown in Reference 3). Such characteristics are to be expected and are not catastrophic for our intended use in a humanoid robot. The hysteresis and crosstalk can only be reduced to a certain extent with the choice of different viscoelastic materials while maintaining the softness of the skin, which is important for the safety property of the skin. The thickness of the layer of pure silicone above the sensor units could be reduced (it is currently about 2mm), but the robustness of the sensor has to be taken into account. The sensitivity/range of the sensor units can be changed by the thickness of the copper-beryllium plates, and thinner plates can be used to adjust the sensitivity/range suitable for the intended use in a humanoid robot. The temperature compensation can suppress the susceptibility to thermal change by approximately 80%.

The performance of the force vector detection depends on several factors that will be further studied; the influence of bending angle of the transducers on the measurement will be explored; the angle and also the silicone layer thickness will be optimized. Also the influence of the contact area will be investigated in future work.

It should be noted that the measurable stress in x and y directions is limited by the characteristic of the test setup used (refer to the test setup in Figure 6.13). The test setup cannot tilt more than 45°; hence, the shear stress of the pusher cannot exceed 5.445 kPa in x and y direction.

In future versions of the sensor, multi-layer, partially-bonded flexible PCBs will enable the placing of the chip below the transducers, so that the whole surface of the robot can be covered with transducers. Ports on all sides will enable an efficient networking structure with minimal wiring, as in Reference 34.

6.4 Three-axis Hall-effect sensors

6.4.1 Concept

The three-axis Hall-effect-based sensors use small-sized chips, i.e., MLX90393 from Melexis, which can measure magnet displacements in three axes and provide digital output over I2C. One small cylindrical magnet is placed above each chip, embedded in a soft material, in our case silicone. Forces acting on the silicone surface will cause a magnet displacement (see Figure 6.26). The magnet displacement can be sensed by the chip below the magnet as changes in the magnetic field in three axes. Several chips (with an equal amount of magnets) can be placed next to each other, in order to get distributed measurements. We have implemented flat and curved sensor modules, as described below.

To the best of our knowledge, this is the first distributed soft skin sensor system that can cover various parts of the robot and measure force in three axes, with a sub-centimeter spatial density, and digital output at the same time. Furthermore, each chip also provides temperature measurements.

6.4.2 Implementation

The sensors can be used to cover various parts of a robot's body. Especially robot hands are often in contact with the environment, and our first goal was to cover a robot hand with sensors. The Allegro hand is a commercially available, relatively inexpensive robot hand used by various group for research on multi-fingered grasping and manipulation. The Allegro hand originally has only limited sensing capabilities

Figure 6.26 Concept of measuring three-axis force or pressure with Hall-effect sensor

(joint angle and motor current for each of the 16 joints), and only the fingertips integrate some soft material (with hardness shore A40, which is relatively hard compared to the human skin). We expect that the robot hand capabilities would greatly benefit from distributed sensing and soft skin. With the sensor modules described below, each finger can measure 72 tri-axis force vectors, which is an unprecedented amount and density of force vector measurements.

Each finger of the Allegro hand is composed of three servo motors and a fingertip. The three motors provide a flat surface, and flat skin sensor printed circuit boards (PCBs) with 16 chips were designed that fit on one motor. The PCBs have a size of 27 × 26 mm, and the 16 chips are arranged in a 4 × 4 matrix. Each chip uses four wire I2C communication and has a 7-bit address, out of which two bits can be set by connecting two pins of the chip to either supply voltage (VDD) or ground (VSS), respectively. Therefore, four chips can share one I2C bus segment (SDA). As a result, only seven wires are required to connect to one flat module with 16 chips [VDD, VSS, SCL (clock), and four SDAs]. The backside of the PCB is flat, and the PCBs are mounted on the servo motors directly with thin double-sided sticky tape.

Regarding the fingertip, the original fingertips of the Allegro hand have a cylindrical body and hemispherical tip. Following our results that a more anthropomorphic shape with a smooth change in curvature is beneficial for manipulation with the hand [9], we opted to use the same anthropomorphic shape as in Reference 10 for the tactile fingertip sensors. Covering such a mulit-curved, high convexity shape with sensors is challenging. The desired shape was shrunk by 4 mm on all sides and then approximated with flat surfaces that are big enough for the sensor chips. The resulting shape was 3D printed, and a flexible PCB was designed to cover it. The flexible PCB with the chips was glued to the 3D printed material. A total of 24 chips are used for one fingertip; therefore nine wires need to be connected to one fingertip [VDD, VSS, SCL (clock), and six SDAs].

Both the flat PCBs and the flexible PCBs were covered with silicone skin, which embeds one magnet above each chip (see Figure 6.27). In particular, the magnets are neodymium (grade N50) with a 1.59-mm diameter and a 0.53-mm thickness. Regarding the strength of the magnetic field, the magnet had an optimal pull of 226.8 g and 729 surface gauss; all magnets are facing their corresponding chip with the south pole. The molding process was done in two steps: For the flat modules, first about 3.1-mm-thick silicone was molded, with holes for the magnets (also space for the chips and other additional components like pull-up resistors was left). After the magnets were placed in the silicone, another layer of silicone was molded, which covered the magnets with about 0.9-mm-thick silicon. For the fingertips, the thickness for the two layers is 2.6 mm (again with holes for the magnet) and 1.4 mm above the magnets. Overall, the silicone is 4 mm thick for both types of modules and the skin surface is smooth; the fingertip silicone was molded in such a way to surround the 3D printed core with the flexible PCB on all but one side (the side that is screwed to the rest of the finger). The silicone for the flat sensor modules was molded in a shape to surround the motor and the sensor PCB. Compared to the parts that they surround, the silicone skin is

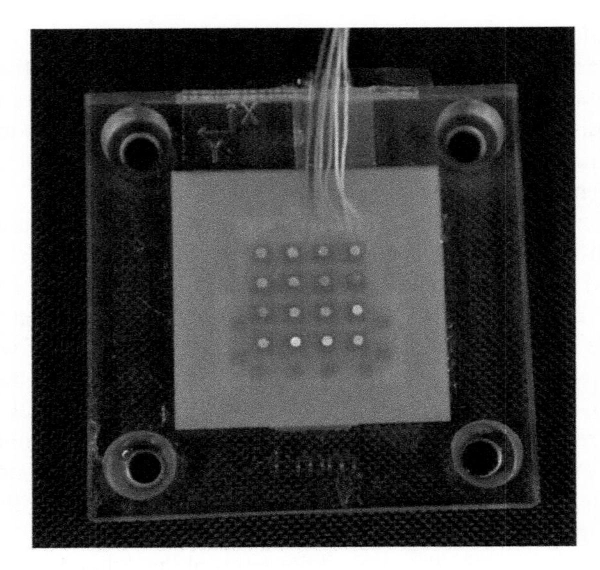

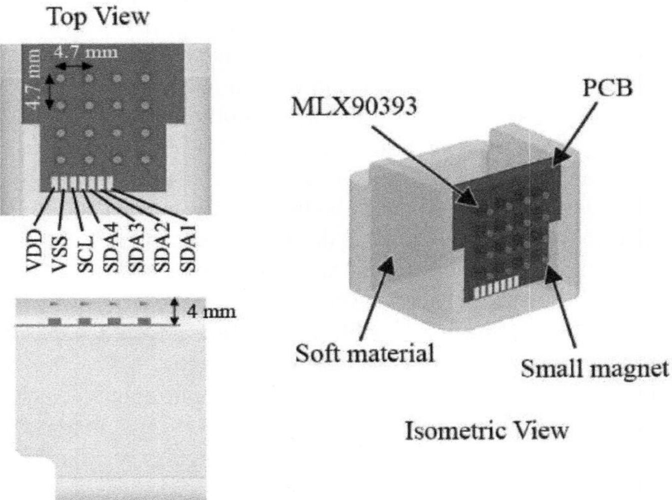

Figure 6.27 *(Top) The complete flat sensor as used for the evaluation experiments. (Bottom) The sensor with the silicone for covering the Allegro hand*

a bit too small, to prevent the silicone from sliding off when operating the hand. Moreover, by surrounding the finger, the silicone provides shock absorbance also to the back of the fingers. The silicone for the flat PCB was Ecoflex 00-30 from Smooth-On. The fingertips used Ecoflex 00-50, a slightly harder Ecoflex, as our

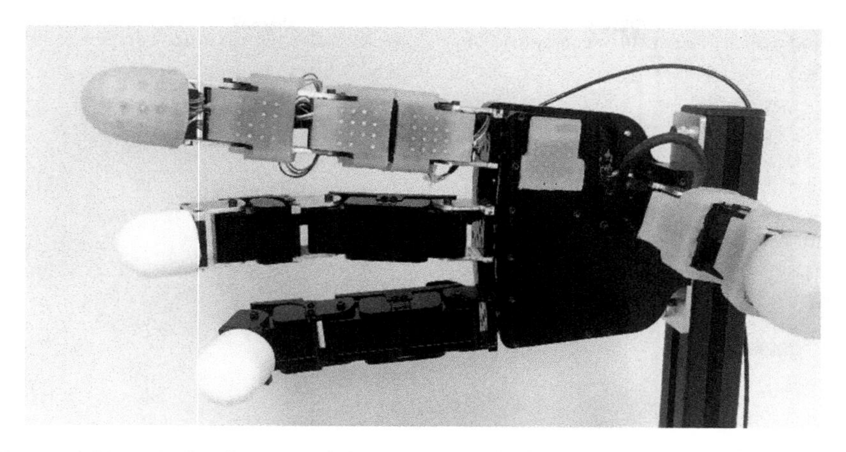

Figure 6.28 Index finger and thumb of Allegro hand covered with skin sensors

experiments showed that with Ecoflex 00-30 the fingertips saturated already at 35 kPa.

To collect the measurements from the sensors, in the experiments described below we use an Arduino Due and a multiplexer (TCA9548A from Adafruit). The Arduino Due is connected to the multiplexer with only one I2C connection, but the multiplexer can collect sensor measurements from four or six SDAs (for the flat module or fingertip, respectively). This setup limits our data collection frequency to around 30 Hz in both cases. In the future, small custom micro-controllers can be used to increase the measurement frequency and reduce the wiring. The index finger covered with one curved and three flat modules can be seen in Figure 6.28.

6.4.3 Experiment

6.4.3.1 Test setup

The test setup used for the tests is nearly identical to the one used for the capacitive sensors (see Figure 6.12), with the only difference that for some experiments a different x–y table was used, which was done only for availability reasons. Furthermore, to apply shear force, the sensors are not tilted; instead, the x–y table is used to slightly displace the sensor after an initial contact with the pusher. An adjustable tilt stage was used so that the voice coil motor applies normal force on the sensor surface (the fingertip has a curved surface). The pusher has a square 12 × 12 mm, except for the experiments for distinguishing the pusher shape (Section 6.4.3.9).

All data shown and used below are unfiltered sensor data. Only an offset (the average of each sensor measurement in the first second, while no force is applied) is calculated for each sensor measurement individually; this offset is used as the zero baseline, i.e., it was subtracted from all measurements.

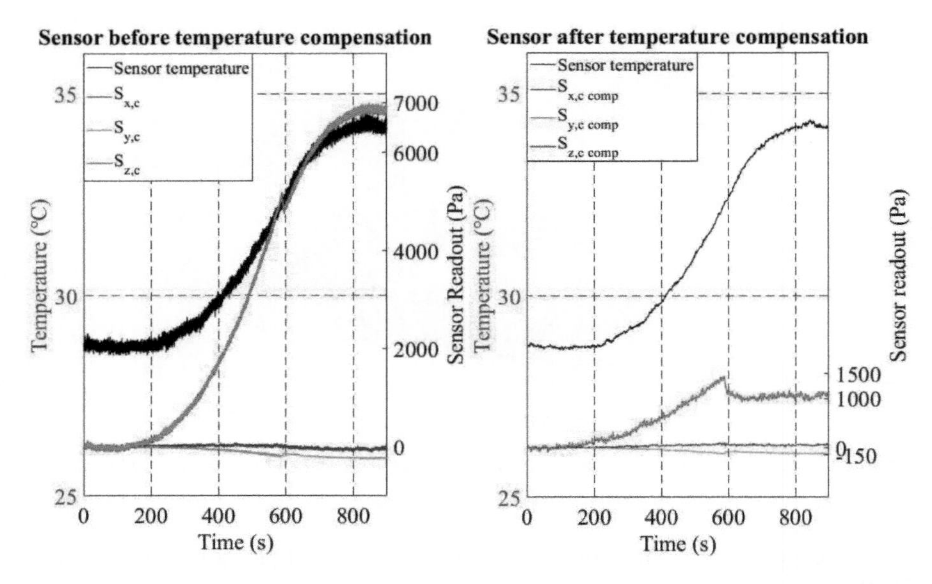

Figure 6.29 *Temperature drift before and after compensation. The right graph shows the filtered temperature measurements*

6.4.3.2 Temperature compensation tests

We expect that the Hall-effect sensor is also sensitive to temperature changes; however, the Hall-effect sensor chip also includes a temperature sensor, and temperature compensation therefore is feasible. Only for this experiment, a prototype version of the sensor was used, with only one three-axis force sensor, and a square magnet [4], but the authors assume that the results are indicative also for the modules with the distributed sensors, with the most relevant difference that the silicone layer between the PCB and the magnet had a thickness of about 4mm. To evaluate the influence of temperature on the measurements, and for temperature influence compensation, a similar method to the one in section 6.3.3.4 is used. In particular, the temperature as measured by the internal temperature sensor multiplied by a gain factor (one for each axis) was subtracted from the raw three-axis measurements. In a first experiment, the sensor was heated from room temperature (about 27°C) to 60°C in about 14 minutes. The z-axis responded strongest, increasing 6 700 digits, corresponding to roughly 12 kPa, using the calibration parameters in Reference 4. From these data, we calculated gain parameters (one for each axis x, y, and z). The measurements before and after the drift compensation with new test data are shown in Figure 6.29. The drift can be reduced from around 6.5 to 1.5 kPa for a temperature increase of 5.4°C. The authors assume that improved temperature compensation is possible, as the trend of the measured temperature and the sensor signal change is similar, if the gain parameters are calculated from training data more similar to the actual use case.

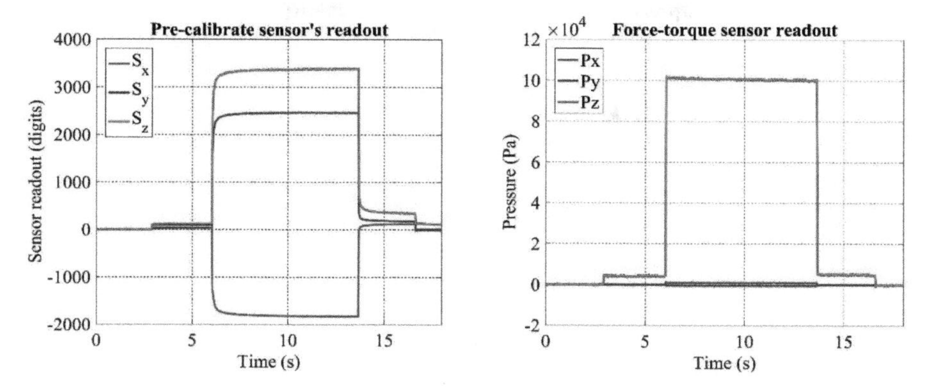

Figure 6.30 Uncalibrated sensor data for flat module

6.4.3.3 Calibration

Uncalibrated sensor responses for the flat and curved modules are shown in Figures 6.30 and 6.31, respectively. There is a non-linear relationship between measured magnet displacement and applied force (as measured by the reference sensor). Furthermore, for both modules, it can be seen that even though only normal force was applied, the Hall-effect sensor measured lateral magnet displacements. This is probably happening because the magnets are not perfectly aligned with the Hall-effect sensors and the silicone is incompressible. Therefore, we use a quadratic model to calculate the calibrated sensor output S_j,c of axis j (x, y, or z) out of the uncalibrated sensor values Sj with the following formula (this was reported wrong in References 6, 7):

$$S_{j.c} = aS_x + bS_y + cS_z + d(S_x * S_y) + e(S_x * S_z) + f(S_y * S_z) + gS_x^2 + hS_y^2 + iS_z^2$$

The calibration parameters a–i are calculated in MATLAB using the Statistics & Machine Learning Toolbox out of independent training data. Details about gathering

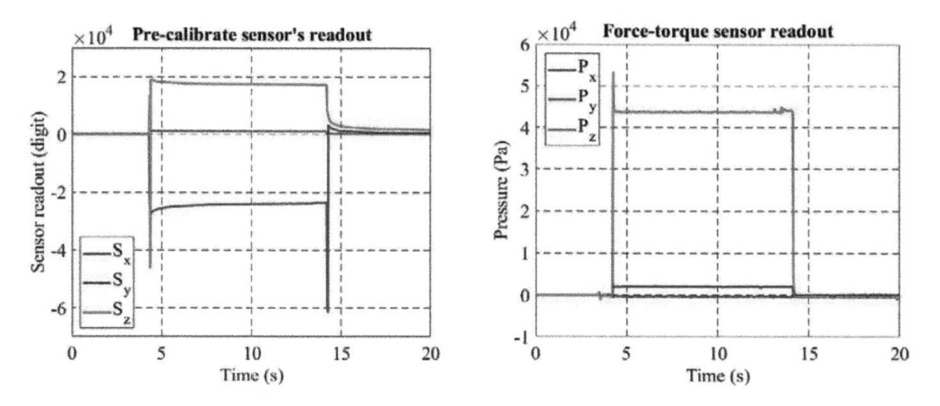

Figure 6.31 Uncalibrated sensor data for fingertip

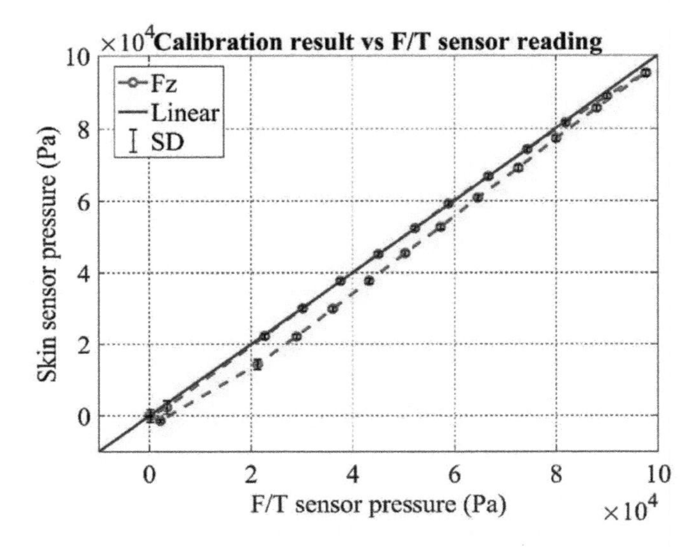

Figure 6.32 *Calibrated sensor in flat module versus reference sensor measurements*

training and test data for the flat module are provided in Reference 7 and for the curved module in Reference 8. After the calibration, the sensors can sense the different axis, shown in Figures 6.32–6.34 for the flat module and Figures 6.35–6.37 for the curved module. The calibrated sensor response is linear and corresponds to

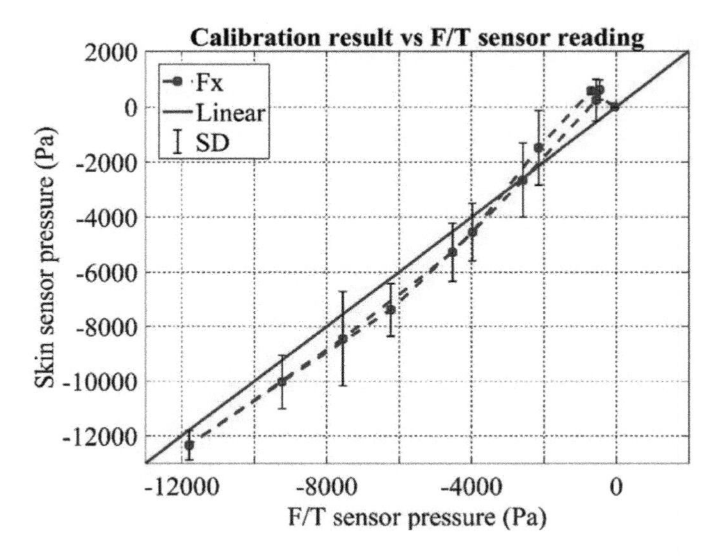

Figure 6.33 *Calibrated sensor in flat module versus reference sensor measurements*

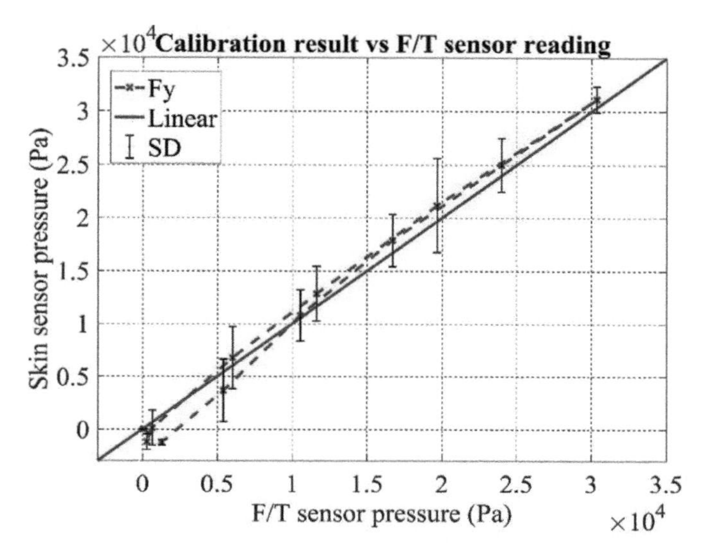

Figure 6.34 *Calibrated sensor in flat module versus reference sensor measurements*

the reference sensor measurements. While each plot focuses on only one measurement axis, it should be noted that the other axis did not respond or respond as the reference sensor. For example, the data for Figure 6.32 and Figure 6.33 were taken in the same experiment.

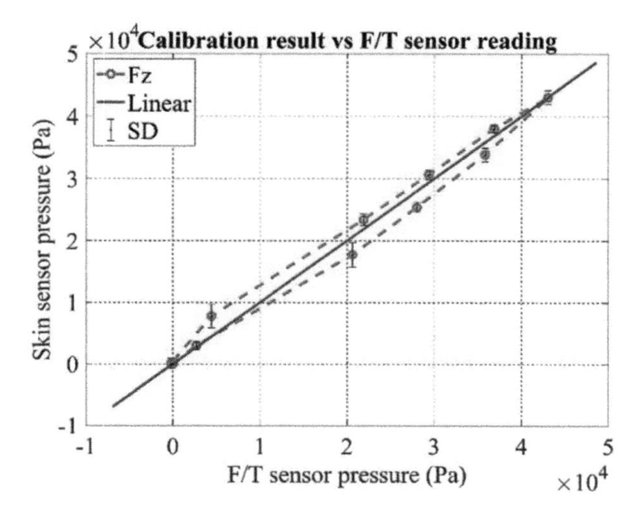

Figure 6.35 *Calibrated sensor in fingertip module versus reference sensor measurements*

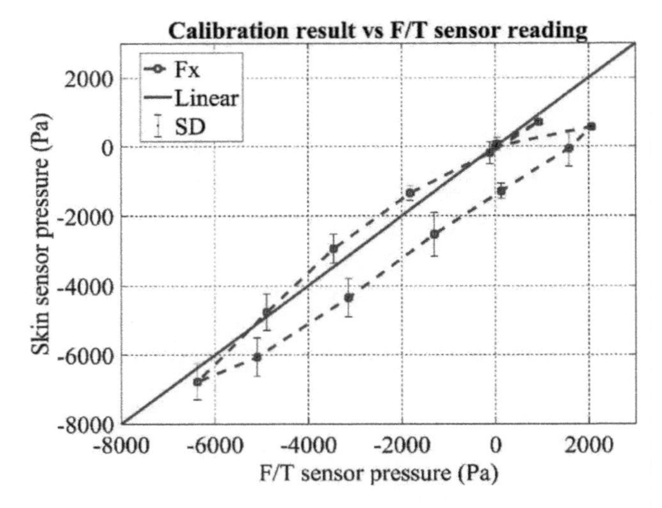

Figure 6.36 *Calibrated sensor in fingertip module versus reference sensor measurements*

6.4.3.4 Response time

When applying the step force for Figure 6.31, Figure 6.35 we could observe that there was a 66 ms (two time steps with 30-Hz sampling frequency) time delay of the skin sensor reply compared to the reference six-axis F/T sensor. The delay could be due to the electronics or the softness of the silicone

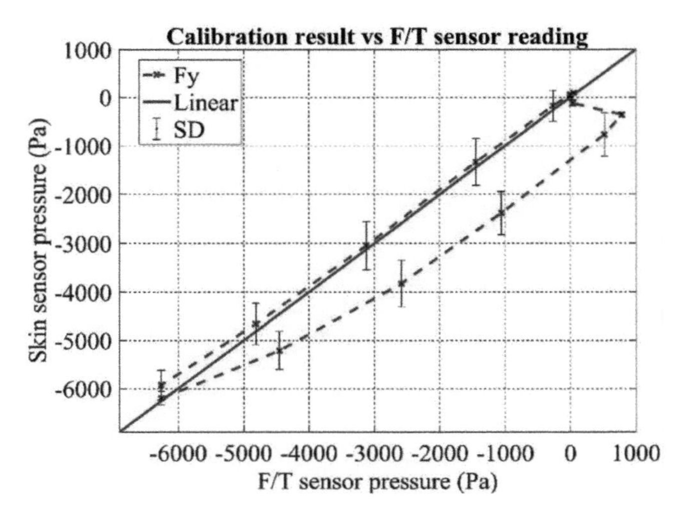

Figure 637. *Calibrated sensor in fingertip module versus reference sensor measurements*

Table 6.5 Minimum detectable force for flat and curved Hall-effect sensor

Axis	Flat module	Curved module
x-axis	± 7 Pa (± 0.001 N)	± 35 Pa (± 0.005 N)
y-axis	± 7 Pa (± 0.001 N)	± 28 Pa (± 0.004 N)
z-axis	± 208 Pa (± 0.03 N)	± 70 Pa (± 0.01 N)

6.4.3.5 Minimum detectable force

We observed the calibrated sensor response of one sensing unit when the sensor is not being pushed (and not being moved). Data were recorded with 30 Hz for 4 seconds before being pushed. The output is not filtered after acquisition; however, the Melexis chip has inbuilt filtering, and we set it to a level that allowed acquiring the measurements at 30 Hz. The sensor output varied within the range given in Table 6.5; those values were set as indicative of the minimum detectable force increase.

6.4.3.6 SNR

We use (6) (6.6) to calculate the SNR values, using the same data as shown in Figure 6.32 and Figure 6.35 (unfiltered, calibrated z-axis measurements). The SNR values for different pressures are shown in Table 6.6.

6.4.3.7 Hysteresis

We calculated the hysteresis with (1) using the data from Figure 6.32 and Figure 6.35. The hysteresis for the z-axis is 5.29% for the flat module and 10% for the fingertip module.

6.4.3.8 Spatial crosstalk test

This test was performed with a flat module. Along a straight line, from one side of the sensor module to the other side, thereby crossing four sensors (only the measurements of those four sensors are shown in Figure 6.38), the sensor module was pushed

Table 6.6 SNR for flat and curved Hall-effect sensor

Pressure (× 10^4 Pa)	SNR (dB)	
	Curved (fingertip)	Flat
0.30	54.66	29.41
1.77	70.02	52.50
2.53	73.14	54.39
3.337	75.63	55.61
4.28	77.70	56.57

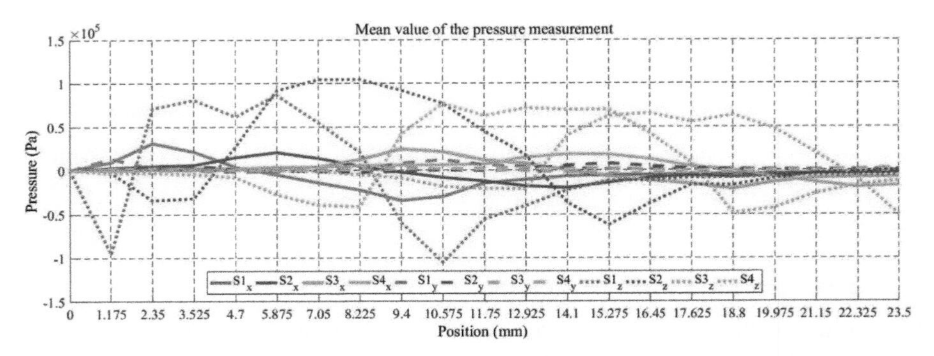

Figure 6.38 *The sensors were pushed at various locations with about 100 kPa. The three-axis response for S1–S4 for all positions is shown. The center of S1 was at about 4.7 mm, S2 at 9.4 mm, S3 at 14.1 mm, and S4 at 18.8 mm*

every 1.175 mm (a quarter of the distance between sensor chips). The response of all four sensors at each location (21 locations in total) when being pushed with 97 kPa is shown in Figure 6.38. In some cases, negative force was measured, which is most probably caused by the incompressibility of the bulk silicone; when the silicone gets pressed in one location, it moves to the surrounding locations. The response is different at each location and it seems possible to extract the pushing location out of those measurements. The different sensors (S1–S4) responded differently, which can at least partially be explained by the fact that they were only calibrated with very limited data (similar to step response data).

6.4.3.9 Contact shape detection

This test was performed with the fingertip module. Pushers of different shapes (shown in the small inlets of Figure 6.39) were pushed against the fingertip, and the response of all sensors in proximity to the pushing location (only those sensors responded) for the different pushers is presented in Figure 6.39. The sensor module was pushed with 10 different forces from 0.5 to 6.3 N, and the sensor responses for all 10 forces are overlaid. The origin of the arrows corresponds to the position of the sensor on the fingertip. The scaling is different for different sensors so that all sensors can be easily seen. Mostly, the blue arrows are scaled down. Overall, it can be seen that the response is different for most pusher shapes. Only the two smallest pushers show a very similar response. Therefore, we argue that the sensor can be used to discriminate shapes, but obviously, there are limitations to the minimum detectable shape differences, as the spatial density is also limited.

6.4.3.10 Measurements on Allegro hand

Both the flat modules and the fingertips were installed on the Allegro hand, as shown in Figure 6.28. The orientation of the sensors matters, due to the

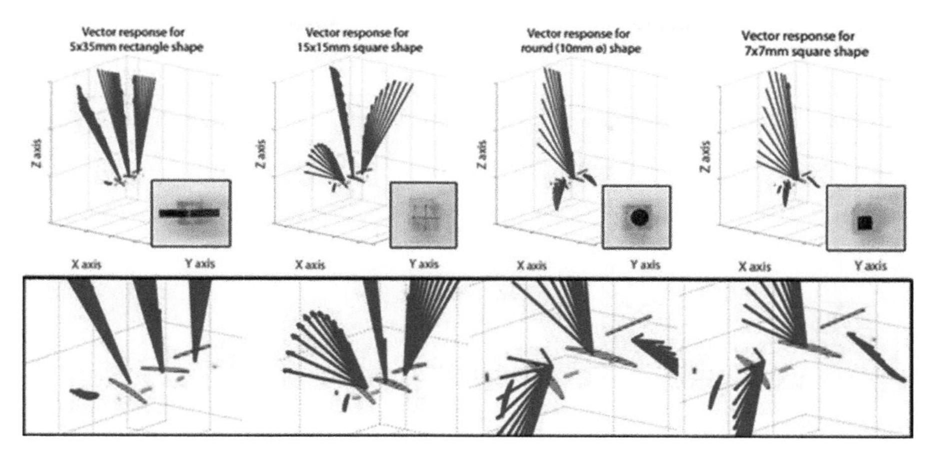

Figure 6.39 The response of the fingertip sensor to different pusher shapes

influence of the earth's magnetic field and the weight of the sensors. In particular, by rotating the flat sensor module in different axis, the calibrated z-axis measurements of the one investigated three-axis sensor changed up to 9 kPa (1.3 N), y-axis up to 2.8 kPa (0.4 N), and x-axis up to 0.7 kPa (0.1 N).

Furthermore, we conducted experiments of the skin sensor measurements when the hand is grasping an object. Figure 6.40 shows the varying reply of one three-axis sensor in the fingertip when the hand is grasping a plastic cup, then weight is added, the cup was moved, and eventually, the hand opens and the cup is released. The hand was controlled with simple joint angle control.

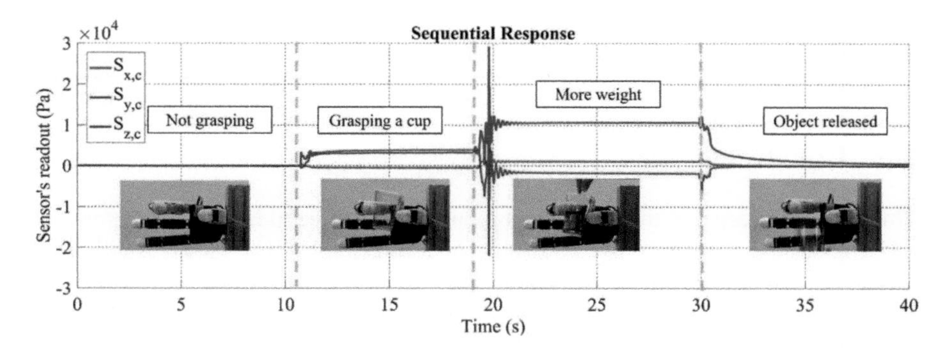

Figure 6.40 The three-axis measurements of one taxel when the hand is grasping an object and when weight is added. Adding weight caused a visible slight vibration of the cup

Table 6.7 The summary of the characteristics of the sensors

Attribute	Capacitive	Hall-effect flat	Hall-effect curved
Size	14 × 14 × 7 mm	4.7 × 4.7 mm center-to-center (coc) distance, 4 mm thick	10 × 6 mm or smaller coc distance, 4 mm thick
Production	Cheap, time consuming	Cheap, fast	
Integration	-	Allegro hand	
SNR (dB)	36.63 at 15.7 kPa	52.50 at 17.7 kPa	70.02 at 17.7 kPa
Hysteresis (z-axis)	3.36%	5.29%	10%
Maximum measured shear stress (kPa)	5.4	30	6
Maximum measured normal stress (kPa)	15.2	100	42
Minimum detectable shear stress (kPa)	2	0.007	0.035
Minimum detectable normal stress (kPa)	2	0.208	0.07
R-squared value for shear stress	0.79 (average of all shear stress data in Table 6.4)	0.99 (average of data for Figures 6.33 and 6.34	0.90 (average of data for Figures 6.36 and 6.37 Figure 36 Figure 37)
R-squared value for shear stress	0.99 (average of all shear stress data in Table 6.4)	0.98 (average of data for Figure 6.32)	0.98 (average of data for Figure 6.35)
Temperature influence after compensation	6.6 kPa for 5° increase (calculated)	1.5 kPa for 5° increase	
Other notable interference	Stray capacitance, can be shielded	Magnetic, difficult to shield	

6.5 Conclusion

This chapter presented the implementation of soft, three-axial tactile sensors for robot skin, based on capacitive and Hall-effect sensors. Both types use small-sized chips that provide digital output in the form of I2C ports, with a 2-bit configurable address, and therefore the amount of necessary wires can be greatly reduced. Table 6.7 compares the properties of the different types. The Hall-effect sensors are smaller and easier to produce and have been successfully integrated in the Allegro hand. Capacitive sensors are sensitive to stray capacitance, but the AD7147 provides an active shield signal, and with careful design, the sensors can be shielded against stray capacitance. The Hall-effect sensor is sensitive to external magnetic fields (including the earth's magnetic field) and is also potentially influenced by the material properties of objects in its vicinity. It is difficult to shield against a magnetic field, especially for our structure. However, reference sensors could be used to compensate for those interferences, or the application could be limited to objects made only from certain materials, which do not influence the sensors. In any case, in our current implementation, the capacitive sensor is less prone to interference.

Some crucial sensor characteristics were investigated, and overall both sensors have state-of-the-art measurement characteristics that make them suitable for robotic skin applications. In particular, the R-squared values for the z-axis are high for both sensors. The Hall-effect sensors have a higher SNR, are more sensitive (lower detectable stress), are more accurate/precise for detecting shear forces (demonstrated by higher R-squared values), and the temperature drift after compensation is lower. However, they also have more hysteresis in their measurements. The SNR and hysteresis have been only evaluated for z-axis measurements so far, and the other axis will be further investigated in future work with a motorized x-y stage. However, all characteristics can be further improved for both sensors with a careful design and the stated values show only the limits of the current implementations of the sensors. The sensors' characteristics can be seen in Table 6.7.

The range of the sensor could not be investigated with our current setup, as the reference sensor and the VCM have force limits. However, the capacitive sensor was tested with up to 5.4 kPa in x-axis and y-axis, and 15.2 kPa in z-axis. The Hall-effect sensors are smaller and could be tested with a smaller pusher, and therefore higher pressures with the same VCM could be achieved; a different way of applying shear stress was used. For the flat module, shear stress of 30 kPa and normal pressures of 100 kPa for the z-axis were applied and successfully measured, and for the curved module with harder silicone 6 and 42 kPa, respectively.

While the sensors proposed in this chapter show several good characteristics, they can be improved in many ways. Potential future work includes further miniaturization, material optimization against hysteresis, and multimodal sensors. Software (filters) and hardware (reference sensors) can be used to make the sensors less prone to interference.

Furthermore, as presented in section 2, tactile sensors can be based on many physical principles, and due to the recent trends of IoT, more and more of those small-sized chips are available. While this chapter shows designs based on capacitive and Hall-effect sensing, employing small-sized chips also for other sensing principles could be a general way for practical implementations of distributed sensors in robots.

References

[1] Somlor S., Schmitz A., Hartanto R.S., Sugano S. 'A prototype force sensing unit for A capacitive-type force-torque sensor'. *2014 IEEE/SICE International Symposium on System Integration (SII)*; Tokyo, Japan, 2015. pp. 684–89.

[2] Somlor S., Schmitz A., Hartanto R.S., Sugano S. 'First results of tilted capacitive sensors to detect shear force'. *Procedia Computer Science*. 2015, vol. 76, pp. 101–06.

[3] Somlor S., Hartanto R.S., Schmitz A., Sugano S. 'A novel tri-axial capacitive-type skin sensor'. *Advanced Robotics*. 2015, vol. 29(21), pp. 1375–91.

[4] Tomo T.P., Somlor S., Schmitz A., Hashimoto S., Sugano S., Jamone L. 'Development of a hall-effect based skin sensor'. *In SENSORS, 2015 IEEE*; 2015. pp. 1–4.

[5] Tomo T.P., Somlor S., Schmitz A, *et al.* 'Design and characterization of a three-axis hall effect-based soft skin sensor'. *Sensors (Basel, Switzerland)*. 2016, vol. 16(4), p. 491.

[6] Tomo T.P., Wong W.K., Schmitz A. 'SNR modeling and material dependency test of a low-cost and simple to fabricate 3D force sensor for soft robotics'. *2016 IEEE/SICE International Symposium on System Integration (SII)*; Sapporo, Japan, 2016. pp. 33–428.

[7] Tomo T.P., Wong W.K., Schmitz A, *et al.* 'A modular, distributed, soft, 3-axis sensor system for robot hands'. Presented at 2016 IEEE-RAS 16th International Conference on Humanoid Robots (Humanoids); Cancun, Mexico.

[8] Tomo T.P., Schmitz A., Wong W.K, *et al.* 'Covering A robot fingertip with uskin: A soft electronic skin with distributed 3-axis force sensitive elements for robot hands'. *IEEE Robotics and Automation Letters*. 2018, vol. 3(1), pp. 124–31.

[9] Or K., Schmitz A., Funabashi S., Tomura M., Sugano S. 'Development of robotic fingertip morphology for enhanced manipulation stability'. *2016 IEEE International Conference on Advanced Intelligent Mechatronics (AIM)*; Banff, AB, Canada, 2016. pp. 25–30.

[10] Or K., Tomura M., Schmitz A., Funabashi S., Sugano S. 'Position-force combination control with passive flexibility for versatile in-hand manipulation based on posture interpolation'. *2016 IEEE/RSJ International Conference on Intelligent Robots and Systems (IROS)*; Daejeon, South Korea, 2016. pp. 2542–47.

[11] [] Or K., MorikuniS., Ogasa S., Funabashi S., Schmitz A., Sugano S. 'A study on fingertip designs and their influences on performing stable prehension for robot hands'. *2016 IEEE-RAS 16th International Conference on Humanoid Robots (Humanoids)*; Cancun, Mexico, 2015. pp. 772–77.

[12] Schmitz A., Bansho Y., Noda K., Iwata H., Ogata T., Sugano S. 'Tactile object recognition using deep learning and dropout'. *2014 IEEE-RAS 14th International Conference on Humanoid Robots (Humanoids 2014)*; Madrid, Spain, 2015. pp. 1044–50.

[13] Kojima K., Sato T., Schmitz A., Arie H., Iwata H., Sugano S. 'Sensor prediction and grasp stability evaluation for in-hand manipulation'. Presented at 2013 IEEE/RSJ international conference on intelligent robots and systems (IROS 2013); Tokyo, 2015.

[14] Funabashi S., Schmitz A., Sato T., Somlor S., Sugano S. 'Robust in-hand manipulation of variously sized and shaped objects'. *2015 IEEE/RSJ International Conference on Intelligent Robots and Systems (IROS)*; Hamburg, 2015. pp. 257–63.

[15] Funabashi S., Sato T., Schmitz A., Sugano S. 'Feature extraction by deep learning for improved in-hand manipulation'. *The Abstracts of the International Conference on Advanced Mechatronics*. 2015, vol. 2015.6, pp. 31–32.

[16] Dahiya R.S., Metta G., Valle M., Sandini G. 'Tactile sensing—from humans to humanoids'. *IEEE Transactions on Robotics*. 1989, vol. 26(1), pp. 1–20.

[17] Siciliano B., Khatib O. *Springer handbook of robotics*. Berlin, Heidelberg: Springer; 2008. pp. 455–76. Available from http://link.springer.com/10.1007/978-3-540-30301-5

[18] Nicholls H.R., Lee M.H. 'A survey of robot tactile sensing technology'. *The International Journal of Robotics Research*. 1989, vol. 8(3), pp. 3–30.

[19] Lee M.H., Nicholls H.R. 'Review article tactile sensing for mechatronics—a state of the art survey'. *Mechatronics*. 1989, vol. 9(1), pp. 1–31.

[20] Jacobsen S.C., McGammon I.D., Biggers K.B., Phillips R.P. 'Design of tactile sensing systems for dextrous manipulators'. *IEEE Control Systems Magazine*. 1989, vol. 8(1), pp. 3–13.

[21] Puangmali P., Althoefer K., Seneviratne L.D., Murphy D., Dasgupta P. 'State-of-the-art in force and tactile sensing for minimally invasive surgery'. *IEEE Sensors Journal*. 2015, vol. 8(4), pp. 371–81.

[22] Howe R.D. 'Tactile sensing and control of robotic manipulation'. *Advanced Robotics*. 1993, vol. 8(3), pp. 245–61.

[23] Argall B.D., Billard A.G. 'A survey of tactile human–robot interactions'. *Robotics and Autonomous Systems*. 2015, vol. 58(10), pp. 1159–76.

[24] Silvera-Tawil D., Rye D., Velonaki M. 'Artificial skin and tactile sensing for socially interactive robots: A review'. *Robotics and Autonomous Systems*. 2015, vol. 63, pp. 230–43.

[25] Göger D., Gorges N., Wörn H. 'Tactile sensing for an anthropomorphic robotic hand: hardware and signal processing'. *In IEEE international conference on robotics and automation*; 2009. pp. 895–901.

[26] Choi B., Lee S., Choi H.R., Kang S. 'Development of anthropomorphic robot hand with tactile sensor: SKKU hand II'. *2006 IEEE/RSJ international conference on intelligent robots and systems*; Beijing, China, 2006. pp. 84.

[27] Jockusch J., Walter J., Ritter H. 'A tactile sensor system for A three-fingered robot manipulator'. *International Conference on Robotics and Automation*; Albuquerque, NM, USA, 2006. pp. 3080–86.

[28] Hosoda K., Tada Y., Asada M. 'Anthropomorphic robotic soft fingertip with randomly distributed receptors'. *Robotics and Autonomous Systems*. 2006, vol. 54(2), pp. 104–09.

[29] Minato T., Yoshikawa Y., Noda T., Ikemoto S., Ishiguro H., Asada M. 'CB2: A child robot with biomimetic body for cognitive developmental robotics'. *2007 7th IEEE-RAS international conference on humanoid robots (humanoids 2007)*; Pittsburgh, PA, USA, 2006.

[30] *ShokacChipTM product outline* [Online]]]. Available from http://www.touchence.jp/en/chip/index.html
[31] Saga S., Morooka T., Kajimoto H., Tachi S. 'High-resolution tactile sensor using the movement of a reflected image'. *In proceedings of eurohaptics*; 2006. pp. 81–86.
[32] Yuan W., Li R., Srinivasan M.A., Adelson E.H. 'Measurement of shear and slip with a gelsight tactile sensor'. *2015 IEEE International Conference on Robotics and Automation (ICRA)*; Seattle, WA, USA, 2014. pp. 304–11.
[33] Ohka M., Tsunogai A., Kayaba T., Abdullah S.C., Yussof H. 'Advanced design of columnar-conical feeler-type optical three-axis tactile sensor'. *Procedia Computer Science*. 2014, vol. 42, pp. 17–24.
[34] Yamada K., Goto K., Nakajima Y., Koshida N., Shinoda H. 'A sensor skin using wire-free tactile sensing elements based on optical connection'. *SICE 2002. 41st SICE Annual Conference*; Osaka, Japan, 2014. pp. 131–34.
[35] Yoshikai T., Hayashi M., Ishizaka Y. 'Development of robots with soft sensor flesh for achieving close interaction behavior'. *Advances in Artificial Intelligence*. 2014, vol. 2012, pp. 1–27.
[36] *optoforce* [Online]. Available from https://optoforce.com/3d-force-sensor-omd
[37] Jamone L., Metta G., Nori F., Sandini G. 'James: A humanoid robot acting over an unstructured world'. *2006 6th IEEE-RAS international conference on humanoid robots*; University of Genova, Genova, Italy, 2006. pp. 50.
[38] Jamone L., Natale L., Metta G., Sandini G. 'Highly sensitive soft tactile sensors for an anthropomorphic robotic hand'. *IEEE Sensors Journal*. 2015, vol. 15(8), pp. 4226–33.
[39] Youssefian S., Rahbar N., Torres-Jara E. 'Contact behavior of soft spherical tactile sensors'. *IEEE Sensors Journal*. 2014, vol. 14(5), pp. 1435–42.
[40] Ledermann C., Wirges S., Oertel D., Mende M., Woern H. 'Tactile sensor on a magnetic basis using novel 3D hall sensor-first prototypes and results'. *2013 IEEE 17th international conference on intelligent engineering systems (INES)*; San Jose, Costa Rica, 2013. pp. 55.
[41] Natale L., Torres-jara E. 'A sensitive approach to grasping'. *In proceedings of the sixth international workshop on epigenetic robotics*; 2006. pp. 87–94.
[42] Iwata H., Sugano S. 'Design of human symbiotic robot TWENDY-ONE'. *2009 IEEE international conference on robotics and automation (ICRA)*; Kobe, 2007. pp. 580–86.
[43] Liu T., Inoue Y., Shibata K. 'Design of low-cost tactile force sensor for 3D force scan'. *2008 IEEE sensors*; Lecce, Italy, 2008. pp. 1513–16.
[44] Iwata H., Sugano S. 'Whole-body covering tactile interface for human robot coordination'. *2002 IEEE International Conference on Robotics and Automation*; Washington, DC, USA, 2002. pp. 3818–24.
[45] Platt R., Ihrke C., Bridgewater L. 'A miniature load cell suitable for mounting on the phalanges of human-sized robot fingers'. *2011 IEEE international conference on robotics and automation (ICRA)*; Shanghai, China, 2007. pp. 5357–62.

[46] Liu H., Meusel P., Seitz N, *et al.* 'The modular multisensory DLR-HIT-hand'. *Mechanism and Machine Theory*. 2007, vol. 42(5), pp. 612–25.

[47] Oddo C.M., Controzzi M., Beccai L., Cipriani C., Carrozza M.C. 'Roughness encoding for discrimination of surfaces in artificial active-touch'. *IEEE Transactions on Robotics*. 2007, vol. 27(3), pp. 522–33.

[48] Takahashi H., Nakai A., Thanh-Vinh N., Matsumoto K., Shimoyama I. 'A tri-axial tactile sensor without crosstalk using pairs of piezoresistive beams with sidewall doping'. *Sensors and Actuators A*. 2013, vol. 199, pp. 43–48.

[49] Noda K., Hoshino K., Matsumoto K., Shimoyama I. 'A shear stress sensor for tactile sensing with the piezoresistive cantilever standing in elastic material'. *Sensors and Actuators A*. 2013, vol. 127(2), pp. 295–301.

[50] Cannata G., Maggiali M. 'An embedded tactile and force sensor for robotic manipulation and grasping'. *5th IEEE-RAS International Conference on Humanoid Robots, 2005*; San Diego, Cali, USA, 2005. pp. 80–85.

[51] Gerardo J., Lanceros-Mendez S. *Sensors: Focus on Tactile Force and Stress Sensors*. 2008, pp. 271–88. Available from http://www.intechopen.com/books/sensors-focus-on-tactile-force-and-stress-sensors

[52] Bridgwater L.B., Ihrke C.A., Diftler M.A. 'The robonaut 2 hand - designed to do work with tools'. *2012 IEEE international conference on robotics and automation (ICRA)*; St Paul, MN, USA, 2007. pp. 3425–30.

[53] Noda K., Matsumoto K., Shimoyama I. 'Flexible tactile sensor sheet with liquid filter for shear force detection'. *2009 IEEE 22nd international conference on micro electro mechanical systems (MEMS)*; Sorrento, Italy, 2013. pp. 785.

[54] Cheng M.-Y., Lin C.-L., Lai Y.-T., Yang Y.-J. 'A polymer-based capacitive sensing array for normal and shear force measurement'. *Sensors (Basel, Switzerland)*. 2010, vol. 10(11), pp. 10211–25.

[55] Hyung- K.L., Jaehoon C., Sun-Il C., Euisik Y. 'Normal and shear force measurement using a flexible polymer tactile sensor with embedded multiple capacitors'. *Journal of Microelectromechanical Systems*. 2008, vol. 17(4), pp. 934–42.

[56] Viry L., Levi A., Totaro M, *et al.* 'Flexible three-axial force sensor for soft and highly sensitive artificial touch'. *Advanced Materials (Deerfield Beach, Fla.)*. 2014, vol. 26(17), pp. 2659–64.

[57] Maiolino P., Maggiali M., Cannata G., Metta G., Natale L. 'A flexible and robust large scale capacitive tactile system for robots'. *IEEE Sensors Journal*. 2013, vol. 13(10), pp. 3910–17.

[58] Schmitz A., Maiolino P., Maggiali M., Natale L., Cannata G., Metta G. 'Methods and technologies for the implementation of large-scale robot tactile sensors'. *IEEE Transactions on Robotics*. 2013, vol. 27(3), pp. 389–400.

[59] Mittendorfer P., Cheng G. 'Integrating discrete force cells into multi-modal artificial skin'. *2012 12th IEEE-RAS international conference on humanoid robots (humanoids 2012)*; Osaka, Japan, 2013. pp. 847–52.

[60] Dobrzynska J.A., Gijs M.A.M. 'Polymer-based flexible capacitive sensor for three-axial force measurements'. *Journal of Micromechanics and Microengineering*. 2013, vol. 23(1), p. 015009.

[61] Devices A. 'AD7147: captouch® programmable controller for single-electrode capacitance sensors (REV.D' in *Data sheet*. pdf; 2011.
[62] Kim K., Sun Y., Voyles R.M., Nelson B.J. 'Calibration of multi-axis MEMS force sensors using the shape-from-motion method'. *IEEE Sensors Journal*. 1992, vol. 7(3), pp. 344–51.
[63] Cheung E., Lumelsky V. 'Development of sensitive skin for a 3D robot arm operating in an\nuncertain environment'. *" in Proceedings, 1989 International Conference on Robotics and Automation*; 1989.
[64] Cheung E., Lumelsky V. 'A sensitive skin system for motion control of robot arm manipulators'. *Robotics and Autonomous Systems*. 1992, vol. 10(1), pp. 9–32.
[65] Ulmen J., Cutkosky M. 'A robust, low-cost and low-noise artificial skin for human-friendly robots'. *2010 IEEE international conference on robotics and automation (ICRA 2010)*; Anchorage, AK, 1992. pp. 4836–41.
[66] Davison B. 'Techniques for robust touch sensing design'. *AN1334 Microchip Technol. Inc*. 2010, pp. 1–28.

Chapter 7

A review of tactile sensing in e-skin, wearable device, robotic, and medical service

Jian Hu[1], Junghwan Back[2], and Hongbin Liu[1,3]

In order to better perceive and manipulate the surrounding environment, various tactile sensing technologies have been developed over the last decades, taking inspiration from the human sense of touch. The tactile sensors have been greatly improved in terms of miniaturization, integration, sensitivity, resolution, etc. However, it is still a huge challenge to integrate them into devices on a large scale with different shapes that require tactile information as feedback. This survey summarizes the mainstream tactile sensing technologies into eight categories in order to discuss the general merits and demerits of each type. An overall picture of the design criteria that can help the researchers to evaluate the performance of a tactile sensing device is presented before an extensive review of the applications, including electrical skins (e-skin), robotics, wearable devices, and medical services. After that, trends of the above fields are presented, such as multifunctional sensing capability, adjustable sensing density in a large area, conformability to complex surfaces, self-powered array, etc. It should be noted that the state-of-art achievements in e-skins will more or less facilitate the development of other fields in which tactile sensing technologies are urgently needed. Finally, challenges and open issues are discussed based on the perspective of mass production, including standardization of the fabrication process, data transmission of a high-density sensing array, fault tolerance and auto-calibration, and the layout of sensing elements on an irregular 3D surface without losing the mechanical and electrical performance of the sensors.

[1]State Key Laboratory of Management and Control of Complex Systems, Chinese Academy of Sciences, Institute of Automation, Beijing, China

[2]School of Biomedical Engineering & Imaging Sciences, Faculty of Life Sciences and Medicine, King's College London, London, UK

[3]School of Biomedical Engineering and Imaging Sciences, King's College London, St Thomas' Hospital, London, UK

7.1 Introduction

Speaking of the sense of touch, it is natural to compare it with sight. Undoubtedly, machine vision is the most widely used sensing technology in the field of robotics. It is the main approach to obtain information about the external environment as well as an important prerequisite to realize the high-precision control, high-reliability human–machine interaction, and intelligent operation of the robots. Actually, more than 80% of the information being received by the brain is from vision [1]. However, other studies have shown that without the sense of touch but only relying on visual information, it is difficult to accurately judge the texture of the surface of an object, thus affecting its fine operation [2]. While physical interaction between objects in the real world is rich [3], and it is also one of the key components to realize force transmission [1]. In addition, the skin, as the largest organ of a human being, can not only feel the slight stimulus and distinguish the type of it (such as stretching, twisting, and compression) but also detect the temperature through interaction with the environment, and even has the perfect characteristic of self-healing.

The density distribution of the mechanoreceptors in different parts of the skin varies greatly, thus spatial resolution that includes the capability of point localization and two-point discrimination threshold varies significantly [4]. Among them, human hands, especially the fingertips, are rich in various mechanoreceptors, which are widely distributed in the layers of epidermis and dermis, as shown in Reference 5. Meissner's corpuscle and Pacinian corpuscle are also named as fast adapting units to detect the low- and high-frequency stimuli, respectively. The former can be used to measure the slip and control the grip, while the latter is sensitive to small motion but only activates at the beginning and end of the stimuli. On the contrary, Merkel cells and Ruffini corpuscles are the slow adapting units, which respond to the low- and high-frequency stimuli, respectively. The former can be used to identify the edge of the objects under static pressure, while the latter has the ability to measure the skin stretch [5–7].

Inspired by the skin's ability to perceive the outside world, tactile sensing technology based on different sensing principles has been developed rapidly [8–10]. The categorization of tactile sensors can also be divided into three types according to the spatial resolution with the corresponding biological body parts in anatomy, which are single-point sensor, high-spatial resolution tactile sensor, and large-area tactile sensor [11]. The research on the first two types of sensors starts early, and the processing methods are relatively mature, so there are commercial products in the market, e.g., the series of Nano sensors for single-point measurement from ATI, the array sensors for multipoint measurement from Pressure Profile Systems (PPS), and Tekscan. In recent years, many achievements have been made in the field of e-skin, but most of the works are mainly focused on the selection of materials and the design of structures based on different sensing principles; meanwhile, the research on practical and commercial tactile sensors is not enough. In addition, the tactile sensing device should be flexible and stretchable enough to be attached to any irregular 3D surface with a long working life. There is no sensing system that can accurately detect the multimode tactile information at the same level as the human being. The development of such technologies remains a technical or even scientific challenge [7].

The perfect e-skin should have the following properties: (1) percept the external stimuli and recognize the types of them simultaneously, (2) mimic the excellent softness of human skin, and (3) transfer the precepted information to the "brain" to realize the human–machine interaction [6]. Therefore, in the field of robot fine operation, medical services, wearable devices, and so on, the tactile sensing system needs to be characterized by flexibility, stretchability, lightweight, high integration, high spatial and temporal resolution, high sensitivity, high reliability, easy expansion, etc. Due to the potential to match the above requirements, e-skin has become a hot topic in the last decade. In summary, tactile is an indispensable source of information for us to explore and operate the external world, as well as a hardware guarantee to assist robots in environmental perception and complete complex operations.

The performance of a tactile sensing system can be improved by selecting the appropriate materials and manufacturing techniques based on different sensing principles, as well as the algorithms of signal processing. Thus, this chapter is organized as follows. In section 2, the current state-of-the-art tactile sensing technologies are divided into eight categories. Meanwhile, the advantages and disadvantages of each type are discussed briefly. After that, the design criteria and performance indexes of the tactile sensing system are given in section 3. Some specific areas that are in urgent need of tactile feedback are introduced in section 4. Future directions and the challenges are discussed in section 5.

7.2 Hardware of various tactile sensing technologies

Based on the sensing principle, tactile sensing technologies can be divided into the resistive, piezoelectric, capacitive, optical, magnetic field, quantum tunneling composite (QTC), triboelectric effect, field-effect transistor (FET) [5, 8–10, 12, 13], as shown in Figure 7.1. A general comparison of the mainstream sensing principles can be found in Table 7.1.

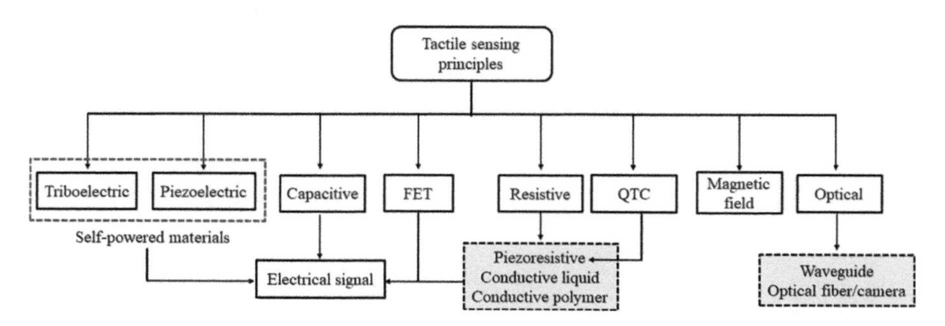

Figure 7.1 Taxonomy of tactile sensing principles

Table 7.1 Comparison of the mainstream sensing principles

Sensing principle		Advantages	Disadvantages
Resistive	Piezoresistive	1. High-spatial resolution 2. High sensitivity	1. Complex machining processes (MEMS may be required) 2. Rigid material
	Soft conductive material	1. Mechanical flexibility 2. Highly conformable to 3D surface	1. Fragile interface between the circuit and the sensing pad 2. Hysteresis effect
Piezoelectric		1. High sensitivity in the dynamic measurement 2. Self-powered material 3. High flexibility (PVDF)	1. Not suitable for static measurement 2. Complex fabrication method (Nanotechnology may be required)
Capacitive		1. High and adjustable sensitivity 2. Fast response time 3. Large dynamic response	1. Flat surface or only one degree of freedom bending compliance 2. Complex machining processes (MEMS may be required)
Optical		1. High humidity tolerance 2. High resistance to electromagnet disturbance 3. Suitable for 3D irregular surface (ESSENT) 4. Simultaneously sensing multimode stimuli (FBG) 5. High-spatial resolution	1. large total package size 2. Easy to be affected by the external light source
Magnetic field		1. Low mechanical hysteresis 2. High sensitivity 3. Wide dynamic range	1. Complex fabrication method 2. High resistance to electromagnet disturbance

(Continues)

Table 7.1 Continued

Sensing principle	Advantages	Disadvantages
QTC	1. Suitable for detecting forces or torques with multiple degrees of freedom 2. Excellent sensitivity	1. Considerable restore time
Triboelectric effect	1. Self-powered material 2. High stretchability (TENG) 3. High sensitivity 4. High resolution 5. A wide selection of materials 6. Easy to be used in combination with other sensing principles	1. Complex fabrication method (nanotechnology may be required)
FET	1. Superior mechanical flexibility (OFET) 2. Biocompatibility (OFET)	1. Complex structure 2. Complex fabrication method

7.2.1 Resistive

In piezoresistive tactile sensors, the applied force is expressed by the resistance change of the pressure-sensitive material. The piezoresistive effect has been found in some metallic [14] and semi-conductive materials [15]. Processing the above materials by micro-electromechanical systems (MEMS) technology, the tactile sensors with high-spatial resolution and sensitivity [16] can be fabricated. By designing various types of the cantilever, the shear and normal force can be distinguished [17, 18]. However, due to the complex micromachining processes, the requirement of mass production in the stage of industrialization is difficult to realize.

Unlike the silicon-based piezoresistive sensors, whose mechanical flexibility is achieved by embedding the sensing elements in an elastomer [12], the other types of resistive tactile sensors have employed a conductive liquid [19] or polymer [20], which is rich in the conductive filler, in order to realize the highly conformability to the 3D surface. Whereas, due to the complexity of the circuit layout, applications are limited to smooth and regular surfaces with small curvature [21]. In addition, the manufacturing processes are relatively easier, but the performance of the sensors may be affected by the hysteresis of the soft materials [22].

7.2.2 Piezoelectric

As for piezoelectric tactile sensing, the conversion of forces to electrical signals depends on the deformation of the piezoelectric materials, which will generate the electric charge. The commonly used material is polyvinylidene fluoride (PVDF) because of its great flexibility, high sensitivity, and chemical stability [23]. Its nanocomposites could further enhance the above capabilities, which are regarded as the most promising materials in the field of wearable devices [13]. Another method that can be used to increase or adjust the sensitivity of a piezoelectric sensor is to combinate it with microstructured polydimethylsiloxane (PDMS) layer [24]. In addition, no requirement for an additional power supply lends this type of sensor to creating a compactable dimension with simple structure. However, due to the limited ability in the static measurement, the disadvantage of piezoelectric sensors is also obvious. To overcome this challenge, a hybrid of more than one sensing principle is conducted to detect static and dynamic interaction tasks. Based on the piezoresistive and piezoelectric effects of ZnO nanowire arrays, two layers with interlocked micropillars are fabricated, which is able to detect the static pressure of 0.6 Pa [25]. William *et al.* proposed a capacitive-piezoelectric tandem stack that has static sensitivity from 0.25 to 0.002 kPa^{-1}, depending on the intensity of pressure [26].

7.2.3 Capacitive

No matter how many layers the capacitive tactile sensor contains during the fabrication process, in theory, this type of sensor usually consists of three functional parts, namely two proximity electrodes and an insulator. When the geometry configuration, such as the thickness of the insulator between the two electrodes, changes under the applied force, the charges are stored which is proportional to the capacitance of the capacitor. The sensitivity of the capacitive tactile sensors can be adjusted by using PDMS as the insulator layer [27]. However, considering that the sensing element should be designed as compact as possible, the size variation of the insulator is limited [28].

Capacitive-based array tactile sensors have been developed with very good sensitivity but are often based on the planar substrates or have only 1D bending compliance [29, 30]. This makes it difficult for these sensors to be attached to the 3D surfaces, especially the surfaces with small radii that are often associated with medical devices. With the development of nanotechnology, Tie *et al.* proposed a flexible capacitive electronic skin with high sensitivity, large dynamic response, and fast response time. The micropatterned PDMS film, which is molded by a lotus leaf together with a substrate of Au, is fabricated as the electrode. It shows great performance in the measurement of bending and stretching [31].

7.2.4 Optical

Polymer-based waveguide was developed by Yun *et al.* [32] to detect multiple contact events simultaneously with a fast responding time, even on curvilinear interfaces, such as the human arm. However, the layout of the waveguide does not consider total reflection condition. It may cause significant light loss during the bending process that limits the adaptability to arbitrary 3D surfaces. Instead of fabricating the waveguide channels, References 33–35 presented a multilayer planer waveguide to achieve a high-spatial resolution based on the pixel size of a Complementary Metal-oxide Semiconductor (CMOS) camera, which is very suitable for detecting the texture of an object. In general, tactile sensing with high-resolution ability has the following characteristics, such as soft contact surface, high resolution of vertical and lateral geometry sensing, slip detection, and shear force estimation based on the deformation of the contact surface [36], non-necessary to consider the effect of surface reflectance function of an object [37], etc.

The photometric stereo was introduced as the sensing principle to measure a 3D topography of a contact surface. To overcome the main drawback of this method, such as the limitation on surface reflectance, controlled illumination, and a large number of images, the GelSight sensor is fabricated with a transparent elastomer membrane with a constant reflectance so that the geometry of the contact surface can be recognized by the distorting of the contact surface [36–38]. If some markers are designed on the membrane, various types of loading, such as the normal force, the shear force, and the torque, can be discriminated based on the pattern of change of the makers [39].

TacTip family is biologically inspired by the Meissner's corpuscles and Merkel cells of the glabrous skin of human hands and feet [40]. Although the calibration of force was not being investigated, the detection of edge showed the sub-millimeter accuracy on a rolling cylinder task [41]. A 40-fold super-resolution can be even achieved based on an active Bayesian perception algorithm [42]. Moreover, in the application of small spatial scales, the sensor with an artificial fingerprint in Reference 43 was found to have significant improvements in positioning perception compared with the one that has a smooth surface. All these advantages make TacTip especially suitable for the field of object exploration [44, 45].

To solve the bottleneck associated with 3D complex surfaces, Back *et al.* [46] have been working on the development of "Embedding Soft material into Structure ENabling Tactile sensing" (ESSENT) technology. The key assumption of this method is that the soft material in a Bernoulli pipe has low compressibility so that the pressure applied to one side can be extruded at the end of the channel in the form of an axial protrusion; therefore, the change in light intensity caused by the deformation of the soft material can be captured by photodiodes, which is used to calibrate the applied force. Federica *et al.* [47] presented an angled tip optical fiber design to measure the intensity modulation of a fluorescence signal under the pressure so that the proposed sensor can be mounted to the surface along a tool body. In addition, optical-based tactile sensors have the potential to realize the simultaneous measurement of multimode stimuli. A hybrid structure fiber was proposed to measure the strain and temperature by the wavelength shift of Fabry–Perot interferometer and fiber Bragg grating (FBG), respectively [48].

7.2.5 Magnetic field

Such sensors are composed of a soft body, an embedded magnet, and magnetoresistive sensing material (such as a Hall sensor) [49]. The intensity of the magnetic field that is proportional to the change in the movement of the magnet can be measured by the Hall effect. Low mechanical hysteresis is the main advantage, as well as the characteristics of high sensitivity (30 mg) and wide dynamic range (6–400 kPa), which are presented by Oh *et al.* [50]. With a 3D Hall sensor, a triaxis tactile sensing device was published that can decouple the shear and normal force [51]. With the development of nanofabrication methods, the nanocomposite cilium successfully reduces the dimension of one sensing element from a few millimeters to 200 μm in diameter [52]. However, these tactile sensors may be affected by the strong external magnetic field and may not work precisely in the environment where ferroalloys are abundant.

7.2.6 Quantum tunneling composite

Technically, QTC shows a piezoresistive effect because its resistance varies with the external stimuli, such as compression, elongation, bending, and twisting [53]. Therefore, QTC is inherently capable of detecting the forces or torques with multiple degrees of freedom, which is suitable to integrate with a robot hand [54]. The composite is prepared by submerging the small metal particles into a nonconducting polymer [55]. QTC behaves like an insulator under normal conditions, but its conductivity increases exponentially within a certain deformation range [56]. During the whole process, there is no contact between the metal particles. However, electrons can transfer from one particle to the others that are close enough based on the quantum tunneling effect. This phenomenon provides excellent sensitivity for QTC, compared with the piezoelectric, piezoresistive, or capacitive materials [57]. However, the disadvantages are equally obvious, e.g., the considerable restore time after removing the applied force, which limits the application of this sensing principle, such as working as a simple switch if there is no other auxiliary structure [58].

7.2.7 Triboelectric effect

Stretchability is a crucial requirement for an e-skin to mimic the properties of human skin, which puts forward the same challenge not only for the sensing part of a tactile device but also for its power supply. E-skin is designed to be constantly exposed to mechanical stimuli. Therefore, the stretchable materials that have the property of converting mechanical energy into electrical signals are ideal for the tactile sensors [59]. To achieve this aim, the triboelectric nanogenerator (TENG) has been investigated in recent years based on a coupling of the triboelectric effect and electrostatic induction, offering the advantages of high sensitivity, high resolution, dynamic tracing, low cost, high efficiency, and a wide selection of materials [9]. The working mechanism of a single-electrode TENG is as follows. The friction layer and electrode layer are fixed and connected to the ground. When friction happens between an active object and the friction layer, equivalent positive and negative charges generate from them respectively. After separating the active object and the friction layer, the electrode layer will induce the positive charges based on the electrostatic induction until the electrical equilibrium is obtained. As the object approaches the friction layer, the process is reversed until the charge is neutralized to a new electrical equilibrium [60]. The contact–separation mode has a similar working principle but consists of two separable electrode layers [61].

Jie *et al.* [62] reported an ultra-flexible skin-conformal tactile sensing device with a micropyramid structure on the friction layer that is used to enhance the density of the charge. Various bending motions of a human body were detected and distinguished by the frequencies and peak-to-peak values of the signals. Another stretchable self-powered e-skin goes a step further in differentiating the multiple mechanical stimuli. Instead of only using either capacitance or resistance signal, the tactile hardware measures them together. However, the architecture was complicated [63]. To improve the output performance of TENG, the combination with other nanogenerators (NGs) is emerging. Junbin *et al.* [64] presented a hybrid piezoelectric–triboelectric sensor so that the incompatible problem of the self-powered sensors between high sensitivity and wide measurement range was broken. Based on the same combination, Wang *et al.* [65] successfully increased the output voltages of a wearable device. Moreover, multimode tactile information including the pressure and temperature can be detected by a single pixel through the triboelectric–piezoelectric–pyroelectric multieffect coupling mechanism [66].

7.2.8 Field-effect transistor

FET is a relatively new sensing principle, relying on an electric field to control a flow of current. The transistor consists of a source, a gate, and a drain as three terminals, which can be used individually to optimize the performance of the FET-based tactile sensors [67]. Compared with FETs, the electronic properties of organic FETs (OFETs) are limited [68], but have superior mechanical flexibility [69], which is suitable for wearing devices. In addition, utilizing graphene, Yogeswaran *et al.* [70] presented a flexible e-skin with a low voltage of 100 mW. Gaining the biocompatibility of parylene as the encapsulation layers, Robert *et al.* [71] fabricated a 300 nm thin e-skin that would not infect human skin. Moreover, the OFET array is the ideal sensing device to achieve the decoupling of multimodal tactile stimuli [72].

7.3 Design criterion and performance index of a tactile sensing system

The performance indexes, including resolution (both spatial and temporal resolution), sensitivity, measuring range, response time, the direction of the force (normal and shear force detection), torque detection, linearity, hysteresis, repeatability, multifunction, fabrication complexity, technological standardization, cost, stretchability, working environment (resistance to electromagnet disturbance, humidity tolerance, etc.), applicable surfaces, modularization, flexibility, dimension, weight, working life, miniaturization, the complexity of data acquisition circuit, etc. [5, 32, 35, 73, 74], should be considered as the basic design criteria for a tactile sensing system so that the interaction between itself and the ambient can achieve high reliability, close to or even beyond human's tactile perception of the environment.

Spatial resolution is determined by the distance between the adjacent sensing elements [73] or the pixel size of a camera for some of the optical-based tactile sensors [35]. Inspired by the super-resolution sensing methods in visual imaging, tactile sensing algorithms that can realize dozens of times the density of the tactile elements have emerged in the last decade [42]. With regards to the temporal resolution, its quality directly affects the ability of slip detection, texture discrimination, as well as stability of operation [6]. In general, the fundamental data of the human hand in these respects are 0.5–10+ mm and 0.4–500+ Hz [5].

The use of soft materials can increase the stretchability and flexibility of a tactile sensor. One of the most famous applications is e-skin, which has the potential to be attached to any 3D irregular surface of the device. Because of the properties of mechanical flexible and chemically stable, PDMS is widely employed in tactile sensors not only as a surface protector but also as an intermediate layer to isolate the electrodes [75]. However, the polymeric materials present a time-dependent highly nonlinear deformation phenomena under the sustaining force [76], in the form of unchanged stain but weakened stress with time. The characteristics of response time, linearity, and hysteresis are closely associated with the temporal resolution, which no longer obeys the linear models. In addition, these three indicators are contradictory to the softness of the sensing materials to some degree. Although the calibration between the applied force and the output of the corresponding sensing technology will give an inverse compensation for nonlinearity and hysteresis, the performance of the repeatability of a sensor is affected by the plastic deformation of the material, which is also called the memory effect.

As for the measuring range, there is usually a trade-off with sensitivity [50]. A PDMS layer with microstructure can be used to further enhance the sensitivity of a piezoelectric or capacitive-based sensor [9, 27]. Although many novel materials are used to fabricate the e-skin that is no less sensitive than human hands, the fabrication processes and data acquisition circuits are complex. It is not easy to modularize and standardize, which means the parts can be replaced and expanded without having to redesign the entire system, let alone reduce the production cost and guarantee the working life.

In order to make the device with the tactile sensor not affecting its original performance, the dimension and weight of the sensor should be as small, thin, and light as possible, which is especially important in perfecting the design of manipulator, minimally invasive surgery (MIS), and wearable devices. The above applications generally involve complex working conditions, such as the disturbance of electromagnet, high humidity, the presence of a variety of electrolytes, fragile targets, etc. To obtain as much tactile information as possible and realize the miniaturization of the sensor, the development of multifunctional tactile sensing element is a natural solution. Besides, the ability to detect shear and normal force, the tactile sensor should also be capable of detecting torques.

7.4 Applications of tactile sensing technologies

7.4.1 Development trend of tactile sensing technologies in e-skin

With the rapid development of computer science, Internet technology, microprocessing technology, and material science, the advent of multifunctional materials that integrate with sensors, processors, and actuators becomes possible to boost a new era of the Internet of things (IoT) [77]. These materials should have the properties of sensing stimuli and performing actions in an interactive and adaptive manner, which naturally enable people to interact and communicate with the surrounding environment [78]. As a stretchable tactile sensing system that can detect various stimuli not limited to the perception of human skin, such as normal force, shear force, lateral strain, bending, temperature, sound, relative humidity, ultraviolet light, magnetic field, and proximity [31, 63, 77, 79], e-skin is a core component and an important data source for robot fine operation, medical surgery, prosthetics, rehabilitation training, continuous health-monitoring, and entertainment [77, 79–81].

When humans manipulate an object, we expected to receive different feedback signals based on different types of external stimuli, including the combination of normal and shear force sensing for grasping control, the combination of vibration and shear force sensing for sliding detection and texture discrimination, the combination of bending and twisting sensing for proprioception, etc. [82]. Inspired by the skin epidermis, the e-skin with a randomly distributed spinosum microstructure is presented by Pang *et al.* [83] to detect numerous human physiological signals. A similar design used to detect the shear force and normal force was reported in Rosenblum [84]. However, one piece of this e-skin can only detect one form of motion at a time. The ability to distinguish among the variety of stimuli based on the limited size of e-skin remains challenging.

To achieve the multimode sensing capability, a stacked structure with different perceptual functions for each layer is proposed [85]. Although the authors claimed that the staggered configuration of the sensing layers and the encapsulated PDMS layers could minimize the crosstalk between different stimuli, and the experiments of a prosthetic hand proved that this e-skin was able to make out various signals corresponding to daily activities. But the tests were carried out separately, it is impossible to tell whether this hierarchy really minimized the crosstalk. Qilin *et al.* [73] did not avoid the issue of simultaneous sensing of multimode information. The performance of a specific sensing signal or a combination of several sensing signals had a good selective detection of the corresponding stimuli, which is able to decouple the interference from other stimuli. However, the layout of the sensing elements is integrated by the mode of stack and distribution. It is difficult to determine whether the distributed layout is the key to decoupling the sensory signals of different stimuli. From the viewpoint of reducing the difficulty of the manufacturing process, the key to fabricate a high-density multifunctional tactile array is that a single sensing element can provide multiple types of noninterference feedback. As we mentioned in section 2, with the rapid development of NG, a single pixel through the multieffect of triboelectric–piezoelectric–pyroelectric successfully realized the detection of pressure and temperature with a degree of flexibility [66].

Another expected property for e-skin that has not been widely studied is the adjustable sensing density from the micro- to the macroscale. The density of mechanoreceptor in human skin varies from place to place [86]. In terms of signal processing, manufacturing, and practicality, the density is not necessary to be consistent everywhere. The configuration of the sensing arrays on a Barrett Hand refers to the density distribution of a human hand, but the number of sensing elements is sparse, and the contact is rigid [87]. An extendible e-skin with microwire is adopted in Reference 88, thousands of sensing nodes can be fabricated on a 100 mm in diameter Kaptor film with up to 1600% linear extension. For more information on how to arrange sensing elements on the 3D surfaces and how to conduct signal processing, please refer to section 5.

7.4.2 Development trend of tactile sensing technologies in a wearable device

Transparency is the development trend of e-skin in the wearable system since the feature of transparency allows the light to pass through so that a photovoltaic cell underneath will harvest the energy accessibly [89]. Networks of nanostructure and graphene sheets can be made into transparent electrodes by maintaining a degree of elasticity without breaking the contiguity of the film [90, 91]. To obtain more superior transparency (>90%), the conductive hydrogel has been investigated to fabricate uniform self-patterned microstructures, but the durability of electrical performance is limited to up to a few hundred of cycles [92]. Moreover, the transparency is affected by the process of dehydration of the hydrogels [93].

Another attractive direction in the field of wearable electronics is the measurement accuracy during the long period of health monitoring or motion tracking.

Because the contact point between the human body and the device changes with the movement of the person [94], to overcome this challenge, one of the solutions is to fabricate the skin-conformal wearable tactile sensor on the irregular surfaces [95]. [64] proposed a hybrid piezoelectric–triboelectric sensor that can be taped to the wrist to sense the weak pulse wave in real time. Based on the sensing technology of triboelectric–photonic, a conformal e-skin was integrated on a robot hand even including the joints so that the bending of the three joints on each finger and the multipoint contact on the rest area of the finger are responded on the photocurrent and voltage, respectively [96]. Furthermore, 3D printing with the novelty characteristics of multimaterial and multiscale provides a flexible method to fabricate complex structures directly and conformally to the freeform surfaces [97]. Apart from the conformability, thinner bending thickness and stronger adhesion to soft tissues also play a significant role in measuring the biological signals accurately [98]. An ultraflexible e-skin presented in Reference 71 has an overall thickness of 300 nm, making it suitable for applications that require extremely light weight. The so-called "ionic gel skin" in Reference 99 shows the super adhesiveness on various surfaces, such as skin, fluting paper, and glass, which was produced by the combination of H-bonding, dynamic covalent bonds, and chemical cross-linkers.

While the above factors are important in designing the e-skin, the safety of interaction with a human is always a critical priority. The materials that are chosen should have the time invariant biocompatibility and stability to avoid the inflammation of human skin [100]. Moreover, due to the requirements of continuous monitoring in some vital situations, the air permeability of a wearable e-skin should be effectively guaranteed. Miyamoto *et al.* [101] presented a nanomesh conductor that can be directly attached to the skin with a very high shape conformability. The permeability had been proved by a one-week skin patch test and water vapor transmission tests. In Reference 102, based on the breath figure method, the silver nanowires are embedded into the porous conductive film so that the function of gas permeable was realized.

7.4.3 Development trend of tactile sensing technologies in robotic

Robot fine operation includes multidegree of freedom force/torque detection, sliding detection, grasping operation, shape perception, material texture perception, material elastic perception, multimodal information fusion, etc. [11]. With the further development of e-skin or the array of tactile sensing technologies, a high density of sensing elements with the ability to perceive the above signals, actuators, and microprocessors will be integrated into it, resulting in a significant increase in energy consumption. However, the acquisition of abundant information should not be at the expense of reducing the requirements for a compact and flexible robot system, making portable, long-term, and compact energy supply an urgent problem to be solved. Accordingly, energy harvesters, energy storage devices, low-power electronics, and efficient/wireless power transfer-based technologies have been fueled to realize the self-powered e-skin [103].

Apart from the triboelectric effect tactile sensors discussed in section 2, the common ways for the self-powered also include photovoltaic, thermoelectric, piezoelectric, pyroelectric, and biochemical reactions, which have benefited from the advances of micro-/nanotechnologies [104]. Among them, transparency is the most basic requirement for a photovoltaic-based self-powered e-skin. An ultralow power consumed sensitive layer, such as single-layer graphene, can be used to reduce the area of the photovoltaic cell [59]. Piezoelectric materials are good candidates for detecting mechanical vibrations but are not suitable to detect static loading because of electricity leakage. Moreover, compared with the triboelectric effect tactile sensors, the electrical outputs are relatively low, so the detection range and sensitivity are limited [28]. In addition, as a large number of circuits and various electrical equipment are integrated into a robotic system, a large amount of heat will be released into the environment during the operation. If the heat energy can be used, the power endurance of the robot will be improved. If the temperature has a gradient, thermoelectric energy will be harvested by the Seebeck effect; otherwise, the pyroelectric effect will be a better solution [105]. Recently, the research about optimizing the material and structure of the pyroelectric NGs was reported in Reference 106.

7.4.4 Development trend of tactile sensing technologies in medical service

In recent years, countries around the world pay more attention to MIS that can reduce the pain of patients, shorten the recovery time, and lessen the cost of treatment. Most the MIS robots currently used in clinical have a visual feedback system. However, one main drawback is the lack of haptic sensing to support palpation because of the limitation of operation space. Correspondingly, the excessive operating force or unexpected slippage may cause damage to the tissues during the surgery [107]. In fact, touch is so vital to acquire environmental information, second only to vision. Although visual feedback can provide some type of haptic information [108], real tactile feedback can make it more convenient for a surgeon to determine the force of MIS devices on the tissues and recognize the characteristics of different organs [109]. Therefore, tactile sensing is not just a supplement to visual but also plays a significant role in detecting vital signs in the medical field.

In medical services, many types of meaningful body parameters, such as blood pressure, blood flow rate, body temperature, heartbeat, elastic of the tissue, texture of the tissue, etc., can be obtained theoretically by the multifunctional tactile sensing devices [103]. Tactile sensors that use strain gauges [110], piezoresistive [111], piezoelectric [112], and capacitive [113] have received early attention in the MIS field because such sensing devices can be fabricated in a thin thickness and have the properties of high sensitivity and wide dynamic range. But the form of electrical signals makes them unsuitable for the environment with multiple electrolytes, especially for MIS and endoscopic surgery. Moreover, if magnetic resonance imaging is used during the surgery not only may these tactile sensing technologies be affected by the magnetic field but also magnetic-based tactile sensors are no longer suitable. Therefore, the typical solutions are to adopt an optical- [114, 115] or acoustic-based sensing principles [116]. Such sensors usually need to measure the deformation of flexible materials. Thus, the phenomenon of stress relaxation may be aggravated, which should be paid extra attention to MIS compared with the applications of tactile sensing technologies in other fields. In addition, considering the size of the medical tools such as the gripper, the density of the sensing elements is relatively sparse with the force detection in only one direction. Accordingly, the procedure of palpation may cost a long time to finish, increasing the likelihood of inconsistent results [117]. Therefore, a distributed tactile sensor is needed to provide both kinesthetic and tactile information. Moreover, barely any of the tactile sensing devices that claimed to have potential in MIS have been validated for performance with an actual surgical robot.

7.5 Challenges and discussion

7.5.1 Standardization of fabrication process

On one hand, mass production is limited by the complex and expensive fabrication process. Currently, commercial tactile sensing devices are only confined to the single-point contact sensors and array tactile sensors, mostly using mature sensing principles such as piezoelectric, piezoresistive, or capacitive. The contact modes between the sensing device and the object are mainly rigid, and the so-called flexible sensor array can only bend into a curve with a small curvature. Moreover, the fastening methods between the sensing device and the surface of a carrier are usually realized by thread or glue. Based on different sensing principles, numerous tactile sensing systems with various advanced properties emerge one after another, taking full advantage of the development of chemistry, material science, and advanced manufacturing technologies. However, most of them are in the laboratory exploration stage. The commercial prospect is not bright by realizing the large-scale production.

On the other hand, the standardization of the fabrication process also determines whether the performance of each sensing element can be consistent in the e-skin or array sensor. For some of the optical-based tactile sensors, the upper limits of their spatial resolution are limited by the number of pixels in the image acquisition equipment [36, 39]. The others are directly related to the number of sensing elements, even the super-resolution detection that can be achieved by algorithms is the same [42]. In addition, the fine operation of the robots not only puts forward the higher requirements for the resolution of the array sensors or e-skin but also promotes the development of the sensors toward a multifunctional and integrated direction. However, most studies are limited to testing the reproducibility of a single sensing element [116] or only provide a qualitative analysis of the output signals [117] or do a simple quantitative analysis but do not involve the calibration [63]. Moreover, the output of each sensing element of a soft and stretchable tactile sensor may get affected by its neighboring taxels and the Euclidean distance to other taxels [12]. It is obvious that there exists a trade-off between the requirement of calibration and spatial resolution. Thus, it is urgent and important to solve the calibration problem of each sensing element via standardization [118].

7.5.2 Data transmission of high-density tactile sensing elements

In addition to vision sensors, robots are usually equipped with array tactile sensors for fine operation, but most of the current research can only collect information on pressure or forces/torques of the joints [119, 120]. In order to accurately perceive the surrounding environment and make appropriate adjustments in real time according to the change of stimuli, it is necessary to expand the number of sensing elements so that the robot has a higher resolution of the sense of touch. Meanwhile, the flexible e-skin with multiple sensing functions should also be considered to achieve the acquisition of multimode information on the object simultaneously [121]. As the amount and types of data increase, in general, the multimodal information is redundant, which has the ability to reduce the uncertainty of the received data and improve the overall performance of the system [122]. However, a series of new challenges such as data transmission, utilization, wiring, heat dissipation, and energy consumption has also been brought.

To overcome the above challenges, a distributed and hierarchical system was identified for a robot with large-scale tactile sensing system, which includes a sensor, module, patch, and region in order of rank from low to high [123]. The serial buses, such as inter-integrated circuit, and the controller area network buses are often used for communication between different functional areas on the same and different hierarchical structures, respectively [74, 124]. Together with local preprocessing, the pressure on the bandwidth of higher levels can be relieved so as to increase the speed of data transmission [125]. Methods of preprocessing usually include feature extraction from an optical-based tactile sensor [126] or just simple filtering. However, it is not a trivial issue of distributing the amount of processing between the local and central units [127].

7.5.3 Fault tolerance and autocalibration

A few ways that can decouple the tactile information have been published. For example, in Reference 116, pressure, shear, and torsion are decoupled based on the unique corresponding relationship between the type of loading and gauge factor. Various tactile information such as pressure, lateral strain, and bending can be distinguished by measuring the capacitance of the sensing structure and the resistance of the electrode of this capacitor without using complicated architecture [63].

The decoupling of the external information helps us accurately judge the influence of different variables, which is a basic way to understand the world, and its importance is self-evident. Moreover, from the above analysis, it is noted that array and integration are the future directions of tactile sensing technologies, but in general, each tactile sensing element still perceives the external environment in an isolated way, separating the internal correlation between the information. The isolated information may not only lose some key features of the physical world but also cannot effectively eliminate the disabled elements, which seriously affects the subsequent decision-making process and intelligence level. Just as it is difficult for a human to accurately recognize and operate an object through the damaged skin, to accurately judge the confidence level of each sensing element in real time and eliminate the information interference caused by the damaged elements are the key issues to improve the safety and accuracy of an operation as well as an important guarantee to extend the service life of the sensor. In the current study, there are few discussions about whether it can be effectively identified by other sensing elements or by the signal processing methods if some sensor elements are damaged.

7.5.4 Layout of sensing elements on an irregular 3D

In order to effectively integrate tactile perception on complex 3D surfaces, the common solution is to increase the flexibility of the sensors so that the electronic components can perfectly fit on the surface of the carrier under any deformation, such as bending, stretching, and shearing, and maintain the mechanical and electrical performance of the sensors. As for the main ways to achieve flexibility, there are usually two types [128]: (i) soft materials such as conductive polymers and fabrics, elastomer composites with conductive filler particles, fluids, etc. [129]; (ii) stretchable structural designs such as serpentine-like structures for interconnects or wires [130], kirigami engineered patterns to impart the electrodes a tunable elasticity [131], serpentine structure [132], etc. However, hysteresis and nonlinearity are the main drawbacks. The manufacturing processes of such sensors are also extraordinarily complex. Generally, technologies, such as spin-coating, spring-coating, photo-patterning, and solution deposition [133], are involved. The corresponding equipment is expensive, and the cost of manufacturing is high. Therefore, how to simplify the production process and cycle of the sensors, reduce the manufacturing cost, and improve productivity through the commonly used materials and mature technologies are an important trend of the future development.

Some researchers also solved the layout of tactile sensing elements on the 3D surface by cutting and splicing, that is, by splicing multiple single [134] or array [135] sensors into the e-skin. Tactile sensors adopted with the flexible printed circuit boards are generally nonstretchable and not suitable for installation on small curved surfaces, but the bendable, multiply capacitance-to-digital converters can be mounted on each printed circuit board [74]. Because of the need to consider wiring and communication issues, it is also not suitable for the carrier with a large area. In addition, the splicing of multiple single sensing points is similar to some wearable haptic devices [88], where the vibrotactile information is provided by the low mass actuators. However, it still comes across the problem of insufficient spatial resolution in addition to the above problems.

References

[1] Jung Y.H., Park B., Kim J.U., Kim T.I. ‘Bioinspired electronics for artificial sensory systems’. *Advanced Materials (Deerfield Beach, Fla.)*. 2019, vol. 31(34), e1803637.

[2] Westling G., Johansson R.S. ‘Factors influencing the force control during precision grip’. *Experimental Brain Research*. 1984, vol. 53(2), pp. 277–84.

[3] Dahiya R.S., Valle M. ‘Tactile sensing for robotic applications’ in *Sensors, focus on tactile, force and stress sensors*; 2008. pp. 298–304.

[4] Silvera-Tawil D., Rye D., Velonaki M. 'Artificial skin and tactile sensing for socially interactive robots: a review'. *Robotics and Autonomous Systems*. 2015, vol. 63, pp. 230–43.

[5] Dahiya R.S., Metta G., Valle M., Sandini G. 'Tactile sensing—from humans to humanoids'. *IEEE Transactions on Robotics*. 2009, vol. 26(1), pp. 1–20.

[6] Chortos A., Liu J., Bao Z. 'Pursuing prosthetic electronic skin'. *Nature Materials*. 2016, vol. 15(9), pp. 937–50.

[7] Konstantinova J., Jiang A., Althoefer K., Dasgupta P., Nanayakkara T. 'Implementation of tactile sensing for palpation in robot-assisted minimally invasive surgery: a review'. *IEEE Sensors Journal*. 2012, vol. 14(8), pp. 2490–501.

[8] Zou L., Ge C., Wang Z.J., Cretu E., Li X. 'Novel tactile sensor technology and smart tactile sensing systems: a review'. *Sensors (Basel, Switzerland)*. 2017, vol. 17(11), p. 2653.

[9] Tao J., Bao R., Wang X, *et al*. 'Self-powered tactile sensor array systems based on the triboelectric effect [online]'. *Advanced Functional Materials*. 2019, vol. 29(41), 1806379. Available from https://onlinelibrary.wiley.com/toc/16163028/29/41

[10] Tiwana M.I., Redmond S.J., Lovell N.H. 'A review of tactile sensing technologies with applications in biomedical engineering'. *Sensors and Actuators A*. 2012, vol. 179, pp. 17–31.

[11] Luo S., Bimbo J., Dahiya R., Liu H. 'Robotic tactile perception of object properties: a review'. *Mechatronics*. 2017, vol. 48, pp. 54–67.

[12] Yousef H., Boukallel M., Althoefer K. 'Tactile sensing for dexterous in-hand manipulation in robotics—a review'. *Sensors and Actuators A: Physical*. 2011, vol. 167(2), pp. 171–87.

[13] Wang X., Sun F., Yin G., Wang Y., Liu B., Dong M. 'Tactile-sensing based on flexible PVDF nanofibers via electrospinning: a review'. *Sensors. 2018.*, vol. 18(2), p. 330.n.d.

[14] Fiorillo A.S. 'A piezoresistive tactile sensor'. *IEEE Transactions on Instrumentation and Measurement*. 1997, vol. 46(1), pp. 15–17.

[15] Barlian A.A., Park W.T., Mallon J.R., Rastegar A.J., Pruitt B.L. 'Review: semiconductor piezoresistance for microsystems'. *Proceedings of the IEEE. Institute of Electrical and Electronics Engineers*. 2009, vol. 97(3), pp. 513–52.

[16] Tian H., Shu Y., Wang X.F, *et al*. 'A graphene-based resistive pressure sensor with record-high sensitivity in a wide pressure range'. *Scientific Reports*. 2015, vol. 5, 8603.

[17] Huang Y.M., Sohgawa M., Yamashita K, *et al*. 'Fabrication and normal/shear stress responses of tactile sensors of polymer/si cantilevers embedded in PDMS and urethane gel elastomers'. *IEEJ Transactions on Sensors and Micromachines*. 2008, vol. 128(5), pp. 193–97.

[18] Noda K., Hoshino K., Matsumoto K., Shimoyama I. 'A shear stress sensor for tactile sensing with the piezoresistive cantilever standing in elastic material'. *Sensors and Actuators A*. 2008, vol. 127(2), pp. 295–301.

[19] Kim S., Lee J., Choi B. 'Stretching and twisting sensing with liquid-metal strain gauges printed on silicone elastomers'. *IEEE Sensors Journal*. 2008, vol. 15(11), pp. 6077–78.

[20] Canavese G., Stassi S., Stralla M., Bignardi C., Pirri C.F. 'Stretchable and conformable metal–polymer piezoresistive hybrid system'. *Sensors and Actuators A: Physical*. 2012, vol. 186, pp. 191–97.

[21] Cheng M.Y., Tsao C.M., Yang Y.J. 'An anthropomorphic robotic skin using highly twistable tactile sensing array'. *5th IEEE Conference on Industrial Electronics and Applications*; IEEE, 2010. pp. 650–55.

[22] Bartolozzi C., Natale L., Nori F., Metta G. 'Robots with a sense of touch'. *Nature Materials*. 2016, vol. 15(9), pp. 921–25.

[23] Goger D., Gorges N., Worn H. 'Tactile sensing for an anthropomorphic robotic hand: hardware and signal processing'. *IEEE International Conference on Robotics and Automation*; IEEE, 2009. pp. 895–901.

[24] Choi W., Lee J., Kyoung Yoo Y., Kang S., Kim J., Hoon Lee J. 'Enhanced sensitivity of piezoelectric pressure sensor with microstructured polydimethylsiloxane layer'. *Applied Physics Letters*. 2014, vol. 104(12), 123701.

[25] Ha M., Lim S., Park J., Um D.S., Lee Y., Ko H. 'Bioinspired interlocked and hierarchical design of zno nanowire arrays for static and dynamic pressure-sensitive electronic skins'. *Advanced Functional Materials*. 2019, vol. 25(19), pp. 2841–49.

[26] Navaraj W., Dahiya R. 'Fingerprint-Enhanced capacitive-piezoelectric flexible sensing skin to discriminate static and dynamic tactile stimuli'. *Advanced Intelligent Systems*. 2019, vol. 1(7), p. 1900051. Available from https://onlinelibrary.wiley.com/toc/26404567/1/7

[27] El-Molla S., Albrecht A., Cagatay E, *et al*. 'Integration of a thin film PDMS-based capacitive sensor for tactile sensing in an electronic skin'. *Journal of Sensors*. 2019, vol. 2016, pp. 1–7.

[28] Yao G., Xu L., Cheng X, *et al*. 'Bioinspired triboelectric nanogenerators as self-powered electronic skin for robotic tactile sensing'. *Advanced Functional Materials*. 2020, vol. 30(6), 1907312. Available from https://onlinelibrary.wiley.com/toc/16163028/30/6

[29] Hyung-Kew L., Jaehoon C., Sun-Il C., Euisik Y. 'Normal and shear force measurement using a flexible polymer tactile sensor with embedded multiple capacitors'. *Journal of Microelectromechanical Systems*. 2020, vol. 17(4), pp. 934–42.

[30] Lee H.K., Chung J., Chang S.I., Yoon E. 'Real-time measurement of the three-axis contact force distribution using a flexible capacitive polymer tactile sensor'. *Journal of Micromechanics and Microengineering*. 2011, vol. 21(3), p. 035010.

[31] Li T., Luo H., Qin L, *et al*. 'Flexible capacitive tactile sensor based on micropatterned dielectric layer'. *Small (Weinheim an Der Bergstrasse, Germany)*. 2016, vol. 12(36), pp. 5042–48.

[32] Yun S., Park S., Park B, *et al.* 'Polymer-waveguide-based flexible tactile sensor array for dynamic response'. *Advanced Materials (Deerfield Beach, Fla.)*. 2014, vol. 26(26), pp. 4474–80.

[33] Kim J.T., Choi H., Shin E., Park S., Kim I.G. 'Graphene-based optical waveguide tactile sensor for dynamic response'. *Scientific Reports*. 2018, vol. 8(1), pp. 1–6.

[34] Yun S., Park S., Park B, *et al.* 'Polymer-waveguide-based flexible tactile sensor array for dynamic response'. *Advanced Materials (Deerfield Beach, Fla.)*. 2014, vol. 26(26), pp. 4474–80.

[35] Lee J.H., Won C.H. 'High-resolution tactile imaging sensor using total internal reflection and nonrigid pattern matching algorithm'. *IEEE Sensors Journal*. 2011, vol. 11(9), pp. 2084–93.

[36] Yuan W., Dong S., Adelson E.H. 'GelSight: high-resolution robot tactile sensors for estimating geometry and force'. *Sensors (Basel, Switzerland)*. 2017, vol. 17(12), p. 2762.

[37] Johnson M.K., Adelson E.H. 'Retrographic sensing for the measurement of surface texture and shape'. *IEEE Computer Society Conference on Computer Vision and Pattern Recognition Workshops (CVPR Workshops)*; Miami, FL, IEEE, 2009. pp. 1070–77.

[38] Johnson M.K., Cole F., Raj A., Adelson E.H. 'Microgeometry capture using an elastomeric sensor'. *ACM SIGGRAPH 2011 papers; Vancouver, B?C, Canada*; New York, NY, 2011. pp. 1–8. Available from http://portal.acm.org/citation.cfm?doid=1964921

[39] Yuan W., Li R., Srinivasan M.A., Adelson E.H. 'Measurement of shear and slip with a gelsight tactile sensor'. *IEEE International Conference on Robotics and Automation (ICRA)*; IEEE, 2015. pp. 304–11.

[40] Chorley C., Melhuish C., Pipe T., Rossiter J. 'Development of a tactile sensor based on biologically inspired edge encoding'. *International Conference on Advanced Robotics*; IEEE, 2009. pp. 1–6.

[41] Ward-Cherrier B., Pestell N., Cramphorn L, *et al.* 'The tactip family: soft optical tactile sensors with 3D-printed biomimetic morphologies'. *Soft Robotics*. 2018, vol. 5(2), pp. 216–27.

[42] Lepora N.F., Ward-Cherrier B. 'Superresolution with an optical tactile sensor'. *IEEE/RSJ International Conference on Intelligent Robots and Systems (IROS)*; Hamburg, Germany, 2018.

[43] Cramphorn L., Ward-Cherrier B., Lepora N.F. 'Addition of a biomimetic fingerprint on an artificial fingertip enhances tactile spatial acuity'. *IEEE Robotics and Automation Letters*; IEEE, 2017. pp. 1336–43.

[44] Lepora N.F., Aquilina K., Cramphorn L. 'Exploratory tactile servoing with active touch'. *IEEE Robotics and Automation Letters*. 2017, vol. 2(2), pp. 1156–163.

[45] Lepora N.F., Church A., de Kerckhove C., Hadsell R., Lloyd J. 'From pixels to percepts: highly robust edge perception and contour following using deep learning and an optical biomimetic tactile sensor'. *IEEE Robotics and Automation Letters*. 2018, vol. 4(2), pp. 2101–07.

[46] Back J., Dasgupta P., Seneviratne L., Althoefer K., Liu H. 'Feasibility study-novel optical soft tactile array sensing for minimally invasive surgery'. *2015 IEEE/RSJ International Conference on Intelligent Robots and Systems (IROS)*; Hamburg, Germany, IEEE, 2018. pp. 1528–33.

[47] Totaro M., Mondini A., Bellacicca A., Milani P., Beccai L. 'Integrated simultaneous detection of tactile and bending cues for soft robotics'. *Soft Robotics*. 2017, vol. 4(4), pp. 400–10.

[48] Zhang X., Peng W., Shao L.Y., Pan W., Yan L. 'Strain and temperature discrimination by using temperature-independent FPI and FBG'. *Sensors and Actuators A*. 2018, vol. 272, pp. 134–38.

[49] Jamone L., Natale L., Metta G., Sandini G. 'Highly sensitive soft tactile sensors for an anthropomorphic robotic hand'. *IEEE Sensors Journal*. 2018, vol. 15(8), pp. 4226–33.

[50] Oh S., Jung Y., Kim S, *et al.* 'Remote tactile sensing system integrated with magnetic synapse'. *Scientific Reports*. 2017, vol. 7(1), pp. 1–8.

[51] Wang H., de Boer G., Kow J, *et al.* 'Design methodology for magnetic field-based soft tri-axis tactile sensors'. *Sensors (Basel, Switzerland)*. 2016, vol. 16(9), p. 1356.

[52] Alfadhel A., Khan M.A., Cardoso de Freitas S., Kosel J. 'Magnetic tactile sensor for braille reading'. *IEEE Sensors Journal*. 2005, vol. 16(24), pp. 8700–05.

[53] Bloor D., Donnelly K., Hands P.J., Laughlin P., Lussey D. 'A metal–polymer composite with unusual properties'. *Journal of Physics D*. 2005, vol. 38(16), pp. 2851–60.

[54] Zhang T., Liu H., Jiang L., Fan S., Yang J. 'Development of a flexible 3-D tactile sensor system for anthropomorphic artificial hand'. *IEEE Sensors Journal*. 2012, vol. 13(2), pp. 510–18.

[55] Azaman N.I.L., Ayub M.A., Ahmad A.A. 'Characteristic and sensitivity of quantum tunneling composite (QTC) material for tactile device applications'. *7th IEEE Control and System Graduate Research Colloquium (ICSGRC)*; Shah Alam, IEEE, 2017. pp. 7–11.

[56] Cirillo A., Cirillo P., De Maria G., Natale C., Pirozzi S. 'Force/tactile sensors based on optoelectronic technology for manipulation and physical human–robot interaction' in *Advanced Mechatronics and MEMS Devices II*. Cham: Springer; 2017. pp. 95–131.

[57] Zhang T., Jiang L., Wu X., Feng W., Zhou D., Liu H. 'Fingertip three-axis tactile sensor for multifingered grasping'. *IEEE/ASME Transactions on Mechatronics*. 2017, vol. 20(4), pp. 1875–85.

[58] Amarasinghe Y.W.R., Kulasekera A.L., Priyadarshana T.G.P. 'Seventh international conference on sensing technology (ICST)'. wellington, new zealand, IEEE, 2017. pp. 1–4.

[59] Chen X., Parida K., Wang J, *et al.* 'A stretchable and transparent nanocomposite nanogenerator for self-powered physiological monitoring'. *ACS Applied Materials & Interfaces*. 2017, vol. 9(48), pp. 42200–09.

[60] Chen S.W., Cao X., Wang N, *et al*. 'An ultrathin flexible single-electrode triboelectric-nanogenerator for mechanical energy harvesting and instantaneous force sensing'. *Advanced Energy Materials*. 2017, vol. 7(1), p. 1601255. Available from http://doi.wiley.com/10.1002/aenm.v7.1

[61] Lin Z.H., Cheng G., Wu W., Pradel K.C., Wang Z.L. 'Dual-mode triboelectric nanogenerator for harvesting water energy and as a self-powered ethanol nanosensor'. *ACS Nano*. 2014, vol. 8(6), pp. 6440–48.

[62] Wang J., Qian S., Yu J, *et al*. 'Flexible and wearable PDMS-based triboelectric nanogenerator for self-powered tactile sensing'. *Nanomaterials (Basel, Switzerland)*. 2019, vol. 9(9), p. 1304.

[63] Park S., Kim H., Vosgueritchian M, *et al*. 'Stretchable energy-harvesting tactile electronic skin capable of differentiating multiple mechanical stimuli modes'. *Advanced Materials (Deerfield Beach, Fla.)*. 2014, vol. 26(43), pp. 7324–32.

[64] Yu J., Hou X., Cui M, *et al*. 'Highly skin-conformal wearable tactile sensor based on piezoelectric-enhanced triboelectric nanogenerator'. *Nano Energy*. 2017, vol. 64, p. 103923.

[65] Wang X., Yang B., Liu J., Zhu Y., Yang C., He Q. 'A flexible triboelectric-piezoelectric hybrid nanogenerator based on P(VDF-trfe) nanofibers and PDMS/MWCNT for wearable devices'. *Scientific Reports*. 2016, vol. 6, 36409.

[66] Ma M., Zhang Z., Zhao Z, *et al*. 'Self-powered flexible antibacterial tactile sensor based on triboelectric-piezoelectric-pyroelectric multi-effect coupling mechanism'. *Nano Energy*. 2016, vol. 66, p. 104105.

[67] Wang J., Jiang J., Zhang C, *et al*. 'Energy-efficient, fully flexible, high-performance tactile sensor based on piezotronic effect: piezoelectric signal amplified with organic field-effect transistors'. *Nano Energy*. 2016, vol. 76, p. 105050.

[68] Trung T.Q., Tien N.T., Seol Y.G., Lee N.E. 'Transparent and flexible organic field-effect transistor for multi-modal sensing'. *Organic Electronics*. 2012, vol. 13(4), pp. 533–40.

[69] Lee Y.H., Jang M., Lee M.Y., Kweon O.Y., Oh J.H. 'Flexible field-effect transistor-type sensors based on conjugated molecules'. *Chem*. 2016, vol. 3(5), pp. 724–63.

[70] Yogeswaran N., Navaraj W.T., Gupta S, *et al*. 'Piezoelectric graphene field effect transistor pressure sensors for tactile sensing'. *Applied Physics Letters*. 2016, vol. 113(1), 014102.

[71] Nawrocki R.A., Matsuhisa N., Yokota T., Someya T. '300-nm imperceptible, ultraflexible, and biocompatible e-skin fit with tactile sensors and organic transistors'. *Advanced Electronic Materials*. 2016, vol. 2(4), p. 1500452.

[72] Kim D.I., Trung T.Q., Hwang B.U, *et al*. 'A sensor array using multi-functional field-effect transistors with ultrahigh sensitivity and precision for bio-monitoring'. *Scientific Reports*. 2015, vol. 5, 12705.

[73] Kappassov Z., Corrales J.A., Perdereau V. 'Tactile sensing in dexterous robot hands—review'. *Robotics and Autonomous Systems*. 2016, vol. 74, pp. 195–220.

[74] Schmitz A., Maiolino P., Maggiali M., Natale L., Cannata G., Metta G. 'Methods and technologies for the implementation of large-scale robot tactile sensors'. *IEEE Transactions on Robotics*. 1996, vol. 27(3), pp. 389–400.

[75] Sun X., Sun J., Li T, *et al.* 'Flexible tactile electronic skin sensor with 3D force detection based on porous cnts/PDMS nanocomposites'. *Nano-Micro Letters*. 2019, vol. 11(1), p. 57.

[76] Holzapfel G.A. 'ON large strain viscoelasticity: continuum formulation and finite element applications to elastomeric structures'. *International Journal for Numerical Methods in Engineering*. 1996, vol. 39(22), pp. 3903–26. Available from 1.0.CO;2-H">http://doi.wiley.com/10.1002/(SICI)1097-0207(19961130)39:22<>1.0.CO;2-H

[77] Hua Q., Sun J., Liu H, *et al.* 'Skin-inspired highly stretchable and conformable matrix networks for multifunctional sensing'. *Nature Communications*. 2018, vol. 9(1), pp. 1–11.

[78] McEvoy M.A., Correll N. 'Materials science materials that couple sensing, actuation, computation, and communication'. *Science*. 2015, vol. 347(6228), 1261689.

[79] Kim J., Lee M., Shim H.J, *et al.* 'Stretchable silicon nanoribbon electronics for skin prosthesis'. *Nature Communications*. 2014, vol. 5(1), pp. 1–11.

[80] Schwartz G., Tee B.C.K., Mei J, *et al.* 'Flexible polymer transistors with high pressure sensitivity for application in electronic skin and health monitoring'. *Nature Communications*. 2013, vol. 4, p. 1859.

[81] Tian L., Zimmerman B., Akhtar A, *et al.* 'Large-area MRI-compatible epidermal electronic interfaces for prosthetic control and cognitive monitoring'. *Nature Biomedical Engineering*. 2014, vol. 3(3), pp. 194–205.

[82] Hammock M.L., Chortos A., Tee B.C.K., Tok J.B.H., Bao Z. '25th anniversary article: the evolution of electronic skin (e-skin): a brief history, design considerations, and recent progress'. *Advanced Materials*. 2014, vol. 25(42), pp. 5997–6038.

[83] Pang Y., Zhang K., Yang Z, *et al.* 'Epidermis microstructure inspired graphene pressure sensor with random distributed spinosum for high sensitivity and large linearity'. *ACS Nano*. 2018, vol. 12(3), pp. 2346–54.

[84] Rosenblum, L.D. *See what I'm saying: the extraordinary powers of our five senses*. New York City, United States: 2011.WW Norton & Company;

[85] Kim J., Lee M., Shim H.J, *et al.* 'Stretchable silicon nanoribbon electronics for skin prosthesis'. *Nature Communications*. 2014, vol. 5(1), pp. 1–11.

[86] Taube Navaraj W., García Núñez C., Shakthivel D, *et al.* 'Nanowire FET based neural element for robotic tactile sensing skin'. *Frontiers in Neuroscience*. 2017, vol. 11, p. 501.

[87] Dang H., Weisz J., Allen P.K. 'IEEE international conference on robotics and automation (ICRA)'. shanghai, china, IEEE, 2014. pp. 5917–22.

[88] Lanzara G., Salowitz N., Guo Z., Chang F.K. 'A spider-web-like highly expandable sensor network for multifunctional materials'. *Advanced Materials (Deerfield Beach, Fla.)*. 2010, vol. 22(41), pp. 4643–48.
[89] Núñez C.G., Navaraj W.T., Polat E.O., Dahiya R. 'Energy-autonomous, flexible, and transparent tactile skin'. *Advanced Functional Materials*. 2017, vol. 27(18), 1606287.
[90] An B.W., Heo S., Ji S., Bien F., Park J.U. 'Transparent and flexible fingerprint sensor array with multiplexed detection of tactile pressure and skin temperature'. *Nature Communications*. 2018, vol. 9(1), pp. 1–10.
[91] Lipomi D.J., Vosgueritchian M., Tee B.C.-K, *et al*. 'Skin-like pressure and strain sensors based on transparent elastic films of carbon nanotubes'. *Nature Nanotechnology*. 2011, vol. 6(12), pp. 788–92.
[92] Ge G., Zhang Y., Shao J, *et al*. 'Stretchable, transparent, and self-patterned hydrogel-based pressure sensor for human motions detection'. *Advanced Functional Materials*. 2017, vol. 28(32), 1802576. Available from http://doi.wiley.com/10.1002/adfm.v28.32
[93] Zhao G., Zhang Y., Shi N, *et al*. 'Transparent and stretchable triboelectric nanogenerator for self-powered tactile sensing'. *Nano Energy*. 2019, vol. 59, pp. 302–10.
[94] Someya T., Yokota T., Lee S., Fukuda K. 'Electronic skins for robotics and wearables'. *IEEE 33rd International Conference on Micro Electro Mechanical Systems (MEMS)*; IEEE, May 2020. pp. 66–71.
[95] Kang S., Cho S., Shanker R, *et al*. 'Transparent and conductive nanomembranes with orthogonal silver nanowire arrays for skin-attachable loudspeakers and microphones'. *Science Advances*. 2018, vol. 4(8), eaas8772.
[96] Bu T., Xiao T., Yang Z, *et al*. 'Stretchable triboelectric-photonic smart skin for tactile and gesture sensing'. *Advanced Materials (Deerfield Beach, Fla.)*. 2018, vol. 30(16), e1800066.
[97] Guo S.Z., Qiu K., Meng F., Park S.H., McAlpine M.C. '3D printed stretchable tactile sensors'. *Advanced Materials (Deerfield Beach, Fla.)*. 2017, vol. 29(27), p. 1701218.
[98] Kim D.-H., Viventi J., Amsden J.J, *et al*. 'Dissolvable films of silk fibroin for ultrathin conformal bio-integrated electronics'. *Nature Materials*. 2010, vol. 9(6), pp. 511–17.
[99] Wang P., Pei D., Wang Z, *et al*. 'Biocompatible and self-healing ionic gel skin as shape-adaptable and skin-adhering sensor of human motions'. *Chemical Engineering Journal*. 2020, vol. 398, p. 125540.
[100] Hassler C., Boretius T., Stieglitz T. 'Polymers for neural implants'. *Journal of Polymer Science Part B*. 2020, vol. 49(1), pp. 18–33.
[101] Miyamoto A., Lee S., Cooray N.F, *et al*. 'Inflammation-free, gas-permeable, lightweight, stretchable on-skin electronics with nanomeshes'. *Nature Nanotechnology*. 2017, vol. 12(9), pp. 907–13.
[102] Zhou W., Yao S., Wang H., Du Q., Ma Y., Zhu Y. 'Gas-permeable, ultrathin, stretchable epidermal electronics with porous electrodes'. *ACS Nano*. 2020, vol. 14(5), pp. 5798–805.

[103] García Núñez C., Manjakkal L., Dahiya R. 'Energy autonomous electronic skin'. *Npj Flexible Electronics*. 2019, vol. 3(1), pp. 1–24.

[104] Wang Z.L., Wu W. 'Nanotechnology-enabled energy harvesting for self-powered micro-/nanosystems'. *Angewandte Chemie (International Ed. in English)*. 2012, vol. 51(47), pp. 11700–21.

[105] Yang Y., Guo W., Pradel K.C, *et al*. 'Pyroelectric nanogenerators for harvesting thermoelectric energy'. *Nano Letters*. 2012, vol. 12(6), pp. 2833–38.

[106] Abad A.C., Ranasinghe A. 'Visuotactile sensors with emphasis on gelsight sensor: a review'. *IEEE Sensors Journal*. 2020, vol. 20(14), pp. 7628–38.

[107] Wagner C.R., Stylopoulos N., Jackson P.G., Howe R.D. 'The benefit of force feedback in surgery: examination of blunt dissection'. *Presence*. 2007, vol. 16(3), pp. 252–62.

[108] Bethea B.T., Okamura A.M., Kitagawa M, *et al*. 'Application of haptic feedback to robotic surgery'. *Journal of Laparoendoscopic & Advanced Surgical Techniques. Part A*. 2004, vol. 14(3), pp. 191–95.

[109] van der Meijden O.A.J., Schijven M.P. 'The value of haptic feedback in conventional and robot-assisted minimal invasive surgery and virtual reality training: a current review'. *Surgical Endoscopy*. 2009, vol. 23(6), pp. 1180–90.

[110] Peirs J., Clijnen J., Reynaerts D, *et al*. 'A micro optical force sensor for force feedback during minimally invasive robotic surgery'. *Sensors and Actuators A*. 2007, vol. 115(2–3), pp. 447–55.

[111] Bandari N.M., Ahmadi R., Hooshiar A., Dargahi J., Packirisamy M. 'Hybrid piezoresistive-optical tactile sensor for simultaneous measurement of tissue stiffness and detection of tissue discontinuity in robot-assisted minimally invasive surgery'. *Journal of Biomedical Optics*. 2017, vol. 22(7), 77002.

[112] Qasaimeh M.A., Sokhanvar S., Dargahi J., Kahrizi M. 'PVDF-based microfabricated tactile sensor for minimally invasive surgery'. *Journal of Microelectromechanical Systems*. 2008, vol. 18(1), pp. 195–207.

[113] Kim U., Lee D.H., Yoon W.J., Hannaford B., Choi H.R. 'Force sensor integrated surgical forceps for minimally invasive robotic surgery'. *IEEE Transactions on Robotics*. 2015, vol. 31(5), pp. 1214–24.

[114] Xie H., Liu H., Luo S., Seneviratne L.D., Althoefer K. 'Fiber optics tactile array probe for tissue palpation during minimally invasive surgery'. Presented at IEEE/RSJ International Conference on Intelligent Robots and Systems; IEEE, Nov 2013, pp. 2539-2544,

[115] Li T., Shi C., Ren H. 'A high-sensitivity tactile sensor array based on fiber bragg grating sensing for tissue palpation in minimally invasive surgery'. *IEEE/ASME Transactions on Mechatronics*. 2018, vol. 23(5), pp. 2306–15.

[116] Ly H.H., Tanaka Y., Fukuda T., Sano A. 'Grasper having tactile sensing function using acoustic reflection for laparoscopic surgery'. *International Journal of Computer Assisted Radiology and Surgery*. 2017, vol. 12(8), pp. 1333–43.

[117] Campisano F., Ozel S., Ramakrishnan A, *et al.* 'Towards a soft robotic skin for autonomous tissue palpation'. *IEEE International Conference on Robotics and Automation (ICRA)*; Singapore, IEEE, 2017. pp. 6150–55.
[118] Iwata H., Sugano S. 'Design of human symbiotic robot TWENDY-ONE'. *IEEE International Conference on Robotics and Automation*; IEEE, May 2009. pp. 580–86.
[119] Pacchierotti, C., Sinclair S., Solazzi M., Frisoli A., Hayward V., Prattichizzo D. 'Wearable haptic systems for the fingertip and the hand: taxonomy, review, and perspectives'. *IEEE Transactions on Haptics*; IEEE, 2017. pp. 580–600.
[120] Kawasaki H., Mouri T. 'Humanoid robot hand and its applied research'. *Journal of Robotics and Mechatronics*. 2019, vol. 31(1), pp. 16–26.
[121] Stiehl W.D., Breazeal C. 'A sensitive skin for robotic companions featuring temperature, force, and electric field sensors'. *IEEE/RSJ International Conference on Intelligent Robots and Systems*; IEEE, 2006. pp. 1952–59.
[122] Felip J., Morales A., Asfour T. 'Multi-sensor and prediction fusion for contact detection and localization'. *IEEE-RAS International Conference on Humanoid Robots*; IEEE, 2014. pp. 601–07.
[123] Youssefi S., Denei S., Mastrogiovanni F., Cannata G. 'A middleware for whole body skin-like tactile systems'. *11th IEEE-RAS International Conference on Humanoid Robots*; IEEE, 2011. pp. 159–64.
[124] Dahiya R.S., Valle M., Metta G. 'System approach: a paradigm for robotic tactile sensing' in *10th IEEE international workshop on advanced motion control*. IEEE; 2008. pp. 110–15.
[125] Mittendorfer P., Cheng G. 'Humanoid multimodal tactile-sensing modules'. *IEEE Transactions on Robotics*. 2011, vol. 27(3), pp. 401–10.
[126] Polic M., Krajacic I., Lepora N., Orsag M. 'Convolutional autoencoder for feature extraction in tactile sensing'. *IEEE Robotics and Automation Letters*. 2019, vol. 4(4), pp. 3671–78.
[127] Oballe-Peinado O., Hidalgo-Lopez J.A., Castellanos-Ramos J, *et al.* 'FPGA-based tactile sensor suite electronics for real-time embedded processing'. *IEEE Transactions on Industrial Electronics*. 2017, vol. 64(12), pp. 9657–65.
[128] Wang C., Wang C., Huang Z., Xu S. 'Materials and structures toward soft electronics'. *Advanced Materials (Deerfield Beach, Fla.)*. 2018, vol. 30(50), e1801368.
[129] Wang B., Facchetti A. 'Mechanically flexible conductors for stretchable and wearable e-skin and e-textile devices'. *Advanced Materials (Deerfield Beach, Fla.)*. 2019, vol. 31(28), e1901408.
[130] Yogeswaran N., Dang W., Navaraj W.T, *et al.* 'New materials and advances in making electronic skin for interactive robots'. *Advanced Robotics*. 2015, vol. 29(21), pp. 1359–73.
[131] Won P., Park J.J., Lee T, *et al.* 'Stretchable and transparent kirigami conductor of nanowire percolation network for electronic skin applications'. *Nano Letters*. 2019, vol. 19(9), pp. 6087–96.

[132] Khang D.Y., Jiang H., Huang Y., Rogers J.A. 'A stretchable form of single-crystal silicon for high-performance electronics on rubber substrates'. *Science*. 2006, vol. 311(5758), pp. 208–12.
[133] Wang S., Xu J., Wang W, *et al*. 'Skin electronics from scalable fabrication of an intrinsically stretchable transistor array'. *Nature*. 2018, vol. 555(7694), pp. 83–88.
[134] Schmitz A., Maggiali M., Natale L., Bonino B., Metta G. 'A tactile sensor for the fingertips of the humanoid robot icub'. *IEEE/RSJ International Conference on Intelligent Robots and Systems*; IEEE, Oct 2010. pp. 2212–17.
[135] Xiao W., Sun F., Liu H., Liu H., He C. 'IEEE international conference on multisensor fusion and integration for intelligent systems (MFI 2012)'. Hamburg, Germany, IEEE, 2016. pp. 52–57.

Chapter 8
Neuroengineering approaches for cognitive hearing technology

Tobias Reichenbach[1]

8.1 Introduction

Hearing impairments affect more than 16% of the adult population worldwide, and more than 5% in children [1–3]. Moreover, hearing impairments progress with age and are hence a particular problem in our aging society: 40% of people above age 50 and 70% of those above age 70 have hearing impairment. This has major effects on a person's personal, social, and economic development [4, 5]. Hearing impairment has also been identified as a potential risk factor for dementia [6, 7].

The most prevalent form of hearing impairment is sensorineural hearing loss, which occurs when the transduction of the mechanical sound vibrations into electrical nerve impulses, accomplished in the inner ear, is impaired. Furthermore, the neural processing of the signals from the inner ear, which occurs in the auditory brainstem and in the auditory cortex, is vulnerable as well [8]. Between 5% and 10% of the population in the United Kingdom, including children, suffer from such auditory processing disorder.

A major difficulty for people with hearing impairment is to understand speech in noisy environments such as when other competing speakers are present. For people with sensorineural hearing loss this difficulty persists even when they wear hearing aids [9]. Patients with auditory processing disorder can benefit from directional microphones, but still have major problems with understanding speech in noise [9].

Despite big advances in the miniaturisation and functionality of hearing aids and their significant cost – the revenues of the hearing industry exceed £5.00 billion per year and even higher revenues are generated in the retail/fitting/dispensing stage – the process of optimizing hearing aids for a particular user and for use in changing real-world acoustic environments has seen little progress. Current developments in

[1]Department of Bioengineering, Imperial College London, South Kensington Campus, London, UK

hearing aids indeed focus on three aspects, none of which incorporates automated user feedback.

First, significant effort has been devoted to noise reduction through engineering of the audio signal, either using single-microphone noise processing such as spectral subtraction, or using the outputs of multiple microphones to create a directional characteristic [10, 11]. State-of-the-art algorithms for speech enhancement can improve the perceived quality of a speech signal in noise and reduce the perceived listening effort of the user, although the benefits in real life have been found to be small. Studies of listening effort have been restricted to the laboratory, and this has prevented the verification, optimization, and adaptation of the algorithms for speech enhancement in real-world listening conditions.

Second, hearing-aid developers seek to integrate the devices into smart environments, via wireless links with smart devices or home entertainment systems. Despite advanced functionality; however, this only tackles limited aspects of user comfort, such as noise reduction in a known environment, and requires additional hardware including microphone arrays and computers.

Third, smartphone-based hearing aid applications aim to give the user some control over the hearing-aid settings and allow the user to change them depending on the acoustic environment [12]. However, due to the many complex settings in a hearing aid and the sparseness of overt user feedback, this approach has produced limited benefits. Moreover, it does not allow for an objective evaluation of hearing-aid performance.

Recently, biofeedback has been proposed as a way to equip hearing aids with cognitive control [13, 14]. In particular, measuring brain activity from scalp electrodes may allow to assess issues relevant to hearing-aid processing such as listening effort, speech comprehension and selective attention to one of several competing speakers. The assessment of these cognitive factors may then be used to feed back to the hearing aid and modify its signal processing. This would equip the user with a way to adjust the settings in a hearing aid in an automatic manner.

Another recent development targets the question of speech enhancement. Stimulating the brain through small electrical currents has been found to modulate the comprehension of speech in background noise, which may provide an avenue for enhancing speech perception [15–18].

In this chapter, we describe different neuroengineering approaches to the cognitive control of a hearing aid as well as to the speech enhancement using transcranial current stimulation.

8.2 General aspects of neurofeedback in a hearing aid

Recent breakthroughs in wearable sensing now make it feasible to use physiological recordings for feedback in hearing aids. In particular, the recently developed in-the-ear (ITE) physiological sensing device records electroencephalographic (EEG) responses from the brain as well as cardiac activity from sensors in the ear canal [19, 20] (Figure 8.1a).

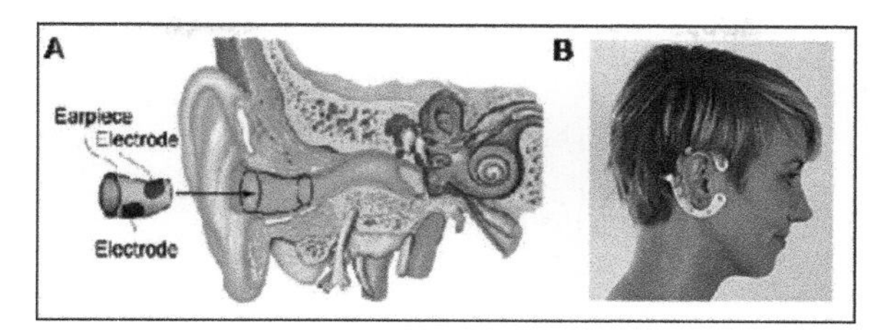

Figure 8.1 *Wearable EEG devices for cognitive control of a hearing aid. (a) The ITE device records EEG and cardiac activity from electrodes within the ear canal. (b) The cEEGrid wraps around the outer ear.*

As an alternative device, the cEEGrid for EEG recordings is transparent, flexible and wraps around the outer ear [21, 22] (Figure 8.1b). Both devices allow to be worn in real-world environments on a 24/7 basis and to be integrated with hearing aids. Moreover, the devices achieve signal-to-noise ratios regarding auditory responses that are comparable to those of scalp EEG despite measuring from a much more limited number of spatial locations. This is due to their fixed position in the ear canal or around the outer ear that provides close contact with the skin, is near the auditory cortex, and is less prone to movement or motion artifacts.

Implementing neurofeedback in a hearing aid requires the analysis of the obtained EEG recordings with respect to cognitive processes. Because the most important function of hearing aids is to enable the user to understand speech, this necessitates the analysis of the neural responses with respect to different cognitive aspects of the processing of naturalistic speech. This type of analysis differs from traditional neuroimaging experiments, which have focused on neural responses to somewhat artificial, highly controlled speech samples such as single syllables, words, or artificial sentences [23–26].

Two types of neural responses to speech have recently been uncovered and employed for cognitive control of hearing aid (Figure 8.2). The first is a neural signal that results mostly from the auditory brainstem. This signal emerges at the temporal fine structure of the voiced parts of speech. These voiced speech segments result from the opening and closing of the glottis, which occurs at the so-called fundamental frequency of speech, between 100 and 300 Hz. This frequency and its many higher harmonics then carry most of the energy of these speech segments. Because some neurons in the auditory brainstem respond at such high frequencies, they cause a signal at the fundamental frequency (and to a lesser extent at the higher harmonics) that can be measured from scalp electrodes [23, 27, 28]. A certain contribution to this neural response results also from the cerebral cortex [29, 30].

As a second main type of neural response, amplitude fluctuations of speech are tracked by neural activity in the cerebral cortex [31–34]. Neurons in the cortex largely respond at much lower frequencies than the ones that carry the energy in speech, below 100 Hz. However, these slower frequencies match the frequencies at which the energy in a speech signal is modulated. Magnetoencephalography (MEG), as well as EEG measurements have found that these amplitude fluctuations,

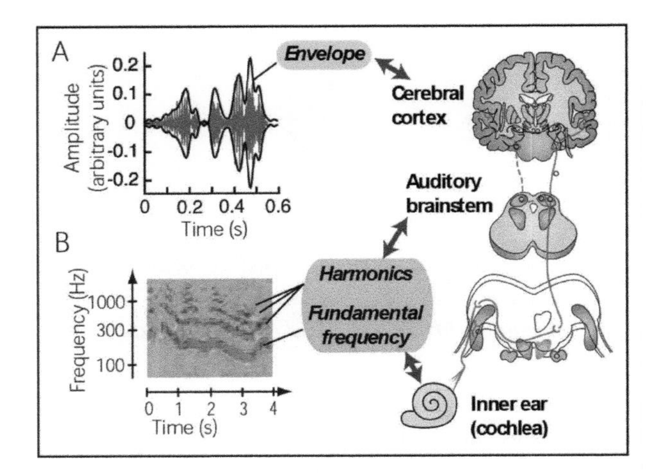

Figure 8.2 *Brain responses to continuous speech. (a) Cortical oscillations can entrain to the envelope (black) of a continuous speech signal (grey). (b) The response of the auditory brainstem tracks the temporal fine structure of speech, and in particular the fundamental frequency.*

which can be quantified through the speech envelope, are tracked by the cortical activity in this frequency range.

8.3 Decoding selective attention to speech from the auditory brainstem response to the temporal fine structure

The temporal fine structure of speech varies over time, making an easy readout of the corresponding neural response difficult (Figure 8.2b). Neuroscientific experiments have therefore used averaging over many repeated short speech tokens, such as single syllables, to increase the signal-to-noise ratio of the neural response and to obtain a measurable signal [23]. However, for applications in a hearing aid, one needs to assess the response to natural, running speech. We have therefore developed a methodology in which we first extract a fundamental waveform from the speech signal (Figure 8.3) [27]. The fundamental waveform is zero whenever there is no voiced part of the speech. For segments that are voiced, the fundamental waveform oscillates, at each time point, at the fundamental frequency. It can be obtained by empirical mode decomposition [35] or by band-pass filtering.

The fundamental waveform can then be related to the EEG recordings using machine learning methods. Broadly, the relationship can be obtained using three different approaches. In the first, the so-called forward model, one seeks to express the EEG recording at a particular channel through the fundamental waveform. This can occur, for instance, by considering different delays of the fundamental waveform, and by seeking an optimal linear superposition of the different obtained signals to reconstruct an EEG channel's time series [36].

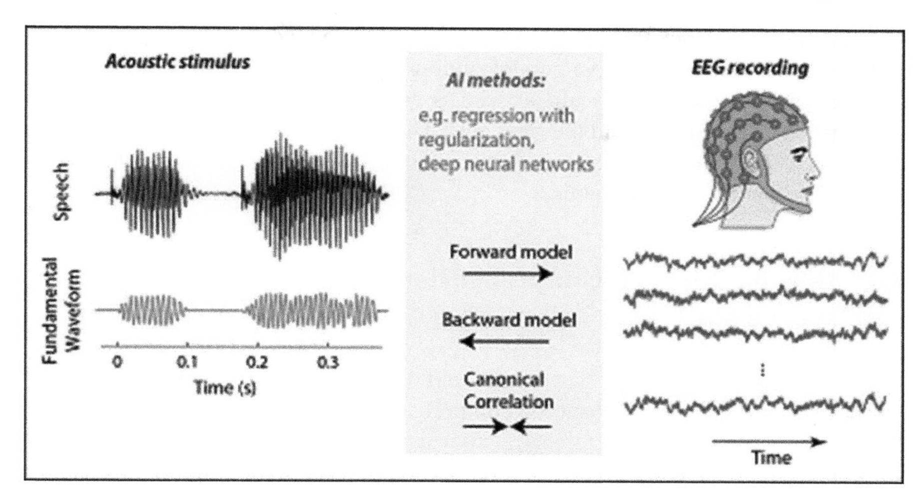

Figure 8.3 *Measuring the neural response to the fundamental frequency of a continuous speech signal. One can extract a waveform (red) from the speech signal (black) that oscillates, at each time point, at the fundamental frequency. This waveform can then be related to EEG recordings using methods from machine learning.*

In the second approach, a backward model, the fundamental waveform is reconstructed from the different EEG channels. Most commonly, this reconstruction proceeds through a linear regression model in which the fundamental waveform is approximated as a linear combination of the different EEG channels, shifted by different delays. Alternatively, the fundamental waveform could be modelled through a deep neural network that takes the EEG recordings as input and has the estimated fundamental waveform as its output.

The third approach, canonical correlation analysis, applies linear analysis to relate the fundamental waveform and the EEG recordings in an intermediate space [37]. This analysis can boost the accuracy of the relationship beyond the ones obtained in either the backward or forward model.

We have shown that linear regression can be used to implement both backward and forward models to measure the brainstem response at the fundamental frequency [27, 36]. The response can be reliably measured in most subjects from a few minutes of EEG recordings.

Importantly, we showed that this neural response is affected by selective attention to one of two competing speakers [27]. To this end, we set up an experiment in which a subject listened to two competing speakers, a male and a female voice. The subject was instructed to sometimes attend the female speaker, and sometimes the male one. When analyzing the neural response, we then found that the neural tracking of the speech's temporal fine structure was larger when the corresponding voice was attended as compared to when it was ignored. Moreover, we showed that the focus of attention could be recorded from short segments of speech, down to

1 second, at an accuracy that was higher than the chance level [36]. In addition, three EEG electrodes sufficed to give good decoding of selective attention: one electrode that was on the center of the scalp, as well as two electrodes that were positioned at the mastoids (near the left and the right ear). These results suggest that the decoding of attention from the brainstem response can contribute to using neurofeedback in a cognitively controlled hearing aid.

8.4 Decoding speech comprehension from cortical tracking of speech features

The most important function that hearing aid users are looking for is to understand speech. Assessing the level of speech comprehension of a wearer in a particular situation can potentially be used to change the setting in a hearing aid, such as aspects of its fitting to a particular hearing loss or the particular noise-reduction algorithms and their parameters.

We, therefore, sought to develop the methodology for decoding speech comprehension from EEG recordings. First, we designed an experiment in which we carefully delineated acoustic processing from higher-level speech comprehension [38]. To this end, we presented native English speakers with English stories in various background noise, while recording their cortical responses through EEG. The different levels of background noise were chosen to alter speech comprehension from excellent to moderate and low. We then also played Dutch stories in the same levels of background noise to the participants, although they did not understand Dutch. Because the acoustics of the two languages English and Dutch are relatively similar, and because we employed the same background noise, the English and the Dutch stimuli had approximately the same acoustics. However, the volunteer's speech comprehension differed hugely between the two sets of stimuli: while it was high, medium, or low for the English stories, depending on the level of background noise, it was uniformly zero for the Dutch narratives.

We then analyzed the EEG data in two steps. First, we assessed the cortical tracking of the amplitude fluctuations in the speech signals through computing backward models to reconstruct the speech envelope from the EEG recordings. Second, we decoded both the level of speech comprehension as well as the clarity of the speech signal, that is, its signal-to-noise ratio, from the coefficients that emerged in the backward model. As a result, we found that even short recordings of 10 seconds in duration allowed us to decode speech comprehension at an accuracy that was above chance level. This suggests that decoding of speech comprehension is feasible in a cognitively controlled hearing aid, and can be achieved on short time scales that make interventions such as the adaption of hearing aid settings and processing feasible.

Our analysis also revealed that the decoding of speech comprehension resulted from a relatively narrow frequency band, the delta band, approximately between 1 and 4 Hz. We verified these results through an encoding approach in which we computed a forward model that expressed the EEG recordings through the speech envelope, shifted by different delays and weighted by the clarity as well as the comprehension level, as well as through the envelope of the background noise. Consistent with our previous analysis using backward models, we found that the delta range of the EEG responses was significantly related to the comprehension level.

The delta frequency range corresponds to the rate of words in speech. The relation of speech comprehension to the neural tracking in the delta band may therefore reflect word-level processing. Recent neuroimaging studies have indeed shown that the cortical tracking of word onsets can reflect both semantic as well as syntactic processing [39, 40].

We, therefore, wondered whether cortical responses to words could indicate speech comprehension in a more direct manner [41]. To assess this question, we quantified both surprisal as well as the associated precision of word sequences. In particular, we trained a recurrent neural network on a large corpus of text to predict the upcoming word from all preceding words. This allowed us to quantify how surprising a particular word was, as well as the precision of the associated prediction.

We then acquired EEG recordings from native English speakers who listened to both stories in English as well as in Dutch. The EEG recordings were related to speech features through linear forward models. For the speech features, we employed words onsets as well as word frequency as control features, since we expected these two features to predict aspects of the neural responses although they do not inform directly speech comprehension. For the latter purpose, we included two further speech features, word surprisal, and word precision.

We found that, when participants listened to the English stories, significant neural responses to both surprisal and precision emerged. These responses were noticeably absent when the subjects were presented with the Dutch narratives, indicating their lack of comprehension. The obtained cortical responses to word surprisal and precision could therefore reliably indicate speech comprehension. However, these responses were tiny and required long recordings of more than 10 minutes to be measurable. While they might be employed to improve the fitting of a hearing aid, they do not appear applicable for real-time neurofeedback.

8.5 Enhancing speech comprehension through transcranial electric stimulation

Neurofeedback in a hearing aid requires not only a readout of cognitive factors such as the focus of attention but also the subsequent control of processes in the hearing aid to improve access to the relevant auditory signals. This involves controlling aspects of the acoustic processing in the aid, such as reduction of background noise as well as aspects of the fitting to the specific hearing loss.

Another process may be to apply electrical stimulation to the auditory cortex to aid the acoustic processing. Recent investigations have indeed shown that transcranial alternating current stimulation, the application of an alternating electrical current over scalp electrodes, can influence speech comprehension (Figure 8.4) [15–17]. The electrical current has thereby been obtained from the speech signal,

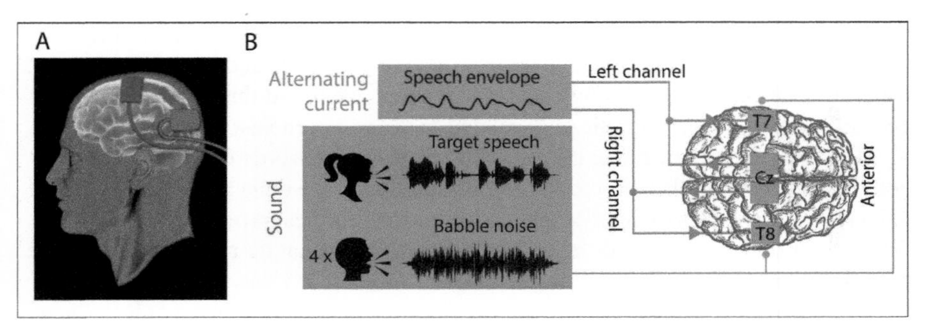

Figure 8.4 Modulation of speech comprehension through transcranial alternating current stimulation. (a) Currents are applied non-invasively through scalp electrodes. (b) When a subject listens to a target signal presented in background noise, and their auditory cortex is stimulated through an alternating current that follows the speech envelope, the speech comprehension can be modulated.

namely following its envelope. Because transcranial alternating current can entrain the brain activity in the corresponding frequency band [42], stimulation with the speech envelope is believed to act on the neural activity in the delta and theta range that matters for speech processing.

We showed that electrical current stimulation can not only modulate, but in fact enhance the comprehension of speech in noise. To this end we first investigated which of the two frequency bands that contribute to the amplitude fluctuations in speech, the delta band and the theta band, matter for the modulation of speech comprehension [18]. We found that the behavioural effect resulted from the theta frequency band, but not from the delta band. This suggests that the electrical current stimulation acts on the lower level acoustic processing, such as the syllable parsing, but not or only little on higher level linguistic processing, such as on the level of words.

We then determined the impact of temporal delays between the acoustic signal and the electrical stimulus on speech comprehension [43]. We found that no delay as well as a short delay of 90 ms of the current with respect to the speech signal resulted in improved speech comprehension. Negative delays as well as longer delays, in contrast, led to a level of speech comprehension that was not significantly different from that under a sham stimulus.

The improvement of speech-in-noise comprehension during the current stimulation was small, about 5%. Application of such a technique in an auditory prosthesis would require the further improvement of this enhancement, such as through modifications of the current waveform that render it yet more efficient to influence the relevant neural processes. A potential avenue to develop such an improved waveform might be the consideration of computational models for speech processing that may allow a faster screening of potential current waveforms [44].

8.6 Summary

Implementing neurofeedback in a hearing aid has recently become feasible due to the development of specialized hardware such as in-ear EEG recordings or the cEEGrid that wraps around the outer ear. The software aspects around the neurofeedback are therefore now beginning to be tackled. They relate to two broader issues. First, neurofeedback requires to obtain information about cognitive aspect of acoustic processing, from the EEG sensors. Second, neurofeedback then necessitates the use of the neural readout to influence the neural processing through altering processes in the hearing aid.

Here we have presented preliminary work on both issues. In particular, regarding the first challenge, we have shown how selective attention can be decoded from fast neural responses to the temporal fine structure of speech. We also presented work on the decoding of speech comprehension from the slower cortical tracking of the amplitude fluctuations in speech. Regarding the second issue, we showed how applying tiny electrical currents in a non-invasive manner can modulate, and in fact enhance, speech comprehension.

Real-world application of these recent advances require; however, considerable further developments. Regarding the readout of cognitive processing, application in a hearing aid requires ideally a real-time analysis. In practice, this means that cognitive factors should be identifiable on a short time scale of a few hundred milliseconds or less. Moreover, these readouts should work reliably under different conditions, such as in different acoustic environment and with different types of background noise. As for the speech enhancement, practical applications would benefit from improving speech comprehension by a considerable margin, say 20% or higher, to make a real difference to the hearing-aid user. This will likely require the optimization of several types of hearing-aid processes, such of the acoustic processing as well as perhaps electrical current stimulation.

References

[1] Davis A.C. 'The prevalence of hearing impairment and reported hearing disability among adults in great britain'. *International Journal of Epidemiology*. 1989, vol. 18(4), pp. 911–17.

[2] Moore D.R., Rosen S., Bamiou D.-E., Campbell N.G., Sirimanna T. 'Evolving concepts of developmental auditory processing disorder (APD): a british society of audiology APD special interest group "white paper"'. *International Journal of Audiology*. 2013, vol. 52(1), pp. 3–13.

[3] Davis A., Wood S. 'The epidemiology of childhood hearing impairment: factor relevant to planning of services'. *British Journal of Audiology*. 1992, vol. 26(2), pp. 77–90.

[4] Bajo V.M., Nodal F.R., Moore D.R., King A.J. 'The descending corticocollicular pathway mediates learning-induced auditory plasticity'. *Nature Neuroscience*. 2010, vol. 13(2), pp. 253–60.

[5] Moore D.R. 'Auditory processing disorder (APD): definition, diagnosis, neural basis, and intervention'. *Audiological Medicine*. 2006, vol. 4(1), pp. 4–11.
[6] Klein C.J., Botuyan M.-V., Wu Y., *et al*. 'Mutations in DNMT1 cause hereditary sensory neuropathy with dementia and hearing loss'. *Nature Genetics*. 2011, vol. 43(6), pp. 595–600.
[7] Frankish H., Horton R. 'Prevention and management of dementia: a priority for public health'. *Lancet (London, England)*. 2017, vol. 390(10113), pp. 2614–15.
[8] Pickles J.O. An introduction to the physiology of hearing. London: Academic Press; 1988.
[9] Chung K. 'Challenges and recent developments in hearing aids. Part I. speech understanding in noise, microphone technologies and noise reduction algorithms'. *Trends in Amplification*. 2004, vol. 8(3), pp. 83–124.
[10] Loizou P.C. Speech enhancement. Boca Raton, FL: CRC Press; 2013. Available from https://www.taylorfrancis.com/books/9781466504226
[11] Hu Y., Loizou P.C. 'Evaluation of objective quality measures for speech enhancement'. *IEEE Transactions on Audio, Speech, and Language Processing*. 2017, vol. 16(1), pp. 229–38.
[12] Amlani A.M., Taylor B., Levy C., Robbins R. 'Utility of smartphone-based hearing aid applications as a substitute to traditional hearing aids'. *Hearing Review*. 2013, vol. 20, pp. 16–18.
[13] Perron M. 'Hearing aids of tomorrow: cognitive control toward individualized experience'. *Hearing Journal*. 2017, vol. 70, pp. 22–23.
[14] Dau T., Maercher Roersted J., Fuglsang S., Hjortkjær J. 'Towards cognitive control of hearing instruments using EEG measures of selective attention'. *The Journal of the Acoustical Society of America*. 2017, vol. 143(3), pp. 1744–1744.
[15] Riecke L., Formisano E., Sorger B., Başkent D., Gaudrain E. ' neural entrainment to speech modulates speech intelligibility '. *Current Biology*. 2018, vol. 28(2), pp. 161–69.
[16] Wilsch A., Neuling T., Obleser J., Herrmann C.S. 'Transcranial alternating current stimulation with speech envelopes modulates speech comprehension'. *NeuroImage*. 2018, vol. 172, pp. 766–74.
[17] Zoefel B., Archer-Boyd A., Davis M.H. 'Phase entrainment of brain oscillations causally modulates neural responses to intelligible speech'. *Current Biology*. 2018, vol. 28(3), pp. 401–08.
[18] Keshavarzi M., Kegler M., Kadir S., Reichenbach T. 'Transcranial alternating current stimulation in the theta band but not in the delta band modulates the comprehension of naturalistic speech in noise'. *NeuroImage*. 2020, vol. 210, p. 116557.
[19] Looney D., Kidmose P., Park C., *et al*. 'The in-the-ear recording concept: user-centered and wearable brain monitoring'. *IEEE Pulse*. 2012, vol. 3(6), pp. 32–42.

[20] Kidmose P., Looney D., Ungstrup M., Rank M.L., Mandic D.P. 'A study of evoked potentials from ear-EEG'. *IEEE Transactions on Bio-Medical Engineering*. 2013, vol. 60(10), pp. 2824–30.

[21] Debener S., Emkes R., De Vos M., Bleichner M. 'Unobtrusive ambulatory EEG using a smartphone and flexible printed electrodes around the ear'. *Scientific Reports*. 2015, vol. 5, 16743.

[22] Mirkovic B., Debener S., Jaeger M., De Vos M. 'Decoding the attended speech stream with multi-channel EEG: implications for online, daily-life applications'. *Journal of Neural Engineering*. 2015, vol. 12(4), p. 46007.

[23] Skoe E., Kraus N. 'Auditory brain stem response to complex sounds: a tutorial'. *Ear and Hearing*. 2010, vol. 31(3), pp. 302–24.

[24] Kutas M., Federmeier K.D. 'Thirty years and counting: finding meaning in the N400 component of the event-related brain potential (ERP)'. *Annual Review of Psychology*. 2011, vol. 62, pp. 621–47.

[25] Frank S.L., Otten L.J., Galli G., Vigliocco G. 'The ERP response to the amount of information conveyed by words in sentences'. *Brain and Language*. 2015, vol. 140, pp. 1–11.

[26] Hagoort P., Indefrey P. 'The neurobiology of language beyond single words'. *Annual Review of Neuroscience*. 2014, vol. 37, pp. 347–62.

[27] Forte A.E., Etard O., Reichenbach T. 'The human auditory brainstem response to running speech reveals a subcortical mechanism for selective attention. 2017'. *ELife*. n.d. Available from https:/doi.org/10.7554/eLife.27203

[28] Sohmer H., Pratt H., Kinarti R. 'Sources of frequency following responses (FFR) in man'. *Electroencephalography and Clinical Neurophysiology*. 1977, vol. 42(5), pp. 656–64.

[29] Coffey E.B.J., Herholz S.C., Chepesiuk A.M.P., Baillet S., Zatorre R.J. 'Cortical contributions to the auditory frequency-following response revealed by MEG'. *Nature Communications*. 2016, vol. 7, 11070.

[30] Bidelman G.M. 'Subcortical sources dominate the neuroelectric auditory frequency-following response to speech'. *NeuroImage*. 2018, vol. 175, pp. 56–69.

[31] Ding N., Simon J.Z. 'Emergence of neural encoding of auditory objects while listening to competing speakers'. *Proceedings of the National Academy of Sciences of the United States of America*. 2012, vol. 109(29), pp. 11854–59.

[32] Giraud A.-L., Poeppel D. 'Cortical oscillations and speech processing: emerging computational principles and operations'. *Nature Neuroscience*. 2012, vol. 15(4), pp. 511–17.

[33] Thwaites A., Nimmo-Smith I., Fonteneau E., Patterson R.D., Buttery P., Marslen-Wilson W.D. 'Tracking cortical entrainment in neural activity: auditory processes in human temporal cortex'. *Frontiers in Computational Neuroscience*. 2015, vol. 9, 5.

[34] Ding N., Simon J.Z. 'Cortical entrainment to continuous speech: functional roles and interpretations'. *Frontiers in Human Neuroscience*. 2014, vol. 8, 311.

[35] Huang H., Pan J. ‘Speech pitch determination based on hilbert-huang transform’. *Signal Processing*. 2006, vol. 86(4), pp. 792–803.

[36] Etard O., Kegler M., Braiman C., Forte A.E., Reichenbach T. ‘Decoding of selective attention to continuous speech from the human auditory brainstem response [online]’. *NeuroImage*. 2019, vol. 200, pp. 1–11. Available from https:/doi.org/10.1016/j.neuroimage.2019.06.029

[37] de Cheveigné A., Wong D.D.E., Di Liberto G.M., Hjortkjær J., Slaney M., Lalor E. ‘Decoding the auditory brain with canonical component analysis’. *NeuroImage*. 2018, vol. 172, pp. 206–16.

[38] Etard O., Reichenbach T. 'Neural speech tracking in the theta and in the delta frequency band differentially encode clarity and comprehension of speech in noise'. *Journal of Neuroscience*. 2019, vol. 39(29), pp. 5750–59.

[39] Ding N., Melloni L., Zhang H., Tian X., Poeppel D. ‘Cortical tracking of hierarchical linguistic structures in connected speech’. *Nature Neuroscience*. 2016, vol. 19(1), pp. 158–64.

[40] Broderick M.P., Anderson A.J., Di Liberto G.M., Crosse M.J., Lalor E.C. 'Electrophysiological correlates of semantic dissimilarity reflect the comprehension of natural, narrative speech'. *Current Biology*. 2018, vol. 28(5), pp. 803–9.

[41] Weissbart H., Kandylaki K.D., Reichenbach T. ‘Cortical tracking of surprisal during continuous speech comprehension’. *Journal of Cognitive Neuroscience*. 2020, vol. 32(1), pp. 155–66.

[42] Helfrich R.F., Schneider T.R., Rach S., Trautmann-Lengsfeld S.A., Engel A.K., Herrmann C.S. ‘Entrainment of brain oscillations by transcranial alternating current stimulation’. *Current Biology*. 2014, vol. 24(3), pp. 333–39.

[43] Keshavarzi M., Reichenbach T. 'Transcranial alternating current stimulation with the theta-band portion of the temporally-aligned speech envelope improves speech-in-noise comprehension [online]’. *Frontiers in Human Neuroscience*. 2020, vol. 14, 187. Available from https:/doi.org/10.3389/fnhum.2020.00187

[44] Hyafil A., Fontolan L., Kabdebon C., Gutkin B., Giraud A.-L. ‘Speech encoding by coupled cortical theta and gamma oscillations’. *ELife*. 2015, vol. 4, e06213.

Chapter 9

Mobile robot olfaction state-of-the-art and research challenges

Lino Marque[1], Hugo Magalhães[1], Rui Baptista[1], and João Macedo[1]

9.1 Introduction

Olfaction is a fundamental sense for most animals, which use it to find mates, food or avoid predators [1]. The literature reporting studies on animal olfaction is extensive. The studies on insect olfaction and the olfactory capabilities of some male moths are particularly interesting. Such moths are known to detect pheromones released from females located several hundred meters away and are able to navigate in natural environments until finding their mate [2]. The notorious olfactory capabilities of several animals are well known and frequently exploited for practical applications. Dogs, pigs and even rats often team up with humans to find landmines [3] or used in airport patrols and other public places sniffing for drugs, explosives or other illegal substances [4]. They also assist search and rescue teams searching for victims [5], and can even detect diseases, such as some types of cancer or COVID-19 [6]. The use of trained animals to augment the limited capacities of human olfaction is acceptable for many applications but has some obvious drawbacks, such as the need to train the animals, lack of quantified detections, limited operating time, the impossibility to operate in hazardous environments and the need for a human supervisor. These drawbacks may be, at least to some extent, mitigated by mobile robots with advanced olfactory capabilities. Such robots would be able to explore the oceans, searching for substances of interest, such as valuable minerals, pollution or organic matter. They may explore the atmosphere, searching for pollution sources, monitor farms while detecting pests and diseases threatening the cultures, monitor forests, detecting wildfires at an early stage and eventually tracking their progress, or operate as advanced scientific instruments, helping scientists in the study of dangerous natural phenomena, such as volcanic eruptions. We may also imagine robot dogs, with the olfactory capabilities of their animal counterparts, but with the ability to operate autonomously, without human supervisors. A common feature to any of

[1]Institute for Systems and Robotics, University of Coimbra, Coimbra, Portugal

these robots is their capacity to smell, i.e., to detect a target substance moving in a fluid, be it water, or air. By analogy with biology, we call this detectable substance an odour and the chemical sensing system, eventually complemented by a flow sensing system, the robot's olfactory system.

Most of the research on mobile robot olfaction (MRO) has addressed the capability of monitoring a given phenomenon, by mapping the concentration of some substance of interest [7], tracking the boundary of a spreading phenomenon, such as an oil spill [8], or searching a substance and finding its origin, also known as the odour source localisation (OSL) problem. This problem is frequently decomposed into three sub-problems, which are searching for traces of the target substance, tracking its odour plume to its origin, and finally, localising its source.

In the past two decades, since the surveys by Marques and Almeida [9] and by Russell [10] about robotic applications for odour-sensing technology, several other surveys related with MRO have been written. Just in the last four years, six interesting surveys addressing different topics on this subject have appeared. Hutchinson *et al.* [11] review methods for source-term estimation (STE), both using static sensor networks or mobile units, such as robots. This chapter also provides a brief overview on boundary-tracking and source-seeking strategies. Bayat *et al.* [12] survey methods for environmental monitoring, focusing their analysis towards works using aquatic platforms and works published mostly between 2011 and 2016. Chen and Huang [13] provide a general overview about OSL works and classified the reviewed works into four categories: (1) gradient-based, (2) bio-inspired, (3) multi-robot and finally, (4) probabilistic and map-based algorithms. Burgues *et al.* [14] surveyed recent works on OSL using aerial vehicles. This chapter also provides an interesting description of the sensing technology usually employed in aerial vehicles. Recently, Jing *et al.* [15] published an interesting survey with a comprehensive analysis on OSL methods, extensively discussing reactive methods, heuristic search, probabilistic inference and learning methods. This chapter highlights the importance of using simulators in OSL research.

This chapter, on MRO, provides a review on the subject, detailing state-of-the-art approaches related with probabilistic and learning methodologies for OSL, and provides our view about the major challenges in the area. The chapter addresses MRO for atmospheric environments. Section 9.2 describes the dispersion mechanisms of substances in moving fluids. Section 9.3 describes the technologies employed in robotic olfactory systems, namely, gas and flow sensors. To limit the chapter, only devices for air environments are treated here. Section 9.4 identifies the main methods and algorithms developed for OSL. Section 9.5 presents machine learning approaches to implement robots able to evolve and improve their OSL capabilities. And finally, section 9.6 identifies the main open-research challenges.

9.2 Odour dispersion

Odour molecules move in the environment driven by two phenomena: advection and diffusion. Advection is the phenomenon resulting from the flow of a fluid, usually in the horizontal direction, while diffusion is the phenomenon responsible to homogenise the concentration of different substances, spreading from regions of higher concentration to regions of lower concentration. Diffusion happens at molecular level, but this is a very slow process at these scales. In natural environments, the interaction of flowing fluids with surfaces and other heterogeneities producing turbulence of different scale eddies, which by its turn, also mixes the fluid, at a much faster rate than molecular diffusion. This process is called turbulent diffusion or dispersion. Small-scale eddies, smaller than the plume width, are responsible for mixing the substance in a plume with the surrounding clean air or water, while large scale eddies, larger than the plume width, distort the plume laterally, producing meandering effects. These effects are shown in Figure 9.1(a,b) where we can see the internal intermittency, caused by the small-scale eddies, and the distortion of the global plume, caused by large-scale eddies. The changes of concentration in space and time caused by these phenomena are modelled by the so-called advection-diffusion equation [17, 18]:

$$\frac{\partial C}{\partial t} + \nabla\left(C\vec{u}\right) = \nabla\left(\boldsymbol{K}\nabla C\right) + Q \tag{9.1}$$

where C represents the odour concentration, $\vec{u}$ is the velocity of the fluid, $\boldsymbol{K}$ is the turbulent diffusion coefficients and Q is the source release rate. The advection phenomenon is represented by the term $\nabla\left(C\vec{u}\right)$ and the turbulent diffusion is represented by the term $\nabla\left(\boldsymbol{K}\nabla C\right)$. The solution of (9.1) for a point source, located at position (0,0,h) in a Cartesian coordinates frame, releasing at constant rate Q, with wind u blowing steadily in the direction of the X coordinates, and assuming an infinite ground plane in XY, is known as the Gaussian plume equation.

$$\bar{c}(x,y,z) = \frac{Q}{2\pi\bar{u}\sigma_y\sigma_z} exp\left(\frac{-y^2}{2\sigma y^2}\right)\left[exp\frac{(-(z-h)^2)}{2\sigma_z^2} + exp\left(\frac{-(z+h)^2}{2\sigma_z^2}\right)\right] \tag{9.2}$$

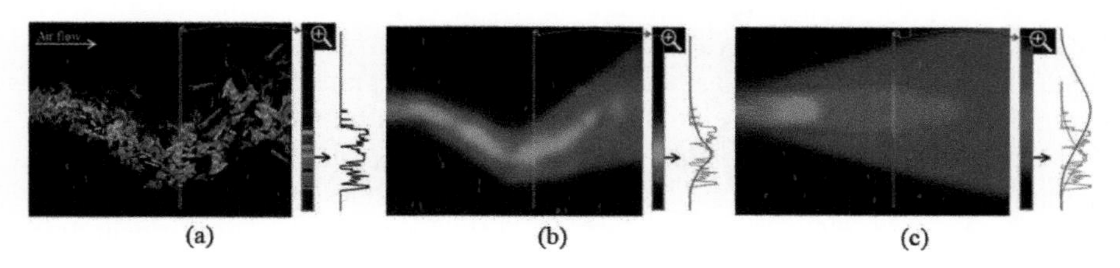

Figure 9.1 *Structure of an odour plume at various timescales [16]. (a) Instantaneous structure, showing the internal concentration intermittency. The black signal at the right shows the instantaneous measurements of a fast gas sensor while crossing the plume at the middle of the image. (b) Short-time scale average, showing how the plume would be perceived by a slow response time chemical sensor (red signal at the right). (c) Long-time scale average, showing the average concentration of the odour plume, as predicted by a Gaussian model.*

The lateral and vertical dispersion coefficients, σ_y and σ_z, respectively, are functions of the downwind distance to the source (x), frequently modelled by the polynomials of equations [9.3 and 9.4], whose coefficients depend on the atmosphere stability [19]:

$$\sigma_y = ax^b \tag{9.3}$$

$$\sigma_z = cx^d \tag{9.4}$$

Despite the strong assumptions that gave rise to (9.2), in practice, it provides a reasonable approximation to environments where the dimensions of the plume are much larger than the dimensions of the nearby obstacles, being frequently employed in atmospheric pollution modelling and in MRO. Figure 9.1(c) shows the average concentration of an odour plume, as predicted by a Gaussian plume model. The coloured area contains the area where the plume may be detected, which is also called "active area" of the plume. For an extensive review about gas dispersion modelling, the reader is encouraged to check References [20, 21].

As we will see in section 9.4, dispersion models are fundamental tools to reason about the olfactory observations that a robot acquires along its mission. These dispersion models are also fundamental tools for research in MRO as they provide the basis for building dispersion simulators. Olfactory experiments in real environments are costly, time-consuming and usually limited in terms of our ability to change the parameters of an experiment. Simulators overcome those difficulties at the cost of missing details from the reality. The usefulness of dispersion simulators has been recognised since the early times of MRO. Marques used CofinBox in his early works [22, 23], a very realistic simulator developed in the framework of the European Project COFIN [24], but only the executable version of this simulator was available and it no longer runs on the current Windows platforms. Farrell *et al.* [25] developed a 2D filament-based simulator that is one of the most widely used simulators in this field to date. This simulator was the basis for several later works, being extended by Sutton to a 3D version [26]. Lochmatter adapted it for use in Webots [27]. And Macedo *et al.* [28] developed a lightweight version optimised for multi-robot evolutionary experiments. More recently, there has been a tendency to integrate dispersion simulators into mobile robots. The first step into that direction was provided by Cabrita *et al.* [29] with a player/stage dispersion simulator which was later integrated into robot operating system (ROS). This simulator supports three modes of operation: Gaussian plume models with different noise levels, data files created by a computational fluid dynamics tool, or data files logged by a sensor network during a real experiment and later used with simulated robots. GADEN is one of the last simulators in this line, supporting a 3D filaments-based model and scene customisation in ROS [30].

9.3 Artificial olfaction

A fundamental feature of any robot's olfactory system is its ability to sense chemicals in the environment. For the case of air environments, that means using some type of gas sensor, animals sense gases either through nasal cavities, where air is sampled and analysed, or through open sampling systems (OSS), such as antennas, where odour molecules pass through and eventually bind to olfactory receptors, providing detections. Robots also follow those approaches, being possible to find works with both types of sampling systems. OSS are simpler and may potentially provide faster response times but leave the sensors mechanically exposed to the environment and susceptible to damages. Sampling nostrils use a chamber containing the transducers, providing mechanical protection and better control on the sampling, and measuring conditions. However, these systems require some type of device to force the air through the sampling chamber, usually a vacuum pump or a small ventilator. Additionally, this approach tends to provide slower response times, as the volume of the chamber and the inlet tube act as a low-pass filter to the whole system.

9.3.1 Gas sensing

There are several sensing technologies that have been used in MRO. Some, such as quartz crystal microbalances (QCM) or conductive polymers, were used in the early works, but later their interest vanished. Others, such as metal oxide semiconductors (MOX or MOS), have been in use since the early beginning and are still some of the most popular ones. Table 9.1 lists some of the most common sensor technologies and their respective characteristics.

The use of signals from insect antennae was demonstrated in the 1950s in studies related with insect pheromones [31]. The procedure to implement and use such sensors is detailed in the literature [32], consisting in cutting both ends of a living antenna from an insect and implanting electrodes in its extremities [33] (see Figure 9.2). When the antenna detects odour, it generates a signal of a few millivolts, which should be amplified to become usable. To keep the sensor alive for longer periods (more than half an hour), the cut extremes should be inserted in a saline solution, such as a Ringer's solution. Motivated by the very high sensitivity, high selectivity and fast response time of biological olfactory systems, some researchers adapted this type of biological chemical sensor to robots and demonstrated olfactory navigation capabilities with them [34]. Myrick *et al.* [35] studied the possibility of using multiple insect antennae to identify different odours emanating from more than one point source, but the short life span of approximately one hour limits the application scenarios for this type of systems. Martinez *et al.* [36] provided a guide on how to integrate insect antennae in a mobile robot. Their proposed approach improved the operation time to approximately one working day. In Reference [37], the authors proposed a biosensor with EAGs capable of being attached to a small drone (named as bio-hybrid drone) that can operate on plume-tracking experiments. Despite the growing interest in this sensing approach, its real impact is not clear yet since all the experiments so far were limited to very short distances lacking further characterisations about the achievable limit of detection.

QCM are made from electronic oscillators, coated with a sensitive layer with affinity to a target gas. When the gas molecules bind to the sensitive layer, the mass of the whole oscillating structure increases, decreasing the resonance frequency.

Table 9.1 Characteristics of common chemical sensors

Technology	Sensitivity (ppm)	LoD	Response time (s)	Selectivity	Lifetime	Price
EAG	10^{-4}				80 min	
AGS	1 – 10,000	2 – 10 ppb	4 – 5	O_2, CO, SO_2 NO_2, NO, O_3	2 y	< 300€
MOX	0.1 – 10	50 ppb	1 – 2	VOCs	> 10 y	> 5€
PID	0 – 40	1 ppb	6 ms – 5	VOCs	< 5 y	> 550€
NDIR	0 – 5,000	2 – 20 ppm	2 – 120	SO_2, CO_2 CH_4	> 15 y	< 300€
TDLAS	40,000 ppm.m	1 ppm.m	<1	CH_4	–	> 10,000€

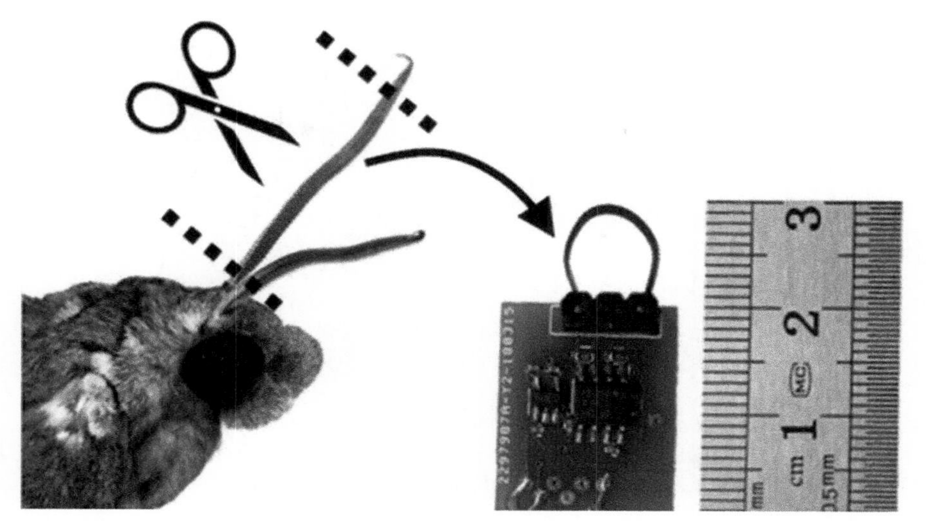

Figure 9.2 The electroantennogram (EAG) is composed of a single excised antenna and a custom signal conditioning circuitry [33]

This type of sensor was used in some of the early works [38], but maybe due to the lack of commercial availability and consequent inconsistency among the used sensors, its utilisation in MRO was abandoned.

A basic amperometric gas sensor (AGS) is composed by an impermeable case containing an electrolyte and two electrodes, called working electrode and counter electrode. The case is sealed by a membrane permeable to a target gas. The electrodes are maintained at a fixed electric potential. Every molecule of the target gas diffusing through the membrane produces a chemical reaction at the surface of the working electrode and a balancing reaction at the surface of the counter electrode, producing an electrical current. If the diffusion at the membrane is small enough to not saturate the chemical reactions at the electrodes, the intensity of the current will be proportional to the rate of molecules passing through the membrane, which is proportional to the concentration of the target gas [39]. These sensors are relatively sensitive (ppb range) and selective, being widely used to monitor the following gases: O_2, CO, NO_2, NO, O_3, H_2S, SO_2, NH_3, HCN, HBr, CS_2, Cl_2, H_2, HCl and HF. Their major drawbacks are a limited lifetime of about two years (the electrolyte is consumed in the chemical reactions) and a relatively high response time (more than 10 s) if we aim to use them to control a robot. Despite these limitations, these sensors are frequently used in environmental monitoring tasks with mobile robots. For example, Villa *et al.* [40] employed AGS on a unmanned aerial vehicle (UAV) to measure CO, NO and NO_2. These are also a regular sensor payload in the study of volcanic plumes [41].

Metal oxide semiconductors (MOX or MOS) are the most used gas sensors in MRO. As represented in Figure 9.3, these sensors are composed by a metal oxide layer (such as tin dioxide, SnO_2) heated at a relatively high temperature (150 to

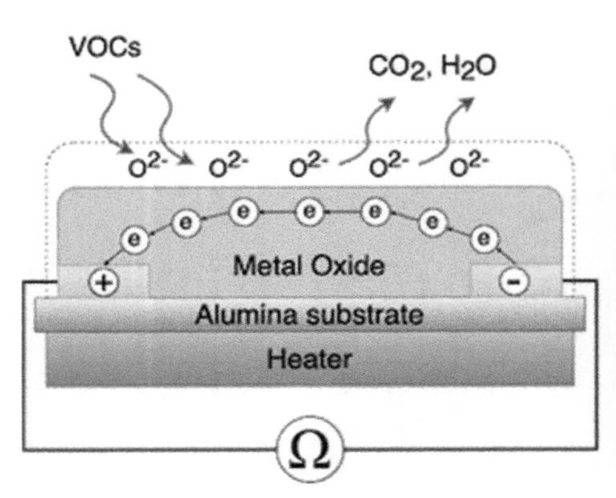

Figure 9.3 Metal oxide semiconductor technology. The left picture presents a diagram with the MOX sensor working principle. The right picture shows a small differential robot equipped with three MOX sensors.

500°C). At this temperature, the oxygen from the air is adsorbed on the crystal surface, reducing the concentration of charge carriers and increasing the electrical resistance. In atmosphere with oxidising gases, this reacts with the adsorbed oxygen and decreases the electrical resistance. The sensor's electrical resistance can be approximated by $R_s = A.C^{-\alpha}$, where A and α are characteristic constants and C is the gas concentration [42]. These sensors are very inexpensive, relatively fast (1-2 s) and sensitive (ppm range), being the type of chemical sensor more widely employed, since the early MRO works [43]. Their main drawbacks are the long-term drift and lack of selectivity. This problem can be mitigated by the implementation of electronic noses, using an array of different transducers, whose output can be processed to classify the sampled odours [44]. The array may be exposed to the atmosphere, in an OSS [45] or it can be arranged inside a sampling chamber [46]. Given the multitude of works on MRO done with these sensors, it would be hard to summarise them all. Lilienthal and his students did extensive work on concentration mapping using MOX sensors. The group studied multiple kernel-based interpolation algorithms to generate efficient concentration maps from the information captured by MOX sensors and ultrasonic anemometers [47]. A combination of MOX sensors and an RGB-D camera attached to an indoor mobile robot is proposed Monroy and his colleagues [48]. Burgues *et al.* [49] implemented MOX sensors on a small nano-quadcopter for OSL. The platform was tested in indoors and the effect of propulsion on the sensors signals was analysed.

All gas molecules exhibit a characteristic absorption spectrum, which may be exploited for their detection. Optical gas sensors explore this interaction of gas molecules with light, being potentially very selective, fast and with very low drift, making them an excellent choice for MRO. There is a wide variety of these sensors, but only three technologies have been used in this field: photoionization detector (PID), non-dispersive infra-red detectors (NDIR), and tunable diode laser absorption spectroscopy (TDLAS).

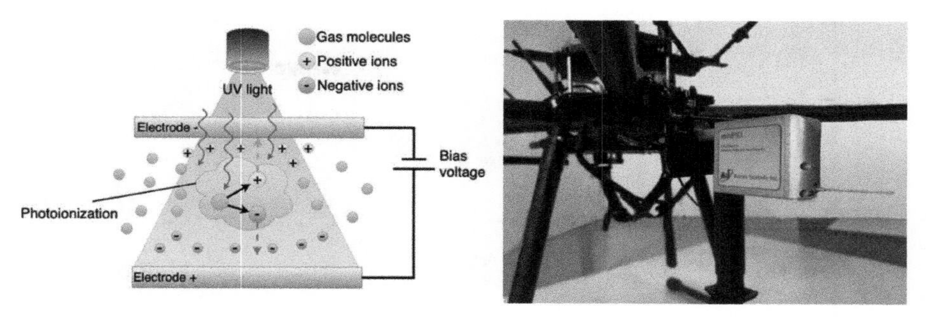

Figure 9.4 *PID sensor technology. The left picture presents the working principle of a PID sensor. The right picture presents a drone, equipped with an Aurora Scientific miniPID.*

A PID is composed by a sampling chamber containing an ion detector and a deep ultraviolet light (see Figure 9.4). When a gas molecule, with ionisation potential lower than the energy of the emitted photons, enters the chamber, it may absorb a photon and become ionised, being detected by the detector. This sensor can be extremely fast (few milliseconds), with a time response limited by the dynamics of the sampling chamber, which depends on its volume. Its major drawback is the lack of selectivity since any ionisable molecule will be detected [50]. Nevertheless, PID sensors are frequently used to characterise the total content of volatile organic compounds (VOCs) in a given site [51], or to study the transport of odours in the atmosphere [52], or in MRO, in works accounting for the plume dynamics [53], including the internal intermittency [54] or to characterise the response of other gas sensors [55].

NDIR uses a sampling chamber containing a broadband light source (usually a miniature incandescent light bulb) and a pair of photodetectors with narrowband spectral filters: one tuned to a spectral band of absorption of the target gas (active channel) and the other (reference channel) tuned to a spectral band where no absorption is expected to occur (see Figure 9.5). The sensor uses the Beer–Lambert law to relate the readings from both channels with the gas concentration [56]. NDIR sensors are very selective and relatively fast (sub-second). The sensitivity of these sensors depends on the length of the sampling chamber, but normal commercial devices are sensitive in the ppm range. These sensors have been used by Zhou *et al.* [57] to verify the vapours expelled from large ships, using an UAV equipped with an AGS sensor, measuring SO_2 , and an NDIR measuring the CO_2 concentration. Watai *et al.* [58] employed NDIR sensors with an UAV kite plane to measure CO_2 levels in the atmosphere, up to 3,000 m of altitude. The system has a response time of approximately 20 s and an accuracy of ± 0.26 *ppm*. Sha *et al.* [59] also made use of a NDIR sensor with an UAV to measure methane emissions from a controlled release. This work aimed to evaluate the performance of the sensor while measuring from an airborne platform.

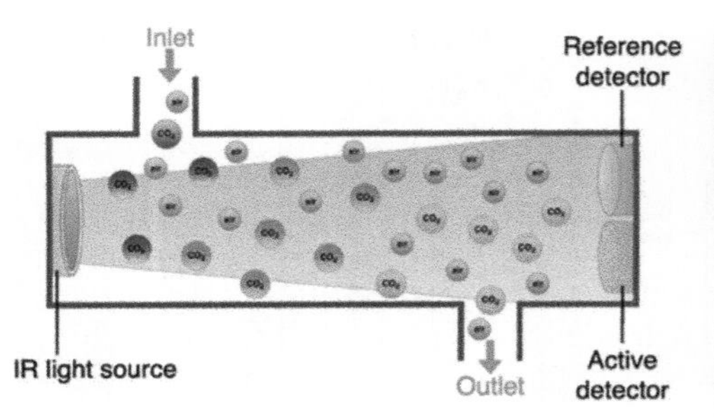

Figure 9.5 NDIR sensor technology. The left picture presents the working principle of a NDIR sensor. The right picture presents a SprintIR-W NDIR sensor.

TDLAS sensors expand the concepts employed by NDIR. These sensors use a tunable wavelength laser source and a photodetector [56]. These sensors operate either with a sampling chamber, where the laser and photodetector are arranged similarly to the NDIR, or in a stand-off arrangement, where both elements probe the environment (see Figure 9.6). The laser scans the emitted light in a band around an absorption line of the target gas and the photodetector provides a signal corresponding to the absorption spectra in the scanning band, providing readings with the peak of absorption and readings with a reference level. For a system using a sampling chamber, the concentration is obtained similarly to the NDIR. For the stand-off system, the sampling length depends on the distance to the reflecting surface and should be accounted to calculate the average concentration in the measuring zone. By providing remote concentration measurements, these sensors are particularly interesting to MRO, but their expensive price limits a wider utilisation in this field. Most of the implementations that use TDLAS for MRO use large platforms, able to operate in industrial environments [60] and outdoor scenarios [61]. Recently, smaller and light-weight TDLAS sensors appeared on the market, driving the interest to adapt this sensor to agile platforms, able to quickly scan the environment from multiple

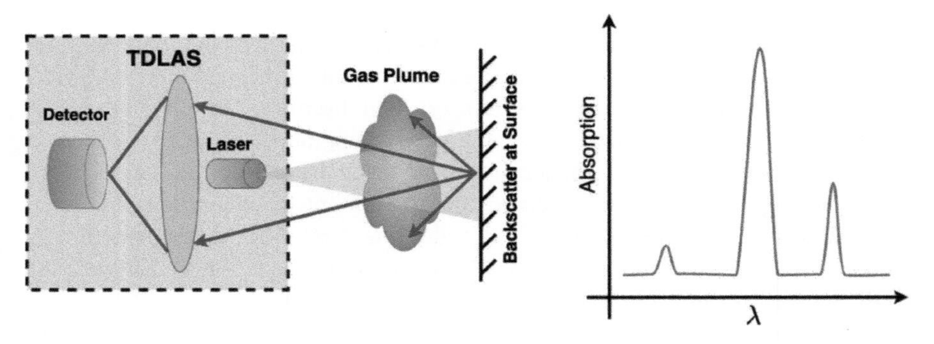

Figure 9.6 Principle diagram of TDLAS sensor

directions and supporting the use of tomography-based methods to estimate gas distributions from multiple integrative measurements [62].

9.3.2 Flow sensing

As seen in section 9.2, flow plays a fundamental role in the dispersion of odour across the environment and consequently, perceiving flow is also fundamental to find odour traces and to track odour plumes to their source. This statement is reinforced by biology, observing that animals highly successful in tracking odour plumes are also those with advanced capabilities to sense flow across their bodies. Common ways to perceive this flow are sensing the deflection in salient hairs, or sensing pressure differentials across their body, as used by some fishes to feel the water flow [63], or feeling increased cooling in areas facing stronger flows.

Mechanical flow sensors explore the drag forces produced by a flowing fluid onto a resisting surface. Examples of such sensors are the rotating cups anemo meter and wind vane used in traditional weather stations to measure wind speed and direction. Another simple example is observing the deflection from vertical in drag resistive loads, such as a light-weight suspended sphere or a helium-filled balloon. This approach served as an inspiration to Chapman *et al.* [64], who implemented anemotaxis navigation with a robot containing an anemometer based on a vertical spring surrounded by four metallic poles, forming an array of electric switches. The frequency of contacts of the spring on each pin provided an estimate of the wind direction. In a more sophisticated approach, Russell [65] implemented an anemometer based on an electric motor actuated, rotating vane. The current consumed by the motor provides a measure of its torque and consequently provides an estimate of the directions of highest and lowest drag, which is related to the wind direction and intensity. More recently, Macedo *et al.* [66] used a wind vane on the top of a small mobile robot to measure the wind direction relative to the robot in a ventilated arena. The major problems with these mechanical wind sensors used in MRO are being bulky and not sensitive to low-intensity flows. There is extensive research on small mechanical anemometers, usually using micro-electro-mechanical systems (MEMS), claiming high performance, but these sensors are only available in research laboratories and have not yet been used in MRO works [67].

Pressure flow sensors use the Bernoulli equation, which relates pressure with velocity in a fluid to measure velocities from measured pressures. A typical example of such sensors are Pitot tubes, which provide pressures at two orifices, one located in an area of null velocity and another in the area where we want to measure the velocity. In this case, the pressure differential provides a direct way to calculate the unknown velocity. This concept is sometimes expanded to arrangements measuring multiple pressures around a structure to estimate multi-dimensional velocity vectors [68]. Differential pressures are appealing to high-speed flows or high-density fluids, such as water. Otherwise, it lacks sensitivity to be used in MRO.

Thermal flow sensors explore the dependence between power dissipation from a heated element and the speed of the surrounding fluid. Power dissipation varies approximately with the square root of the velocity, so these sensors are very sensitive at low speeds [69]. Additionally, although they can be easily purchased on the market, they are also easy to implement, and several researchers have built their own anemometers using this principle. Thermal anemometers are frequently implemented with thin hot wires or with bead-shaped thermistors. Hot wires are sensitive to the flow direction, so they can be used to build a directional anemometer arranging multiple wires in orthogonal directions. On the contrary, bead-shaped thermistors are approximately spherical and consequently insensitive to the flow direction. In this case, a common approach to extract the wind direction consists in arranging an array of such elements around a wind deflecting pillar and processing the signals provided by that array to estimate the wind intensity and direction, as can be seen in Figure 9.7 [70]. Thermal anemometers have been used since the early days of MRO. Ishida *et al.* [72] and Marques *et al.* [70] combined gas sensors and directional anemometers to implement what was called odour compasses.

Ultrasonic anemometers explore the dependence on the speed of propagation of acoustic waves in a medium with the velocity of that medium. These sensors measure changes in the propagation time across a fixed length path to calculate the speed of the fluid in that path. Each measuring dimension needs a pair of ultrasonic transducers, and the sensitivity of the measurement is proportional to the length of propagation; consequently, these sensors are bulky and relatively expensive (€500 to €3000). Despite these problems, they are very accurate in the range normally employed in the environment, being frequently used in MRO. Osório *et al.* used a 2D anemometer to track 3D plumes in an indoor environment [71].

Figure 9.7 Thermal and ultrasonic directional anemometers employed in MRO. On the left a Nomadic Super Scout robot with a MOX-based electronic nose and four self-heated thermistors around a wind-deflecting structure [70]. On the right, an Erratic robot with an array of four gas sensors at different levels and a 2D ultrasonic anemometer, used for 3D plume tracking [71].

9.4 Odour source localisation

OSL is the process of finding the position of an odour source. As shown in Figure 9.8, this process is usually carried out with one or multiple robots searching a given space for traces of an odour and tracking the corresponding plume to its source [38].

A related process, called STE, uses a plume model and environmental measurements taken inside a plume's active area to estimate the parameters of its source, including its position. These measurements may be acquired by a fixed sensor network [73, 74] or by one or multiple mobile agents actively sampling the plume [11]. In the latter case, the STE can be integrated in the plume-tracking phase to speed up the localisation process. Some STE methods are probabilistic and provide the uncertainty of the estimation and, consequently, the uncertainty about the location of the odour source.

Mapping odour concentrations is another class of methods that can be used to estimate the location of odour sources [49, 53, 75]. This approach exploits the nature of dispersion phenomena and its implications in terms of odour distribution, which implies that a source is located in the place of higher odour concentration. As a result, a concentration map can be analysed to find regions with high probability of containing sources. A mapping process requires extensive sampling of the environment. Consequently, it takes longer than implementing a process which explicitly aims to find the odour source by sampling only the regions of high expected interest.

Regardless of the approach, from a robotics standpoint, any robot will be always performing the tasks represented in Figure 9.9, consisting in sensing the environment at their current position and evaluating the acquired data to decide if the goal was reached (i.e., if the source was found) or if it is necessary to move to another position that may provide better insight about the source location.

9.4.1 Searching odours

The first phase in an OSL process is searching for traces of the target odour across the environment. In a typical scenario, the searching agent needs to cover a significant part of the environment before it comes into contact with an odour plume.

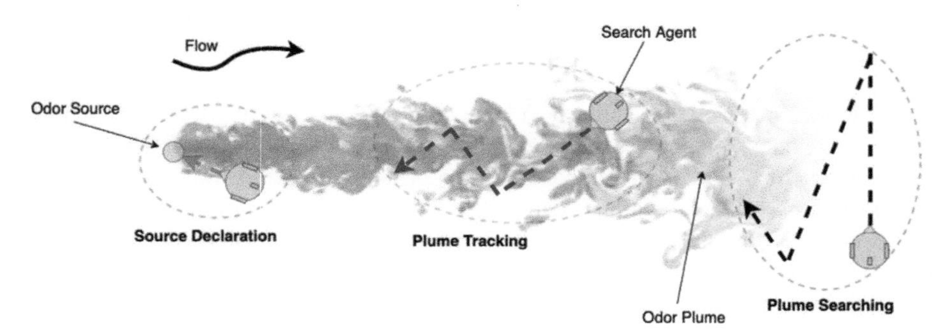

Figure 9.8 Phases of an OSL process: plume searching, plume tracking and source declaration

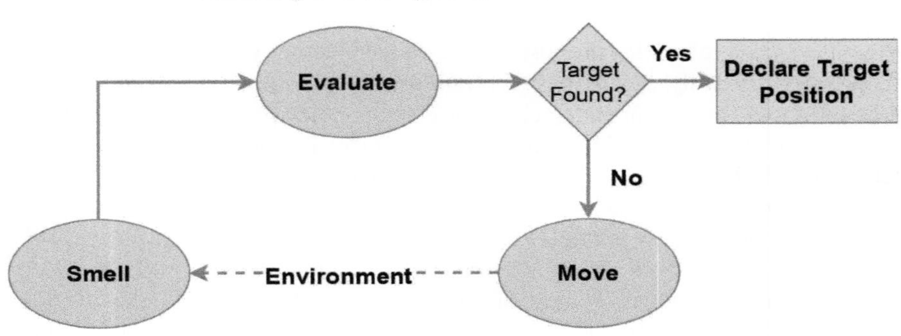

Figure 9.9 Task sequence for MRO

In such situations, the fundamental question is: *What is the best strategy to sample the environment while no odour clue was detected?* This type of problem was first studied during World War II, to plan strategies for finding submarines in the sea and later to plan rescue missions, resulting in several textbooks on the theory of optimal search [76, 77]. This theory considers the properties of the used sensor, namely, its probability of detection (PoD) and detection width (DW) to find optimal trajectories to search targets across a given area. These optimal trajectories are usually sequential sweeping patterns, separated by a distance related by the PoD and DW of the employed sensor. In the case of olfactory search, it would make sense to consider the active area of an odour plume to determine the optimal searching trajectories. Marjovi and Marques formalised this problem and proposed optimal formations to detect odour plumes, both with a network of stationary sensors [78] and with a moving robot swarm [79]. The problem of searching odour plumes has been extensively studied in biology and ecology to understand and model foraging behaviours [80, 81]. This community observed intermittent searching patterns with random long runs mediated by periods of short random motions around small clustering areas. It has been proposed the hypothesis that these patterns follow a Lévy distribution, being modelled by what is called Lévy-walks or Lévy-flights. These pseudo-random motion patterns found in nature may be justified by the distribution of natural resources, which are typically arranged in clusters, or they may be due to the very low PoD inside the active area of an odour plume, resulting from the mechanisms of odour spreading across the environment. In such searching conditions, the receiver benefits from staying longer at a given place, as the probability of detecting an odour patch increases with the sampling time. The small PoD of these processes also means that the increased performance of optimal search patterns against random ones decrease, meaning that random trajectories, without the need for global localisation or mapping capabilities, may be a good option for simple searching agents. These patterns have been employed in the searching phase of previous OSL works, including during the initialisation of meta-heuristic-based methods, such as References [23, 75]. Pasternak *et al.* [82] adapted elements from both Lévy-walks and correlated rndom walks to model the search and plume-tracking phases and proposed an OSL algorithm called Lévy-taxis. The method assumes spatial correlation between the chemicals and the fluid flow, and exploits this correlation to guide the agent in the upwind direction when no chemical is detected. This approach was later improved by Emery *et al.* [83] in what they called Adaptive Lévy-Taxis (ALT) by adding short-memory capabilities to the agent. The actual decision is a combination of the previous action with the current flow and concentration gradient measurements, providing a more robust search strategy.

9.4.2 Tracking odour plumes

The tracking stage is the most challenging of the three tasks, and that is why it is also the one to which scholars dedicate more efforts. The absence of smooth gradients makes it difficult to rely solely on different concentration levels. The intermittent nature of the plume results in a constant change between detection and non-detection events, while the unstable nature of the flow adds further complexity to the task as it leads to a highly dynamic environment that is constantly changing.

9.4.2.1 Reactive strategies

Due to the ability of animals to successfully locate odour sources, most of the existing plume tracing methods are inspired by their behaviours, which are designed to work under specific conditions. One of the most important environmental conditions is the strength and stability of the medium flow. In environments where the flow is weak or non-existent, biological organisms employ chemotactic strategies, which use information regarding the chemical gradient. One such behaviour is the biased random walk observed in bacteria and other small organisms [84]. In environments containing strong winds, animals typically take advantage of the medium flow information to guide their search process [85]. A popular anemotactic approach is inspired by the behaviours of the male silkworm moth while tracking a trail of pheromone [45].

9.4.2.2 Meta-heuristic search

Meta-heuristics are optimisation algorithms that have been used extensively for MRO. The OSL problem is formulated as finding the best solution according to an objective function. The chemical measurements are used as observations to evaluate the predicted concentrations, generated from a dispersion model such as the Gaussian plume model, and minimise the cost function. A typical cost function is the minimisation of the sum of squared residuals of the observation and predicted values:

$$\mathrm{R} = \sum_{i=1}^{N} \left(c_i - c_i' \right)^2 \tag{9.5}$$

where c is the measured concentration and c' is the predicted concentration value. Popular meta-heuristics such as the genetic algorithm (GA) have been proposed to locate multiple odour sources scattered across large areas [22]. The simulated annealing algorithm (SAA) has been employed to estimate the parameters of a Gaussian plume model from observations values obtained with a single mobile agent [86]. While the GA draws inspiration from natural selection and genetics, the SAA takes inspiration from the annealing process of metals. Different algorithms can also be combined into a single method to improve the inference by tackling their individual limitations [87].

Population-based algorithms are often considered for swarm applications where typically each robot is modelled as a hypothesis of the overall algorithm and the measured concentrations are used to compute the fitness values that guide the search process towards the locations with high odour concentrations [23]. The performance of popular population-based algorithms such as particle swarm optimisation (PSO), bacterial foraging optimisation (BFO) and ant colony optimisation (ACO) has been evaluated by Turduev *et al.* [75] with a group of mobile robots operating on controlled scenarios.

9.4.2.3 Probabilistic methods

Probabilistic methods use a model of odour dispersion and iteratively estimate the source parameters by minimising the error between the environmental observations and those obtained with the candidate parameters. One of the advantages of probabilistic-based methods lies in the estimation of the uncertainty value that provides an assessment of how certain the results are.

One of the first probabilistic approaches for OSL with mobile agents was proposed by Farrel *et al.* [88]. It considered hidden Markov models with a discrete grid of the search space and binary detection events. This algorithm generates two distribution maps: an odour source distribution map from which the source location is inferred and a plume distribution map to plan the optimal path.

Later, Infotaxis was proposed by Vergassola *et al.* [89] as a new search method based on information-theoretic concepts that considers detection rates to approximate the posterior probability of the odour source location according to the observations. This information-driven method was designed to search for an odour source in complex environments, with sporadic odour encounters and relies on Bayesian estimation to approximate a belief of the source location, considering the history of binary detections during a time interval. A movement decision-making process is employed with the objective to minimise the uncertainty level quantified by the Shannon entropy.

As new information becomes available, the entropy level is attenuated, eventually reaching zero when the source location is found. Using a grid-based probability map can be computational demanding for large spaces as probability updates need to be computed for every grid cell. Different approaches have been proposed to tackle this limitation. One of such approach consists of approximating the estimated posterior probability with different Gaussian terms instead of using a grid map, along with the free energy criterion instead of the entropy on planning stage [90], as to efficiently minimise the entropy level, a precise dispersion model is required. Other approaches using sequential Monte Carlo methods have been gaining increased relevance (e.g., using a particle filter) to approximate a belief map of the source location from a tractable amount of particles, where each particle is a hypothesis of a source location [91]. This method was later enhanced [92] to account for multi-robot scenarios instead of sharing the full set of particles, each robot only shares the mean and covariance matrix of the Gaussian density and uses the Kullback–Leibler divergence to measure the difference between the shared and local data. This results in a more robust strategy that could potentially be used in a scenario considering multiple sources.

Partially observable Markov decision processes (POMDPs) are planning frameworks for uncertain environments with partial observability. They provide an adequate platform for MRO as the source location is usually unknown, and the environment as well as the robot localisation have a substantial uncertainty associated. A POMDP requires three main components to operate: an information state obtained from a Bayesian framework that estimates the source parameters, a set of admissible actions guiding the agent to the next location and a reward function corresponding to the expected knowledge gain from the possible actions. An interesting feature is the flexibility that POMDPs offer allowing different belief representations, reward functions or encoding additional source parameters besides the location [93]. A proposed POMDP method called Entrotaxis [94] employs a particle filter to update the posterior distribution of the source parameters and adopts the concept of maximum entropy for the navigation reward by choosing actions that guide the searcher to the locations with highest uncertainty. Other approaches employ a Metropolis–Hastings algorithms to estimate the source parameters and the relative entropy also known as the Kullback–Leibler divergence to reward the selected actions [95]. This results in a motion guided by the weighted sum of the highest information value (given from the Kullback–Leibler divergence) and the maximum a posteriori value of the source position. A higher exploratory behaviour is achieved in the beginning of the search, and an increasingly higher exploitative behaviour as the agent moves towards the source location. POMDPs can also be used in a distributed fashion where each unit individually processes a POMDP and with some cooperation strategy (e.g., policy auctions [96]) shares data with the others to improve the search process. Since exchanging information is the main source of cooperation, it is extremely relevant to determine what data should be exchanged between the team to optimise the time and effort needed in pinpointing the source location [97].

While frequently used as a tool within other probabilistic search algorithms (e.g., to represent a belief of the odour source), particle filters can also model OSL algorithms. Previous works employed a particle filter estimation algorithm in parallel with a bio-inspired plume tracing strategy that ends when the estimation converges to a certain position [98]. Other approaches combine the results from the particle filter algorithm with reactive behaviours of bio-inspired tracking methods to improve the agents' guidance towards the source location [99]. A great amount of research work focus on estimating the reduced number of parameters of the chemical plume and often rely on binary data. In Reference [100], a particle filter-based method was proposed to estimate a great number of parameters of the Gaussian plume model. Instead of initially spreading the hypothesis throughout the search space, the filter is initialised with a prior estimation process, which roughly estimate the parameters of the dispersion model from a set of environmental measurements, collected with a bio-inspired navigation strategy. These rough estimates restrict the particles' search space, improving the quality and speed of the estimation.

9.4.3 Source declaration

The last stage of an OSL process consists in pinpointing the position of the source. In biology, this is frequently accomplished with the searcher moving sufficiently close to the source, to be able to identify it with complementary sensors, such as vision. In the case of simpler organisms, such as bacteria, the source declaration can be made through its motion to the area of highest concentration, which happens to be the source. In robotics, all the previous approaches are valid. Additionally, we may also consider STE as localisation methods, since these provide an estimate for the source position (usually with a given uncertainty associated).

Most plume tracking works disregard this last phase of OSL, usually considering that the source is found when the searching robot passes closer than a given distance from the source, which in this case would be identified by a complementary sense, such as vision.

The works addressing specifically this phase are scarce and usually rely on the use of the divergence operator. In Reference [22], Marques *et al.* updated a concentration grid map and classified a given cell as an odour source when it contained some concentration, and all of its upwind neighbours have no odour. Later, this approach evolved from the application of the divergence operator to an odour concentration map [101]. Figure 9.10. shows the concentration map measured by a group of small robots exploring an area with an odour source. The intermittency observed on that map difficult the localisation of the source, but after smoothing the measurements by Kriging [102] and applying the divergence operator, the resulting map clearly shows the area of higher divergence (Figure 9.10.), which corresponds to the location of the source. This work was later extended with the concept of virtual cancellation plume (VCP) to deal with regions containing multiple sources, possibly with overlapping active regions [103]. Developing on VCP [104] proposed another multiple source localisation method, based on ACO. This method was validated in simulated scenarios containing multiple sources in challenging relative positions.

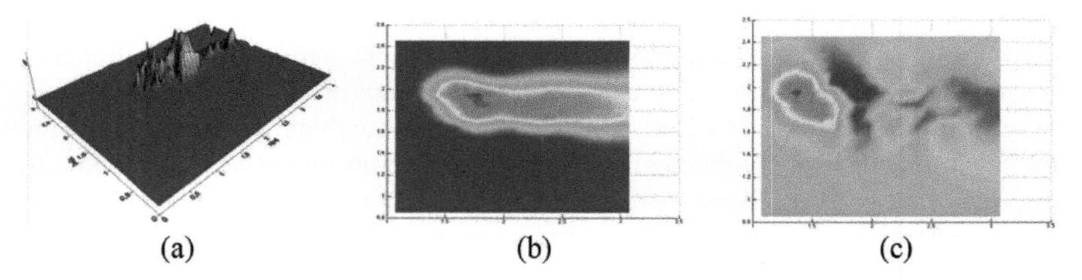

Figure 9.10 Use of the divergence operator to localise an odour source in a concentration map [101]

In a completely different line of thought, some researchers combine olfaction with vision in order to locate odour sources. Ishida *et al.* [105] use these modalities in a small mobile bot controlled by a behaviour-based subsumption architecture to search for coloured odour sources (bottles) inside a homogeneous, almost empty room. When no odour is detected, the robot wanders through the environment. Conversely, when an object is detected, the robot approaches it to verify its smell. If an odour is sensed, the robot tracks the corresponding plume until finding the coloured object releasing the gas. In a more elaborated, but somehow contradictory approach, if we consider olfaction as a long-distance sense, Monroy *et al.* [48] used vision to detect potential odour sources in an environment. After identifying a potential source, the robot moves closer and smells it to check if it is leaking a gas.

9.5 Learning in mobile robot olfaction

Animal behaviours can be divided into two types: innate and learned. Innate behaviours are not acquired but inherited. The migration of birds can be considered as an example of innate behaviour. On the other hand, learned behaviours are acquired by observing others and through experience. Robots may adapt their behaviours through artificial intelligence (AI) methods. Machine learning (ML) and evolutionary computation (EC) are the two main fields of AI that aim to produce knowledge from data, differing in the methods used. ML relies on statistical techniques while EC is composed of search heuristics loosely inspired by the principles of evolution by natural selection and Mendel's genetics. Despite the different methods, both produce knowledge that is used to map inputs into outputs. In robotics, the knowledge produced often takes the form of controllers, which map the states of the robot into appropriate actions. Due to the similarity of the goals of both fields, in this document, we shall often use the terms learning and adaptation to refer to both EC and ML approaches.

There are many applications for learning in robotics. At a lower level, the sensors provide information that must be perceived, and knowledge should be extracted from it in the form of features. At higher levels, the robot may learn strategies that use the perceptions of the environment, together with the current and past states of the robot to decide which action is the most advantageous for achieving its goals. In the case of MRO, learning may be applied to evolve policies for the robots or to estimate the parameters of gas distribution models (among which there is typically the position of the odour source).

9.5.1 Source-term estimation

In recent years, the popularity of deep learning has greatly increased and so has its application to various domains. An example of such applications is gas source localisation, where deep long short-term memory neural networks (LSTM) are used to predict the most probable location for an odour source [106, 107]. These works rely on discretising the environment, and based on the samples from sensors placed over the environment, the LSTMs are trained to output the most probable location for the odour source. Contrarily to these approaches, Thrift *et al.* [108] do not assume that the source is inside a mesh of sensors. As a result, they attempt to predict the direction to the odour source, rather than its location. The authors developed an odour compass based on surface-enhanced Raman scattering sensors, whose output is given to various machine learning models to compute the direction of the odour source. Their results showed that convolutional neural networks (CNN) and support vector machines (SVM) attained the best results.

9.5.2 Policy search

Most OSL works involving learning focus on producing policies, i.e., controllers for the robots. This section overviews ML and EC approaches to learn robotic policies.

9.5.2.1 Machine learning

ML is a sub-field of AI that uses several methods to draw knowledge from data, creating models that map inputs into outputs [109]. These techniques have several applications in robotics, such as computer vision for navigation, localisation and goal recognition, motion control and pattern recognition from sensor arrays, including arrays of chemical sensors employed to build electronic noses (eNoses), or systems composed by chemical and flow sensors, which output the direction towards an odour source as an odour compass [110]. One area of ML is reinforcement learning (RL) [111], which enables an agent to adapt its behaviour by interacting with its environment and receiving feedback from it. The goal of RL is to find the optimal policy π that selects the action that maximises the reward for each state, leading to the maximum cumulative reward over an entire episode. Many RL works in robotics use artificial neural networks (ANNs) as the controllers [112]. Ducket *et al.* [113] carried out one of such works, where a recurrent neural network (RNN) was trained to act as an odour compass. The RNN took as input the time-series of gas sensor readings and outputted the probable direction to the odour source. A small network was manually designed, having a single hidden layer with ten neurons. The network was trained with back-propagation and the training data were acquired in the real world, in 10° intervals. Later, Farah and Ducket [114] improved this work, performing source localisation experiments with a mobile robot. More recently, Hwangbo *et al.* [115] trained an ANN to perform reactive control of a real quadcopter. They provided the network with information about the pose of the quadcopter, as well as its angular and linear velocities, and the network outputs four real values, corresponding to the velocities of the propellers. The authors performed the training in simulation, outperforming other methods from the state-of-the-art. The best policy found was able to transfer well to the real quadcopter, being able to stabilise the robot after being manually thrown by a human operator and to perform way-point tracking.

Q-Learning [116] is a popular reinforcement learning algorithm used for developing robotic controllers for Markov decision processes. One of the problems of applying Q-Learning to robots is related to the state space. The states of traditional Q-Learning approaches are discrete, whereas the states provided by the robots' sensors are usually best expressed by vectors of continuous values. A common approach to cope with continuous state spaces is to approximate the value function with an

ANN. Chen *et al.* [117] trained a deep neural network to approximate the value function of Q-learning to act as the robotic controller to locate an odour source. The network takes a stack of historical frames, containing odour detection and wind information, and outputs the expected future cumulative reward for each possible action. Despite the novelty of this method, further experiments need to be carried out, as it was only tested in a 10\:m x 10\:m simulation environment, discretised into a grid with 1\:m x 1\:m cells. Duisterhof *et al.* [118] also proposed to train a Deep Q-Network to tackle a source localisation task with a nano-drone. They perform online and on-board training of a Deep Q-Network that controls the drone in exploring the environment and seeking for a light-source while avoiding obstacles. The network receives as inputs the signals of four proximity sensors (with a maximum range of 5\:m) and two source-term sensors which provide estimates for source gradient and strength. The network outputs one of three actions: moving forward or rotating left or right.

The previous approaches use no memory. However, using memory for OSL may be beneficial, for instance, for remembering where and when odour traces were detected. RNNRNN and LSTM networks are special types of ANNs that make use of memory. Hu *et al.* [119] propose to train LSTMs through a reinforcement learning algorithm known as deterministic policy gradient (DPG). Their goal is for the LSTM to guide an autonomous underwater vehicle (AUV) tasked with tracking a chemical plume. They model the problem as a POMDP with continuous state and action spaces and trained the LSTMs in a purpose-built simulator. The method consists of an actor-critic approach [120], where the actor network takes as input the state perceived by the robot and outputs the action to be performed. In turn, the critic is trained to approximate the value function, for predicting the expected reward for the action chosen by the actor in the current state. Wang *et al.* [121] also model the OSL problem as a POMDP. More specifically, their approach to plume tracking can be split into two processes: modelling and planning. In the modelling process, the belief states of the POMDP are used as a probability map for the location of the odour source. In the planning process, the robot selects its trajectory based on a reward function that takes into account both the source probability map as well as a plume distribution map. A fuzzy controller is employed to identify the current search condition and balance the weights of the two maps to encourage either exploration or exploitation. A path planning method is then applied to find the optimal route, i.e., the route that maximises the information about the odour source.

9.5.2.2 Evolutionary computation

Evolutionary algorithms (EA) [122] are a family of stochastic search heuristics from the area of AI, inspired by the principles of evolution by natural selection and genetics. They can be divided into GA and genetic programming (GP), depending on whether they evolve the solutions for a problem or computer programs that produce those solutions. EAs have been successfully applied over the years to solve problems from the classes of optimisation, learning and design, which have no analytic solution, or where that solution would be too hard to be found. OSL is one of such problems, as it is impossible to completely and accurately model the entire environment and the interactions taking place within it, which influence the way odour propagates. Moreover, using robots, there is a constant presence of noise, both from their sensors and actuators, that the controllers must deal with.

Evolutionary robotics (ER) is an area of AI that focuses on using EC techniques for robotic applications. Examples of applications are the evolution of the shape [123] and of the controllers for one or more robots [124]. GAs have already been used in robotic odour search for estimating intermediate goals for the robots [125]. The method works by continuous evolving a population of candidate locations for the source. Those locations are encoded as vectors of continuous values (x,y). The evaluation of the candidate solutions is made by having the robots navigate to them, being the concentration measured at the target location used as fitness. They only used crossover, as the assignment of the locations to each robot and their navigation was considered to introduce enough randomness into the process. This method was able to perform well in simulation, in an environment without wind flow, being the odour dispersion dictated only by molecular diffusion. Later [22], evolutionary strategies (ES) were also applied to create the intermediate goals for the robots. ES are EAs that create new candidate solutions mainly through mutation. In the case of Reference [22], a directed mutation operator was proposed, which biased the new location with the upwind or crosswind directions depending on whether odour was sensed at the current location.

Infotaxis (Section 9.4.2.3) is a probabilistic method that uses a gas dispersion model to compute the probability map for the location of the odour source, which in turn is used in the decision-making process. Most works assume that the parameters of this gas distribution model are known, even though it is not trivial to select a set of values that accurately match a given environment. Macedo *et al.* [126] tackle this task by having a GA evolve the parameters of Infotaxis' gas distribution model for two different environments, showing that inaccurately modelling the environment leads to a significant loss in performance.

Genetic programming has also been used to evolve search strategies for OSL in the form of decision trees. Villareal *et al.* [127] evolved decision trees for controlling a robot, tasked with locating an odour source in a small, indoor environment. The controllers were evolved in simulation, being only the best individual tested in the real world. Later, Macedo *et al.* [128] proposed to evolve controllers for a group of mobile robots that attempt to locate a single odour source in a large, simulated environment. They use a hierarchical approach, where GP evolves decision trees that provide goal vectors for each robot, based on its perceptions and on those of its teammates. A potential field is used, at a lower level, to combine the goal vector with the obstacle avoidance behaviour. This method was compared to the silkworm moth algorithm, being faster and converging more accurately to the location of the source.

One of the main advantages of evolving decision trees is the ability to analyse the search strategies created. However, GP is often affected by bloat, meaning that there is an uncontrolled growth of the decision trees without a corresponding gain in performance, hindering a human's ability to inspect them. Macedo *et al.* tackled this problem with geometric syntactic genetic programming (GSynGP) [129]. This GP variant uses a novel crossover operator that performs geometric operations between two trees, implicitly controlling bloat without restricting the algorithm's search ability. GSynGP was later applied to evolve search strategies for the OSL problem with a single robot [66]. That work extended GSynGP to evolve decision trees with multiple symbols per node, i.e., a main symbol (a perception or action to perform) and a list of parameters (e.g., distance, termination criteria). The results showed that GSynGP evolved significantly smaller strategies than a standard GP algorithm, while having equivalent performance. Moreover, it was able to outperform the bio-inspired strategies from the literature. More recently [126], the performance of GSynGP was compared to Infotaxis. The results of that study showed that GSynGP can produce strategies with equivalent success rates to those of Infotaxis, while requiring significantly less time. Moreover, computational effort required by GSynGP is also significantly less than that of Infotaxis.

ANNs have also been used as robotic controllers in many works. The simplest approaches rely on having the experimenter define the topology of the neural networks and evolving only the connection weights. Such an approach was proposed by Heinerman *et al.* [130] who distributed an evolutionary algorithm over a population of six robots for performing online evolution of a foraging behaviour. All robots use the same network topology that is pre-determined by the experimenter. Thus, only the connection weights are evolved, in the form of real-valued arrays. Their approach consists of a hybrid algorithm, where each robot can evolve its controllers individually but also perform social learning by sharing its best controller with other agents. Croon *et al.* [131] followed a similar approach, where continuous-time RNNs were used to control a single robot to locate an odour source in a simulated environment with a turbulent plume. The topology of the network is hand-designed, being the connection weights and time constants evolved with a GA.

Although evolving fixed-topology ANNs require a simple representation, it relies heavily on the practitioners' experience to design the networks appropriately. Unfortunately, defining the optimal structure of ANNs is a hard task, and for that reason, some works have been proposed to simultaneously evolve the topology and connection weights of ANNs. Macedo *et al.* [132] followed that approach, evolving the topologies and connection weights of ANNs to control a robotic swarm in an OSL task. Each robot had its own ANN which took as input a set of attractive and repulsive forces and outputted the direction of the robot's motion. The proposed method was tested in simulation, being able to consistently move the swarm closer to the odour source than an *E. coli*-inspired chemotaxis algorithm.

9.5.2.3 Swarm approaches

Swarm robotics consist of employing groups of many simple robots that are unable to perform a given task on their own. However, working in cooperation, complex behaviours emerge [133]. The difficulty with using these systems lies in their design. They can either be hand-designed, where the designer engages in a trial and error cycle, specifying the individual behaviours and evaluating if the desired global behaviour is achieved, or they can be designed automatically, resorting to RL techniques or evolutionary algorithms [134]. The robots are often guided using potential fields that can be combined by ANNs. The adaptation generally focuses on the connection weights, being the architectures hand-defined with no quality guarantees. Macedo *et al.* [132] used a different approach, where an EA was used to simultaneously evolve the connection weights and topologies of the ANNs that combine the forces of the potential field.

Another work aimed at using a robotic swarm guided by DAPSO to locate multiple odour sources [103]. The decentralised asynchronous particle swarm optimisation algorithm (DAPSO) uses a population of particles, where each one represents a robot in the environment. DAPSO works by iteratively measuring the chemical concentration in the location of each robot and using that information, along with the perceptions of its neighbours, to move it closer to the source. This method is based on the original particle swarm optimisation algorithm [135]; however, it has been modified to take into account the restrictions imposed by real robots, such as the time spent moving between way-points.

9.6 Open challenges

Despite all the advancements in locating odour source, there are still plenty of issues to tackle. This section reports the open OSL challenges.

9.6.1 Artificial olfaction

Even with many new technologies emerging, the capabilities of mobile robots to sense the environment are still far behind to those of living organisms. Improving the sensitivity and response time of sensors is of utmost importance, while smaller and lower cost sensors would allow to enhance the robots' performance and aid in understanding how animals respond to certain stimulus.

Response time dictates the maximum speed at which a mobile platform can move while still obtaining spatially resolved measurements. Some studies demonstrated that odours have fluctuations widely distributed in time with a duration of 0.1 s and separated by 0.5s [136]. While sensors that sample at higher frequencies exist (e.g., PID or EAG), the lower cost and more frequently adopted MOX sensors cannot fulfil these requirements. Improving the response time of these sensors can enhance the performance of lower cost units, commonly adopted on multi-robot or swarm approaches.

Sensitivity levels vary according to the target odour and application at hand. As an example, humans can sense 110 ppt concentrations of amyl acetate, whereas dogs can detect as little as 11.4 ppt [137]. At larger scale scenarios such as environmental pollution, concentration levels are at ppm orders of magnitude, although for search and rescue applications the levels vary between 1ppm and sub levels of ppb [138]. Regarding the sensors, the ones with higher sensitivity ranges tend to have higher limits of detection (LOD) due to the larger structural dimensions. For OSL missions, the LOD is of high relevance since it determines the maximum downwind distance at which the chemical plume can be detected. The challenge is to employ micro-structures to reduce the detection limits and combining different materials to increase the sensitivity range.

Aside from measuring concentration levels of a certain odorant, identifying a particular chemical within a mixture of substances present in the environment is also relevant. The human olfactory system can express as many as 1,000 different olfactory receptors cells (ORCs) [139] while dogs have more than 220 million ORCs in the nasal cavity [140]. Dogs' immensely higher amount of ORCs is one of the reasons why they are so effective in search and rescue missions. Identifying different odorants requires a chemical decomposition process which takes a considerable amount of time. As such, improving sensors selectivity is a necessity and still an open subject. Recognition methods that consider temperature and humidity values or improving the sensors sampling chamber can contribute towards improving the identification of odours. Considering that sensors respond differently to chemicals, another alternative would be combining the signals of different sensors with a pattern recognition algorithm.

9.6.2 Odour source localisation

Compared with probabilistic and learning methods, research regarding bio-inspired strategies seems to be fading out, but the capacity of robots to search for odour sources is still very limited as opposed to the capacity of animals. There are still a great number of behaviours and techniques employed by animals that are not fully comprehended. Understanding those could lead to the development of improved strategies for mobile platforms.

Another subject is the growing interest in OSL strategies that consider a 3D space. Emitting sources of airborne plumes are usually on the ground. In marine environments, they are either at the bottom of the sea or near the water surface. So, it is still not clear if there is any practical interest in 3D OSL strategies or if they are just an academic challenge. Additionally, the negative effect caused by the propellers from mobile platforms such as UAVs, autonomous surface vehicles (ASVs) or AUVs needs to be overcome. The odour dispersion from these propulsion systems requires further understanding, and eventually, the adoption of less intrusive platforms may be a path to follow. Platforms such as blimps [14, 141] have already been tested and show promising results.

Most of the research is still focused on single-odour sources emitting at a constant rate. Dealing with dynamic emission rates or moving sources are still challenges that need to be overcome. One way would be to incorporate knowledge from multiple domains by combining terrestrial of aquatic platforms with aerial units. A surface vehicle could provide in-situ measurements while an airborne platform observes a broader field of view, searching for hints or regions that could aid the ASV to locate the source position. This combination of multi-domain platforms was already employed to track the front of riverine plumes [142].

A great number of the existing methods may not be robust enough for real-world scenarios as they make assumption that do not hold (e.g., the presence of static or mobile obstacles). These should be considered more often for future approaches.

Another limitation is the assumption of steady-flow conditions. While for terrestrial robots, the presence of flow sensors to measure the wind is common; for aerial or aquatic robots, it is not so frequent. This is due to UAVs having limited payload, requiring smaller and more expensive sensors, while Doppler velocity logs for aquatic platforms are also expensive.

9.6.3 Learning to locate odour sources

The main difficulty of applying learning techniques for OSL is the definition of a reward function, as it tells the learning algorithm what it should learn. Most works only evaluate whether their approaches reach the chemical source and on how much time they took. These are sparse reward functions, meaning that the robot needs to perform the entire task before receiving any feedback. Conversely, in traditional learning tasks, a controller is evaluated using dense reward functions, where a reward value can be assigned to each state-action pair. Dense reward functions construct smooth fitness landscapes, being easier to measure slight improvements in the controllers. Moreover, the dynamics inherent to the OSL environments lead to noisy and unfair fitness evaluations. This is particularly true when considering online learning, where the controllers are typically evaluated by having controlled the robot for a pre-defined time slot. As a result, consecutive controllers are not evaluated under the same conditions, being hard to compare their quality. A recent work [143] made the first step in solving this issue by designing and comparing various evaluation methods, with the goal of producing an evaluation function that accurately measures how well a given controller searched for a chemical source. Another way around this problem would be to convert it into a supervised learning approach, using a data set of desirable state actions to train the controllers.

Another difficulty in designing learning methods for OSL is devising a proper representation of the state of the robot and the world. Many of the existing methods for agent learning are designed for discrete states, which reduce the size of the state-space and in the creation of higher level policies. However, methods must be devised to convert the continuous signals produced by a robot's sensors to discrete states. The simplest approaches consist of discretising the features [144]. However, more powerful representations may be achieved by resorting to autoencoders [145] or CNNs [117].

The third learning challenge that we highlight is related with where the learning is performed. Due to the time and risks involved in learning on the real world, simulators are typically employed. However, simulators are abstractions of the real world, containing many simplifications that often lead to different results when moving simulation to reality. This phenomenon is known as the Reality Gap [146]. While there is yet no optimal solution to the reality gap, the best alternative is to evolve a suitable controller in simulation and carry on adapting it on the real robot [147]. While not ideal, such approach avoids spending too much time evaluating poor controllers, while tailoring a solution to the real hardware.

References

[1] Wyatt T.D. Pheromones and animal behavior [online]. 2nd ed. Cambridge University Press; 2014 Jan 23. Available from https://www.cambridge.org/core/product/identifier/9781139030748/type/book

[2] Cardé R.T. 'Navigation along windborne plumes of pheromone and resource-linked odors'. *Annual Review of Entomology*. 2021, vol. 66, pp. 317–36.

[3] Prada P.A., Chávez Rodríguez M. 'Demining dogs in colombia – a review of operational challenges, chemical perspectives, and practical implications '. *Science & Justice*. 2016, vol. 56(4), pp. 269–77.

[4] Jezierski T., Adamkiewicz E., Walczak M., *et al.* 'Efficacy of drug detection by fully-trained police dogs varies by breed, training level, type of drug and search environment'. *Forensic Science International*. 2014, vol. 237, pp. 112–18.

[5] Greatbatch I., Gosling R.J., Allen S. 'Quantifying search dog effectiveness in a terrestrial search and rescue environment'. *Wilderness & Environmental Medicine*. 2015, vol. 26(3), pp. 327–34.

[6] Jendrny P., Twele F., Meller S., *et al.* 'Scent dog identification of SARS-cov-2 infections in different body fluids'. *BMC Infectious Diseases*. 2021, vol. 21(1), pp. 1–14.

[7] Dunbabin M., Marques L. 'Robots for environmental monitoring: significant advancements and applications'. *IEEE Robotics & Automation Magazine*. 2001, vol. 19(1), pp. 24–39.

[8] Hwang J., Bose N., Nguyen H.D., Williams G. 'Oil plume mapping: adaptive tracking and adaptive sampling from an autonomous underwater vehicle'. *IEEE Access: Practical Innovations, Open Solutions*. 2001, vol. 8, pp. 198021–34.

[9] Marques L., de Almeida A.T. 'Apapplication of odor sensors in mobile robotics' in *Autonomous robotic systemrobotic systems*. London: Springer; 1998.

[10] Russell R.A. 'Survey of robotic applications for odor-sensing technology'. *The International Journal of Robotics Research*. 2001, vol. 20(2), pp. 144–62.

[11] Hutchinson M., Oh H., Chen W.-H. 'A review of source term estimation methods for atmospheric dispersion events using static or mobile sensors'. *Information Fusion*. 2017, vol. 36, pp. 130–48.

[12] Bayat B., Crasta N., Crespi A., Pascoal A.M., Ijspeert A. 'Environmental monitoring using autonomous vehicles: a survey of recent searching techniques'. *Current Opinion in Biotechnology*. 2017, vol. 45, pp. 76–84.

[13] Chen X.-X., Huang J. 'Odor source localization algorithms on mobile robots: a review and future outlook'. *Robotics and Autonomous Systems*. 2017, vol. 112, pp. 123–36.

[14] Burgués J., Marco S. 'Environmental chemical sensing using small drones: a review'. *The Science of the Total Environment*. 2020, vol. 748, 141172.

[15] Jing T., Meng Q.-H., Ishida H. 'Recent progress and trend of robot odor source localization' [online]. *IEEJ Transactions on Electrical and Electronic Engineering*. 2021, vol. 16(7), pp. 938–53. Available from https://onlinelibrary.wiley.com/toc/19314981/16/7

[16] Marjovi A., Marques L. 'Multi-robot odor distribution mapping in realistic time-variant conditions'. *2014 IEEE International Conference on Robotics and Automation (ICRA)*; Hong Kong, China, IEEE, 2014.

[17] Arya S.P. *Air pollution meteorology and dispersion*. New York: Oxford University Press; 1999.

[18] Seinfeld J., Pandis S. *Atmospheric Chemistry and Physics: from Air Pollution to Climate Change*. 3rd ed. New York: John Wiley and Sons; 2016.

[19] Carrascal M.D., Puigcerver M., Puig P. 'Sensitivity of Gaussian plume model to dispersion specifications'. *Theoretical and Applied Climatology*. 1993, vol. 48(2–3), pp. 147–57.

[20] Stockie J.M. 'The mathematics of atmospheric dispersion modeling'. *SIAM Review*. 2011, vol. 53(2), pp. 349–72.

[21] Leelőssy Á., Molnár F., Izsák F., Havasi Á., Lagzi I., Mészáros R. 'Dispersion modeling of air pollutants in the atmosphere: a review'. *Open Geosciences*. 2014, vol. 6(3), pp. 257–78.

[22] Marques L., Nunes U., Almeida A.T.D. 'Odour searching with autonomous mobile robots: an evolutionary-based approach'. *The 11th International Conference on Advanced Robotics*; University of Coimbra, Portugal, 2003.

[23] Marques L., Nunes U., de Almeida A.T. 'Particle swarm-based olfactory guided search'. *Autonomous Robots*. 2006, vol. 20(3), pp. 277–87.

[24] Nielsen M., Chatwin P.C., Jørgensen H.E., Mole N., Munro R.J., Ott S. *Concentration Fluctuations in Gas Releases by Industrial Accidents. Final report* [Technical Report]. Roskilde; 2002.

[25] Farrell J.A., Murlis J., Long X., Li W., Cardé R.T. 'Filament-based atmospheric dispersion model to achieve short time-scale structure of odor plumes'. *Environmental Fluid Mechanics*. 2002, vol. 2(1/2), pp. 143–69.

[26] Sutton J., Li W. 'Development of CPT_m3d for multiple chemical plume tracing and source identification'. *Seventh International Conference on Machine Learning and Applications*; San Diego, California, USA, 2008.

[27] Lochmatter T. 'Bio-inspired and probabilistic algorithms for distributed odor source localization using mobile robots' in *Odor source localization*. EPFL; 2010.

[28] Macedo J., Marques L., Costa E. 'A comparative study of bio-inspired odour source localisation strategies from the state-action perspective'. *Sensors (Basel, Switzerland)*. 2019, vol. 19(10), pp. 1–34.

[29] Cabrita G., Sousa P., Marques L. 'Player/stage simulation of olfactory experiments'. IEEE/RSJ International Conference on Intelligent Robots and Systems; IEEE, 2010.

[30] Monroy J., Hernandez-Bennets V., Fan H., Lilienthal A., Gonzalez-Jimenez J. 'GADEN: a 3D gas dispersion simulator for mobile robot olfaction in realistic environments'. *Sensors (Basel, Switzerland)*. 2017, vol. 17(7), p. 1479.

[31] Wu C., Du L., Zou L. 'Smell sensors with insect antenna' in PingW., QingjunL., Chunsheng W., JimmyH.K. (eds.). *Bioinspired smell and taste sensors*. Springer; 2015. pp. 77–102.

[32] Yamada N., Ohashi H., Umedachi T., Shimizu M., Hosoda K., Shigaki S. 'Dynamic model identification for insect electroantennogram with printed electrode'. *Sensors and Materials*. 2021, vol. 33(12), p. 4173.

[33] Anderson M.J., Sullivan J.G., Talley J.L., Brink K.M., Fuller S.B., Daniel T.L. 'The "smellicopter," a bio-hybrid odor localizing nano air vehicle' [online]. *IEEE/RSJ International Conference on Intelligent Robots and Systems (IROS)*; Macau, China, 2019. Available from https://ieeexplore.ieee.org/xpl/mostRecentIssue.jsp?punumber=8957008

[34] Kuwana Y., Shimoyama I., Miura H. 'Steering control of a mobile robot using insect antennae'. Proceedings 1995 IEEE/RSJ International Conference on Intelligent Robots and Systems. Human Robot Interaction and Cooperative Robots; 1995.

[35] Myrick A.J., Park K.C., Hetling J.R., Baker T.C. 'Detection and discrimination of mixed odor strands in overlapping plumes using an insect-antenna-based chemosensor system'. *Journal of Chemical Ecology*. 2009, vol. 35(1), pp. 118–30.

[36] Martinez D., Arhidi L., Demondion E., Masson J.-B., Lucas P. 'Using insect electroantennogram sensors on autonomous robots for olfactory searches'. *Journal of Visualized Experiments*. 2014, vol. 90, e51704.

[37] Terutsuki D., Uchida T., Fukui C., Sukekawa Y., Okamoto Y., Kanzaki R. 'Real-time odor concentration and direction recognition for efficient odor source localization using a small bio-hybrid drone '. *Sensors and Actuators B*. 2021, vol. 339, p. 129770.

[38] Russell R.A., Thiel D., Deveza R., Mackay-Sim A. 'A robotic system to locate hazardous chemical leaks'. *Proceedings of 1995 IEEE International Conference on Robotics and Automation*; 1995.

[39] Stetter J.R., Li J. 'Amperometric gas sensors – a review'. *Chemical Reviews*. 2008, vol. 108(2), pp. 352–66.

[40] Villa T.F., Salimi F., Morton K., Morawska L., Gonzalez F. 'Development and validation of a UAV based system for air pollution measurements'. *Sensors (Basel, Switzerland)*. 2016, vol. 16(12), p. 2202.

[41] James M., Carr B., D'Arcy F., *et al*. 'Volcanological applications of unoccupied aircraft systems (UAS): developments, strategies, and future challenges'. *Volcanica*. 2002, pp. 67–114.

[42] Ihokura K., Watson J. The Stannic Oxide Gas Sensor: Principles and Applications. Boca Raton, FL: CRC Press; 1994.

[43] Rozas R., Morales J., Vega D. '91 ICAR'. Fifth International Conference on Advanced Robotics 'Robots in Unstructured Environments; Pisa, Italy, 2002.

[44] Röck F., Barsan N., Weimar U. 'Electronic nose: current status and future trends '. *Chemical Reviews*. 2008, vol. 108(2), pp. 705–25.

[45] Marques L., Nunes U., de Almeida A.T. 'Olfaction-based mobile robot navigation'. *Thin Solid Films*. 2002, vol. 418(1), pp. 51–58.

[46] Larionova S., Almeida N., Marques L., de Almeida A.T. 'Olfactory coordinated area coverage'. *Autonomous Robots*. 2006, vol. 20(3), pp. 251–60.

[47] Lilienthal A.J., Reggente M., Trincavelli M., Blanco J.L., Gonzalez J. 'A statistical approach to gas distribution modelling with mobile robots-the kernel dm+ V algorithm'. *IEEE/RSJ International Conference on Intelligent Robots and Systems (IROS 2009)*; St. Louis, MO, 2009.

[48] Monroy J., Ruiz-Sarmiento J.-R., Moreno F.-A., Melendez-Fernandez F., Galindo C., Gonzalez-Jimenez J. 'A semantic-based gas source localization with a mobile robot combining vision and chemical sensing'. *Sensors (Basel, Switzerland)*. 2018, vol. 18(12), p. 4174.

[49] Burgués J., Hernández V., Lilienthal A.J., Marco S. 'Smelling nano aerial vehicle for gas source localization and mapping'. *Sensors (Basel, Switzerland)*. 2019, vol. 19(3), p. 478.

[50] Coelho Rezende G., Le Calvé S., Brandner J.J., Newport D. 'Micro photoionization detectors'. *Sensors and Actuators B*. 2019, vol. 287, pp. 86–94.

[51] Calderón J.B. 'Signal processing and machine learning for gas sensors: gas source localization with a nano-drone'. *TDX (Tesis Doctorals En Xarxa)*. 2021.

[52] Justus K.A., Murlis J., Jones C., Cardé R.T. 'Measurement of odor-plume structure in a wind tunnel using a photoionization detector and a tracer gas'. *Environmental Fluid Mechanics*. 2002, vol. 2(1/2), pp. 115–42.

[53] G. Monroy J., Blanco J.-L., Gonzalez-Jimenez J. 'Time-variant gas distribution mapping with obstacle information'. *Autonomous Robots*. 2002, vol. 40(1), pp. 1–16.

[54] Bailey J.K., Quinn R.D. 'A multi-sensory robot for testing biologically-inspired odor plume tracking strategies'. *Proceedings, 2005 IEEE/ASME International Conference on Advanced Intelligent Mechatronics*; Monterey, California, USA, 2005.

[55] Vincent T.A., Xing Y., Cole M., Gardner J.W. 'Investigation of the response of high-bandwidth MOX sensors to gas plumes for application on a mobile robot in hazardous environments'. *Sensors and Actuators B*. 2002, vol. 279, pp. 351–60.

[56] Hodgkinson J., Tatam R.P. 'Optical gas sensing: a review'. *Measurement Science and Technology*. 2012, vol. 24, p. 012004.

[57] Zhou F., Pan S., Chen W., Ni X., An B. 'Monitoring of compliance with fuel sulfur content regulations through unmanned aerial vehicle (UAV) measurements of SHIP emissions'. *Atmospheric Measurement Techniques*. 2019, vol. 12(11), pp. 6113–24.

[58] Watai T., Machida T., Ishizaki N., Inoue G. 'A lightweight observation system for atmospheric carbon dioxide concentration using a small unmanned aerial vehicle'. *Journal of Atmospheric and Oceanic Technology*. 2006, vol. 23(5), pp. 700–10.

[59] Shah A., Pitt J., Kabbabe K., Allen G. 'Suitability of a non-dispersive infrared methane sensor package for flux quantification using an unmanned aerial vehicle'. *Sensors (Basel, Switzerland)*. 2019, vol. 19(21), p. 4705.
[60] Bonow G., Kroll A. Presented at 2013 IEEE International Conference on robotics and automation (ICRA). Karlsruhe, Germany; 2006.
[61] Bennetts V.H., Schaffernicht E., Stoyanov T., Lilienthal A.J., Trincavelli M. 'Robot assisted gas tomography—localizing methane leaks in outdoor environments'. IEEE International Conference on Robotics and Automation (ICRA); Hong Kong, China, 2014.
[62] Neumann P.P., Kohlhoff H., Hüllmann D., *et al.* 'Aerial-based gas tomography – from single beams to complex gas distributions'. *European Journal of Remote Sensing*. 2019, vol. 52(sup3), pp. 2–16.
[63] Northcutt R.G. 'Swimming against the current'. *Nature*. 1997, vol. 389(6654), pp. 915–16.
[64] Chapman T., Hayes A.T., Tilden M. 'Reactive maze solving with a biologically inspired wind sensor'. *Proceedings of the 6th International Conference on Simulation of Adaptive Behavior SAB-00*; Paris, France, 2000.
[65] Russell R.A., Kennedy S. 'A novel airflow sensor for miniature mobile robots'. *Mechatronics*. 2000, vol. 10(8), pp. 935–42.
[66] Macedo J., Marques L., Costa E. 'Locating odour sources with geometric syntactic genetic programming'. *Proceedings of the International Conference on the Applications of Evolutionary Computation (Part of EvoStar)*; Online, 2020. Available from https://www.evostar.org/2020/evoapps/
[67] Tao J., Yu X. (Bill 'Hair flow sensors: from bio-inspiration to biomimicking—a review'. *Smart Mater Struct*. 2012, vol. 21(11), p. 113001.
[68] Prudden S., Fisher A., Marino M., Mohamed A., Watkins S., Wild G. 'Measuring wind with small unmanned aircraft systems'. *Journal of Wind Engineering and Industrial Aerodynamics*. 2018, vol. 176, pp. 197–210.
[69] Bruun H.H. 'Hot-wire anemometry: principles and signal analysis'. *Measurement Science and Technology*. 1996, vol. 7(10).
[70] Marques L., Almeida N., de Almeida A.T. 'Olfactory sensory system for odour-plume tracking and localization'. IEEE SENSORS 2003 (IEEE Cat. No.03CH37498); Toronto, Canada, 1999.
[71] Osório L., Cabrita G., Marques L. 'Mobile robot odor plume tracking using three dimensional information' ECMR; 2011.
[72] Ishida H., Kobayashi A., Nakamoto T., Moriizumi T. 'Three-dimensional odor COMPASS'. *IEEE Transactions on Robotics and Automation*. 1999, vol. 15(2), pp. 251–57.
[73] Murray-Bruce J., Dragotti P.L. 'Estimating localized sources of diffusion fields using spatiotemporal sensor measurements'. *IEEE Transactions on Signal Processing*. 1999, vol. 63(12), pp. 3018–31.
[74] Boubrima A., Bechkit W., Rivano H. 'On the deployment of wireless sensor networks for air quality mapping: optimization models and algorithms'. *IEEE/ACM Transactions on Networking*. 1999, vol. 27(4), pp. 1629–42.

[75] Turduev M., Cabrita G., Kırtay M., Gazi V., Marques L. 'Experimental studies on chemical concentration MAP building by a multi-robot system using bio-inspired algorithms'. *Autonomous Agents and Multi-Agent Systems*. 2014, vol. 28(1), pp. 72–100.
[76] Stone L.D. Theory of optimal search. Amsterdam, Netherlands: Elsevier; 1976.
[77] Koopman B.O. *Search and screening: general principles with historical applications*. Pergamon Press; 1980.
[78] Marjovi A., Marques L. 'Optimal spatial formation of Swarm robotic gas sensors in odor plume finding'. *Autonomous Robots*. 2014, vol. 35(2–3), pp. 93–109.
[79] Marjovi A., Marques L. 'Optimal Swarm formation for odor plume finding'. *IEEE Transactions on Cybernetics*. 2014, vol. 44(12), pp. 2302–15.
[80] Viswanathan Gandhimohan M., da Luz M.G.E., Raposo E.P., Stanley H.E. The Physics of Foraging. Cambridge University Press; 2011. Available from https://www.cambridge.org/core/product/identifier/9780511902680/type/book
[81] Bénichou O., Loverdo C., Moreau M., Voituriez R. 'Intermittent search strategies'. *Reviews of Modern Physics*. 2011, vol. 83, 81.
[82] Pasternak Z., Bartumeus F., Grasso F.W. 'Lévy-taxis: a novel search strategy for finding odor plumes in turbulent flow-dominated environments'. *Journal of Physics A*. 2009, vol. 42(43), p. 434010.
[83] Emery R., Rahbar F., Marjovi A., Martinoli A. 'Adaptive Lévy taxis for odor source localization in realistic environmental conditions'. IEEE International Conference on Robotics and Automation (ICRA); Singapore, Singapore, 2017.
[84] Adler J. 'Chemotaxis in bacteria'. *Annual Review of Biochemistry*. 1975, vol. 44, pp. 341–56.
[85] Kennedy J.S. 'Zigzagging and casting as a programmed response to wind-borne odour: a review'. *Physiological Entomology*. 1983, vol. 8(2), pp. 109–20.
[86] Cabrita G., Marques L. 'Estimation of gaussian plume model parameters using the simulated annealing algorithm'. *ROBOT2013: First Iberian Robotics Conference*; Madrid, Spain, 2014.
[87] Wang Y., Chen B., Zhu Z., *et al.* 'A hybrid strategy on combining different optimization algorithms for hazardous gas source term estimation in field cases'. *Process Safety and Environmental Protection*. 2020, vol. 138, pp. 27–38.
[88] Farrell J.A., Pang S., Li W. 'Plume mapping via hidden Markov methods'. *IEEE Transactions on Systems, Man, and Cybernetics. Part B, Cybernetics*. 2003, vol. 33(6), pp. 850–63.
[89] Vergassola M., Villermaux E., Shraiman B.I. '"Infotaxis" as a strategy for searching without gradients'. *Nature*. 2007, vol. 445(7126), pp. 406–09.
[90] Masson J.-B. 'Olfactory searches with limited space perception'. *Proceedings of the National Academy of Sciences of the United States of America*. 2013, vol. 110(28), pp. 11261–66.

[91] Song C., He Y., Lei X. 'Autonomous searching for a diffusive source based on minimizing the combination of entropy and potential energy'. *Sensors.* 2020, vol. 19(11), p. 2465.

[92] Song C., He Y., Ristic B., Lei X. 'Collaborative infotaxis: searching for a signal-emitting source based on particle filter and Gaussian fitting'. *Robotics and Autonomous Systems.* 2020, vol. 125, 103414.

[93] Ristic B., Skvortsov A., Gunatilaka A. 'A study of cognitive strategies for an autonomous search'. *Information Fusion.* 2020, vol. 28, pp. 1–9.

[94] Hutchinson M., Oh H., Chen W.H. 'Entrotaxis as a strategy for autonomous search and source reconstruction in turbulent conditions'. *Information Fusion.* 2020, vol. 42, pp. 179–89.

[95] Rahbar F., Marjovi A., Martinoli A. 'An algorithm for odor source localization based on source term estimation'. International Conference on Robotics and Automation (ICRA); Montreal, QC, 2019.

[96] Capitan J., Spaan M.T.J., Merino L., Ollero A. 'Decentralized multi-robot cooperation with auctioned pomdps'. *International Journal of Robotics Research.* 2013, vol. 32(6), pp. 650–71.

[97] Rahbar F., Martinoli A. 'A distributed source term estimation algorithm for multi-robot systems ' [online]. IEEE International Conference on Robotics and Automation (ICRA); Paris, France, 2020. Available from https://ieeexplore.ieee.org/xpl/mostRecentIssue.jsp?punumber=9187508

[98] Li J.G., Meng Q.H., Wang Y., Zeng M. 'Odor source localization using a mobile robot in outdoor airflow environments with a particle filter algorithm'. *Autonomous Robots.* 2011, vol. 30(3), pp. 281–92.

[99] Bourne J.R., Pardyjak E.R., Leang K.K. 'Coordinated bayesian-based bioinspired plume source term estimation and source seeking for mobile robots'. *IEEE Transactions on Robotics.* 2019, vol. 35(4), pp. 967–86.

[100] Magalhães H., Baptista R., Macedo J., Marques L. 'Towards fast plume source estimation with a mobile robot'. *Sensors (Basel, Switzerland).* 2020, vol. 20(24), E7025.

[101] Cabrita G., Marques L. 'Divergence-based odor source Declaration'. *2013 9th Asian Control Conference, ASCC 2013*; 2013.

[102] Hawkins D.M., Cressie N. 'Robust kriging: a proposal'. *Journal of the International Association for Mathematical Geology.* 1984, vol. 16(1), pp. 3–18.

[103] Cabrita G., Marques L., Gazi V. 'Virtual cancelation plume for multiple odor source localization'. *Intelligent Robots and Systems (IROS), 2013 IEEE/RSJ International Conference*; Tokyo, Japan, 2013.

[104] Cao M.L., Meng Q.H., Wang X.W., Luo B., Zeng M., Li W. IEEE International Conference on robotics and biomimetics (ROBIO). Shenzhen, China, 2013.

[105] Ishida H., Tanaka H., Taniguchi H., Moriizumi T. 'Mobile robot navigation using vision and olfaction to search for a gas/odor source'. *Autonomous Robots.* 2006, vol. 20(3), pp. 231–38.

[106] Kim H., Park M., Kim C.W., Shin D. 'Source localization for hazardous material release in an outdoor chemical plant via a combination of LSTM-RNN and CFD simulation'. *Computers & Chemical Engineering*. 2006, vol. 125, pp. 476–89.

[107] Bilgera C., Yamamoto A., Sawano M., Matsukura H., Ishida H. 'Application of convolutional long short-term memory neural networks to signals collected from a sensor network for autonomous gas source localization in outdoor environments'. *Sensors (Basel, Switzerland)*. 2018, vol. 18(12), p. 4484.

[108] Thrift W.J., Cabuslay A., Laird A.B., Ranjbar S., Hochbaum A.I., Ragan R. 'Surface-enhanced Raman scattering-based odor COMPASS: locating multiple chemical sources and pathogens'. *ACS Sensors*. 2019, vol. 4(9), pp. 2311–19.

[109] Drugan M.M. 'Reinforcement learning versus evolutionary computation: a survey on hybrid algorithms'. *Swarm and Evolutionary Computation*. 2019, vol. 44, pp. 228–46.

[110] Baptista R., Magalhaes H., Macedo J., Marques L. '2D thermal wind sensor for mobile robot anemotaxis: design and validation'. *IEEE International Symposium on Safety, Security, and Rescue Robotics (SSRR)*; Abu Dhabi, UAE, 2020.

[111] Tan M. 'Multi-agent reinforcement learning: independent vs. cooperative agents'. *Proceedings of the Tenth International Conference on Machine LearningProceedings of the tenth international conference on machine learning*; University of Massachusetts, Amherst, 1993.

[112] Arulkumaran K., Deisenroth M.P., Brundage M., Bharath A.A. 'Deep reinforcement learning: a brief survey'. *IEEE Signal Processing Magazine*. 2017, vol. 34(6), pp. 26–38.

[113] Duckett T., Axelsson M., Saffiotti A. 'Learning to locate an odour source with a mobile robot'. IEEE International Conference on Robotics and Automation (Cat. No. 01CH37164); Seoul, South Korea, 2001.

[114] Farah A.M., Duckett T. 'Reactive localisation of an odour source by a learning mobile robot'. *Proceedings of the Second Swedish Workshop on Autonomous Robotics*; Stockholm, Sweden, 2002.

[115] Hwangbo J., Sa I., Siegwart R., Hutter M. 'Control of a quadrotor with reinforcement learning'. *IEEE Robotics and Automation Letters*. 2017, vol. 2(4), pp. 2096–103.

[116] Watkins C.J.C.H., Dayan P. 'Q-learning'. *Machine Learning*. 1992, vol. 8(3–4), pp. 279–92.

[117] Chen X., Fu C., Huang J. 'A deep Q-network for robotic odor/gas source localization: modeling, measurement and comparative study'. *Measurement*. 1992, vol. 183, p. 109725.

[118] Duisterhof B.P., Krishnan S., Cruz J.J, *et al*. 'Learning to seek: autonomous source seeking with deep reinforcement learning onboard a nano drone microcontroller'. *ArXiv Preprint ArXiv:1909.11236*. 2019.

[119] Hu H., Song S., Chen C.L.P. 'Plume tracing via model-free reinforcement learning method'. *IEEE Transactions on Neural Networks and Learning Systems*. 2019, vol. 30(8), pp. 2515–27.

[120] Sutton R.S., Barto A.G. Reinforcement learning: an introduction. MIT press; 2018.

[121] Wang L., Pang S., Li J. 'Olfactory-based navigation via model-based reinforcement learning and fuzzy inference methods'. *IEEE Transactions on Fuzzy Systems*. 2020, vol. 29(10), pp. 3014–27.

[122] Eiben A.E., Smith J.E. 'Introduction to evolutionary computing' in *Introduction to Evolutionary Computing*. 2nd ed. Berlin, Heidelberg: Springer-Verlag; 2015.

[123] Murata S., Kurokawa H. 'Self-reconfigurable robots'. *IEEE Robotics & Automation Magazine*. 2007, vol. 14(1), pp. 71–78.

[124] Siciliano B., Khatib O. 'Springer Handbook of robotics' in *Evolutionary robotics, Springer Handbook of Robotics*. Cham: Springer; 2016. pp. 2035–68.

[125] Marques L., Nunes U., de Almeida A.T. 'Cooperative odour field exploration with genetic algorithms'. Proceedings 5th Portuguese Conference on Automatic Control (CONTROLO 2002); 2002.

[126] Macedo J., Marques L., Costa E. 'Evolving infotaxis for distinct environments'. *Submitted to the International Conference on Intelligent Robots and Systems (IROS2021)*; Prague, Czech Republic, 2021.

[127] Villarreal B.L., Olague G., Gordillo J.L. 'Synthesis of odor tracking algorithms with genetic programming'. *Neurocomputing*. 2016, vol. 175, pp. 1019–32.

[128] Macedo J., Marques L., Costa E. 'Robotic odour search: evolving a robot ' S brain with genetic programming'. IEEE International Conference on Autonomous Robot Systems and Competitions (ICARSC); Coimbra, Portugal, 2017.

[129] Macedo J., Fonseca C.M., Costa E. 'Geometric crossover in syntactic space'. *European Conference on Genetic Programming*; Parma, Italy, 2018.

[130] Heinerman J., Zonta A., Haasdijk E., Eiben A. 'On-line evolution of foraging behaviour in a population of real robots'. *European Conference on the Applications of Evolutionary Computation*; Porto, Portugal, 2016.

[131] de Croon G.C.H.E., O'Connor L.M., Nicol C., Izzo D. 'Evolutionary robotics approach to odor source localization'. *Neurocomputing*. 2013, vol. 121, pp. 481–97.

[132] Macedo J., Marques L., Costa E. 'Evolving neural networks for multi-robot odor search'. International Conference on Autonomous Robot Systems and Competitions (ICARSC); Bragança, Portugal, 2016.

[133] Trianni V. ' evolutionary Swarm robotics ' in Evolutionary Swarm robotics: evolving self-organising behaviours in groups of autonomous robots. Vol. 108. Berlin, Heidelberg: Springer; 2008.

[134] Brambilla M., Ferrante E., Birattari M., Dorigo M. 'Swarm robotics: a review from the Swarm engineering perspective'. *Swarm Intelligence*. 2013, vol. 7(1), pp. 1–41.

[135] Kennedy J. 'Particle swarm optimization' in *Encyclopedia of machine learning*. Boston, MA: Springer; 2011. pp. 760–66.

[136] Murlis J., Jones C.D. 'Fine-scale structure of odour plumes in relation to insect orientation to distant pheromone and other attractant sources'. *Physiological Entomology*. 1981, vol. 6(1), pp. 71–86.

[137] Wackermannová M., Pinc L., Jebavý L. 'Olfactory sensitivity in mammalian species'. *Physiological Research*. 2016, vol. 65(3), pp. 369–90.

[138] Pearce T.C. 'Computational parallels between the biological olfactory pathway and its analogue "the electronic nose": Part II. sensor-based machine olfaction'. *Bio Systems*. 1997, vol. 41(2), pp. 69–90.

[139] Rawson N.E., Gomez G., Cowart B., *et al.* 'Selectivity and response characteristics of human olfactory neurons'. *Journal of Neurophysiology*. 1997, vol. 77(3), pp. 1606–13.

[140] Jenkins E.K., DeChant M.T., Perry E.B. 'When the nose doesn't know: canine olfactory function associated with health, management, and potential links to microbiota'. *Frontiers in Veterinary Science*. 2018, vol. 5, p. 56.

[141] Ishida H., Pardo M., Sberveglieri G. 'Blimp robot for three-dimensional gas distribution mapping in indoor environment'. Olfaction and Electronic Nose; Brescia, Italy, 2009.

[142] Pinto J., Mendes R., da Silva J.C.B., Dias J.M., de Sousa e J.B. 'Multiple autonomous vehicles applied to plume detection and tracking'. OCEANS - MTS/IEEE kobe techno-ocean (OTO); Kobe, 2018.

[143] Macedo J., Marques L., Costa E. 'Designing fitness functions for odour source localisation' [online]. GECCO '21; New York, NY, 2021. Available from https://dl.acm.org/doi/proceedings/10.1145/3449726

[144] Cobo L.C., Isbell C.L., Thomaz A.L. 'Automatic task decomposition and state abstraction from demonstration'. *Proceedings of the 11th International Conference on Autonomous Agents and Multiagent Systems-Volume 1*; Valencia, Spain, 2012.

[145] Tang H., Houthooft R., Foote D, *et al.* 'Exploration: a study of count-based exploration for deep reinforcement learning' in Advances in neural information processing systems. Morgan Kaufmann Publishers Inc; 2017.

[146] Jakobi N., Husbands P., Harvey I. 'Noise and the reality gap: the use of simulation in evolutionary robotics'. *European Conference on Artificial Life*; Granada, Spain, 1995.

[147] Pollack J.B., Lipson H., Ficici S.G., Funes P., Hornby G., Watson R.A. 'Evolutionary techniques in physical robotics'. *International Conference on Evolvable Systems*; Edinburgh, Scotland, UK, 2000.

Chapter 10
Vision sensors for robotic perception

Shan Luo[1], Daniel Fernandes Gomes[2], Jiaqi Jiang[2], and Guanqun Cao[2]

In this chapter, we will introduce vision sensors for robotics applications. It first briefly introduces the working principles of the widely used vision sensors, i.e., RGB cameras, stereo cameras and depth sensors, and also the off-the-shelf vision sensors that have been widely used in robotics research, particularly robot perception. As one of the most widely used sensors to be equipped with robots and thanks to its low cost and high resolution, vision sensors have also been used in other sensing modalities. In recent years, there has been rapid development in embedding vision sensors in optical tactile sensors. In such sensors, visual cameras are placed under an elastomer layer and used to capture its deformation while being interacted with objects. The vision sensors enable robots to sense and estimate the properties of objects, e.g., their shapes, appearances, textures and mechanical parameters. We will cover various aspects of vision sensors for robotic applications, including the various technologies, hardware, integration, computation algorithms and applications in relation to robotics.

10.1 Introduction

Our eyes are crucial for seeing the world around us. With our eyes, we can see the colour of a cat, the texture of a carpet, the face of a person and appearance of a building. Similarly, vision sensors have also been developed in the past decades and can be equipped with robots to enable them to have the sense of sight. Vision sensors provide robots with vital information about the surroundings and vision has been the sensing modality robots rely on most.

Compared to other sensing modalities such as touch sensing, hearing, smell and taste, vision has a much larger field of view (FoV) and is able to capture the view of a scene at a glance, and multiple objects can be observed in a single view. The

[1]Department of Engineering, King's College London, London, UK
[2]SmartLab, Department of Computer Science, University of Liverpool, Liverpool, UK

properties of objects in the scene, e.g., colour, textures, appearances and shapes can be obtained from one single camera image, and it is remarkably easy to collect data with vision sensors.

On the other hand, processing visual data requires high computational resources. More than 50% of the human brain [1] is devoted to processing visual information either directly or indirectly, and therefore visual information has been a key for humans to understand the world. For robots, much of the processing power is also devoted to extracting information from the visual data. One reason is that the abundant data can be accessed via the vision sensors. The other reason is that there are fluctuation factors in visual data that affect the extraction of useful information from visual data.

Such factors include scaling, rotation, translation and illumination. The scaling problem is caused by the projection of the observed objects to the 2D visual sensing panels, making vision an ill-posed problem. The rotation and translation problem arises from that the different positions and orientations of objects may result into different appearances of the objects in the view. The illumination problem is caused by different light conditions in the environment, and visual observations of objects may suffer from occlusions and shadows posed by the robot itself, particularly robot hands in grasping, and other objects in the scene.

In the past decades, vision sensors of different sensing principles have been proposed and commercialised, and many of them have been applied to robotics research. The most widely used vision sensors are the RGB cameras. They are usually equipped with a standard CMOS sensor through which the colour images of persons and objects are acquired. The acquisition of static photos is usually expressed in megapixels that means one million pixels, e.g., 2 MP (1,920 × 1,080 = 2,073,600 pixels, also known as full HD resolution or 1080p), 12 MP and 16 MP. Compared to static images, videos captured by the RGB cameras can reveal the temporal information of the objects in the view, e.g., recognising human actions, tracking moving vehicles and localising a robot in a map.

To enable processing the visual events efficiently, event cameras, also known as neuromorphic cameras or dynamic vision sensors (DVS), emerge in recent years. An event camera responds to local changes in brightness, instead of capturing images using a shutter as conventional cameras do. Pixels inside an event camera operate independently and asynchronously: each pixel reports changes in brightness as they occur and stays silent otherwise. Event cameras demonstrate better temporal resolution in order of millions frames per second compared to conventional cameras in order of hundreds frames per second.

Apart from 2D information extracted from images by the RGB cameras, 3D information can also be captured by vision sensors. One natural way to obtain the 3D information is to simulate the binocular vision of humans that derives information about how far away objects are based on solely relative positions of the object in the two eyes. A stereo camera simulates the human binocular vision by having two image sensors and therefore gives it the ability to perceive depth. There are also other ways to obtain the depth information based on different techniques, e.g., time-of-flight (ToF), structured light, and light fields. These depth sensors are usually

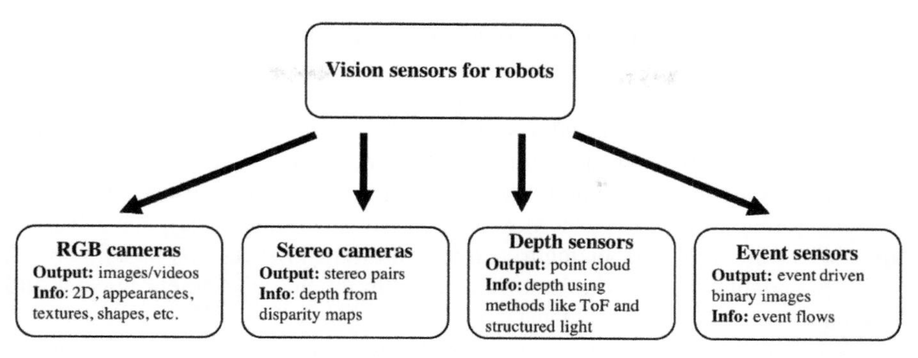

Figure 10.1 *There are different types of vision sensors for robotics applications. RGB cameras have been one of the most widely used sensors in robotics, from robot grasping to visual SLAM. With the images or videos captured by the RGB cameras, rich information of the objects in the scene can be obtained, e.g., appearances, textures and shapes. Stereo cameras can be used to obtain the depth from the object to the camera from the obtained stereo pairs. Other depth sensors include ones based on ToF and structured light. Other types of vision sensors have also emerged and have been applied to robotics like event sensors that output event flows.*

used with the RGB cameras to form the RGB-D cameras so that both 2D appearances cues and depth can be obtained at the same time.

In the recent years, vision sensors have also been used in other sensing modalities. There is a rapid development of embedding vision sensors in optical tactile sensors. Such sensors usually consist of a visual camera at the base of the sensor and an elastomer layer on the top to interact with objects, and the visual camera can capture the deformation of the elastomer in the interaction. The optical tactile sensors bridge the gap between vision and tactile sensing to create cross-modal perception. As visual cameras are used to capture the tactile interactions, the outputs of the optical tactile sensors are essentially camera images. This crossover has enabled techniques developed for computer vision, e.g., convolutional neural networks, to be applied to tactile sensing, connecting the look and feel of objects being interacted with. Recently, there is also development that can transform between or match visual and tactile data from such sensors.

In this chapter, we introduce different vision sensors, i.e., RGB cameras, stereo cameras and depth sensors, that have been used in the robotics research, with an overview shown in Figure 10.1. We will then introduce how vision sensors can be used in other sensing modalities. Various aspects of vision sensors for robotic applications will be covered, including the hardware, integration, computation algorithms and applications to robotics.

10.2 RGB cameras for robotic perception

The projections of real 3D scenes onto 2D planes, generated when light (rays) real objects and filtered through a small cavity has always amused and served as a practical tool to humans, as pointed out by speculative theories about how pre-historic man produced cave paintings and the usage of camera obscura* in ancient and more modern civilisations. More recently, this working principle has been at the core of modern, first analogue and then digital cameras. This basic working principle enables cameras to capture a scene in an energy-efficient manner and instantly.

While the real light phenomenon generates inverted projections, we can conceptually solve this by considering a virtual plane between the observed scene and the camera plane. Given the similarity between the two triangles, the image projected on the virtual plane is proportionally equivalent to the one projected on the real plane. By making the usual thin lens assumptions, the optical sensor can be modelled as a pinhole camera. The projective transformation that maps a point in the world space P into a point in a camera image P' can be defined using the general camera model [2] as:

$$P' = K\left[\boldsymbol{R}|\boldsymbol{t}\right]P$$

$$K = \begin{bmatrix} fk & 0 & c_x \\ 0 & fl & c_y \\ 0 & 0 & 1 \end{bmatrix}$$

where $P' = \left[x'z,\ y'z,\ z\right]^T$ is an image pixel and $P = \left[x,\ y, z,\ 1\right]^T$ is a point in space, both represented in homogeneous coordinates here, $\left[\boldsymbol{R}|\boldsymbol{t}\right]$ is the camera's extrinsic matrix that encodes the rotation R and translation t of the camera, K is the camera intrinsic matrix (f is the focal length; k and l are the pixel-to-meters ratios; c_x and c_y are the offsets in the image frame). If the used camera produces square pixels, i.e., $k = l$, fk and fl can be replaced by α, for mathematical convenience. From the above equations, a point in the world space P can be mapped into a point in an image P' which is a 'well posed' problem, i.e., has a uniquely determined solution. However, the mapping from P' to P usually does not have a uniquely determined solution, i.e., an 'ill-posed' problem. It results into the fact that a camera image may be resulted from different real-world settings. As a result, it is challenging for a robot to understand its ambient world from one single camera image.

*https://en.wikipedia.org/wiki/Camera_obscura

10.3 Stereo cameras

Given that cameras reduce 3D geometry into 2D, this creates one problem: the loss of depth perception and/or size ambiguity. To mitigate this, two cameras and projections can be considered instead, to form a stereo camera. One of the commercially available stereo cameras is the ZED sensor.†

By performing triangulation between the real point and the two corresponding projections, the depth of the object can be inferred. This construction is commonly referred as epipolar geometry. By comparing information about a scene from two corresponding points in left and right cameras that are projected from the same real-world point, 3D information can be extracted by examining the relative positions of objects in the two panels of the left and right cameras. The challenge in forming the epipolar geometry falls in matching the corresponding points in the two cameras, i.e., stereo matching. A large number of algorithms have been proposed for stereo correspondence using convolutional neural networks in recent years [3, 4]. It has great potential to have stereo vision for robotics tasks as well, e.g., grasping.

10.4 Event cameras

10.4.1 Hardware

Compared with conventional cameras, the event cameras offer a number of advantages, including lower latency, less power, microsecond temporal resolution and larger dynamic range. Different from conventional cameras, the event camera employs a DVS, which is able to capture the changes of brightness for each pixel asynchronously [5]. As a result, the event camera provides an asynchronous stream of brightness changes including the location, time information and polarity ('On' and 'Off'), i.e., events. In DVS, each pixel memorises the statement of brightness as a reference when an event is triggered, and compares it with the current statement. If there exits an obvious variation that surpasses the threshold, an event is triggered and the reference is updated by the pixel.

On the other side, due to the use of 'On' and 'Off' polarity, it is difficult to construct a clear and detailed description of a scene. To address this problem, asynchronous time-based image sensor (ATIS) [6] and dynamic and active pixel vision sensor (DAVIS) [7] have been proposed for event cameras. The ATIS includes the DVS pixels for the brightness change detection and another subpixel to measure the absolute values of brightness. As a result, ATIS can capture not only the motion but also the background of static scene. The DAVIS consists of a DVS and a conventional active pixel sensor (APS). This combination makes it able to generate a

†https://www.stereolabs.com/zed/

colourful and detailed static background. However, the APS image usually suffers from motion blur and it is difficult to synchronise with DVS in high-speed motion scenes.

10.4.2 Applications in robotics

In recent years, event cameras have been widely used in many robotic applications, such as object tracking [8, 9], optical flow estimation [10, 11], 3D reconstruction [12, 13] and recognition tasks [14–16]. Compared to conventional vision sensors, event cameras demonstrate nice features low latency, less power and temporal resolution. These features make event sensors highly suitable for tasks that have strong requirements for efficiency in visual processing.

There are also advancements in the algorithmic development of the event-based visual processing in recent years. The spiking neural networks (SNNs) have become a popular method for processing event signals. In the neurons of SNNs, the input signals, i.e., events, are received by the neurons and accumulated in the internal state, named as the 'membrane potential'. When it exceeds a threshold, the neuron generates a spike for the neurons of the next layer, and the internal state of the current neuron resets. Variants of SNNs have been developed, such as leaky integrate and fire model (LIF) and spike response model (SRM) [17], inspired by the properties of human neurons. Thanks to this event-based property, the SNNs have been widely applied with the event camera in many applications, such as in References [18–20].

10.5 Depth cameras

Like its name suggests, RGB-D cameras are able to augment the RGB image with depth information, i.e., the distance from each point in the real scene to the cameras. With the ability to measure object depth, RGB-D cameras have been widely used for object pose estimation, 3D reconstruction and robotic grasping. In order to adapt to different application scenarios, many consumer-grade depth cameras have been developed in recent years.

According to the sensor types used in the cameras, RGB-D cameras can be divided into two categories, i.e., optical depth cameras and non-optical depth cameras. Optical depth cameras occupy a major part of the market thanks to its mature technology, low price and compact size. The most common techniques currently being employed for optical RGB-D cameras are based on structured lights, ToF, and active infrared (IR) stereo methods.

Structured-lights-based RGB-D cameras have a pair of near-IR laser transmitters and receivers. It uses the transmitter and the receiver to project light with certain structural features onto the object and collect the reflected light signals, respectively. Then it calculates the depth information based on the changes in the reflected light signals caused by different depth areas. In the early stages of RGB-D camera development, structured-lights-based RGB-D cameras attracted attention due to its mature technology, low cost and low resource consumption. Some iconic examples of structured-lights-based RGB-D cameras are Intel RealSense R200[‡] and Microsoft Kinect V1.[§] However, they are easily affected by the ambient light and long perception distance, which makes it not suitable for outdoor and large scenes.

Thanks to the increasing processing power, Microsoft successively launched two ToF-based RGB-D cameras, the Kinect V2 in 2014 and the Azure Kinect[¶] in 2018. Different from estimating depth with light signal changes in the structure-lights-based method, ToF-based RGB-D cameras obtain the distance of the target by detecting the round-trip time of the light pulse. Through this way, it can work for long distance detection and reduce the interference of the ambient light. Nonetheless, the larger size of the ToF-based RGB-D cameras limits their use on small mobile platforms like in many robotics applications.

In addition to the cameras mentioned above, RGB-D cameras based on the active IR stereo principle have also played an important role in the development of reliable depth sensors. Different from the naive block-matching methods that are widely used in stereo vision, the active IR stereo cameras use an IR laser projector to generate texture for the stereo cameras, which significantly improves the accuracy. Moreover, the projector can be used as an artificial source of light for night-time or dark situations. There are different series of RGB-D cameras based on active IR stereo technology such as Intel RealSense D415,[**] D435,[††] D435i[‡‡] and D455.[§§]

10.6 Vision sensors for other modalities

Cameras (and depth-cameras) can be used to assess large areas instantly; however, they suffer from occlusions and variances in the scene illumination, shadows and other sources of ambiguity. In contrast, tactile sensors offer a local assessment of the scene that is robust to such problems and, given the fact that precise sensing is more critical near contact, tactile sensors become a crucial sensing modality to consider, that is complementary to vision. Nonetheless, the fabrication of tactile skins is widely challenging due to complicated electronics, cross-talk problems and consequently have traditionally produced low resolution of tactile signals [21–24]. On the other hand, cameras are now ubiquitous and consequently have become extremely cheap while being able to capture high-resolution images. As a consequence, a wide range of works have focused on exploiting such high-resolution cameras to produce optical tactile sensors.

The optical tactile sensors can be grouped into two main groups: marker-based and image-based, with the former being pioneered by the TacTip sensors [25] and the latter by the GelSight sensors [26]. As the name suggests, marker-based sensors

[‡]https://software.intel.com/content/www/us/en/develop/articles/realsense-r200-camera.html
[§]https://en.wikipedia.org/wiki/Kinect
[¶]https://azure.microsoft.com/en-gb/services/kinect-dk/
[**]https://www.intelrealsense.com/depth-camera-d415/
[††]https://www.intelrealsense.com/depth-camera-d435/
[‡‡]https://www.intelrealsense.com/depth-camera-d435i/
[§§]https://www.intelrealsense.com/depth-camera-d455/

exploit the tracking of markers printed on a soft domed membrane to perceive the membrane displacement and the resulted contact forces. By contrast, image-based sensors directly perceive the raw membrane with a variety of image recognition methods to recognise textures, localise contacts and reconstruct the membrane deformations, etc. Because of the different working mechanisms, marker-based sensors measure the surface on a lower resolution grid of points, whereas image-based sensors make use of the full resolution provided by the camera. Some GelSight sensors have also been produced with markers printed on the sensing membrane [27], enabling marker and image-based methods to be used with the same sensor. Both families of sensors have been produced with either flat sensing surfaces or domed/finger-shaped surfaces.

10.6.1 Marker-based sensors

The first marker-based sensor proposal can be found in Reference [28]; however, more recently an important family of marker-based tactile sensors is the TacTip Family of sensors described in Reference [29]. Since its initial domed shaped version [25], different morphologies have been proposed: including the TacTip-GR2 [30], a smaller fingertip design, TacTip-M2 [31], mimicking a large thumb for in-hand linear manipulation experiments, and TacCylinder to be used in capsule endoscopy applications. With its miniaturised and adapted design [30, 31], have been successfully used as fingers (or finger tips) in robotic grippers. Although each TacTip sensor introduces some manufacturing improvements or novel surface geometries, the same working principle is shared: white pins are imprinted onto a black membrane that can then be tracked using computer vision methods.

There are also other optical tactile sensors that track the movements of markers. In Reference [32], an optical tactile sensor named FingerVision is proposed to make use of a transparent membrane, with the advantage of gaining proximity sensing. However, the usage of the transparent membrane makes the sensor lack the robustness to external illumination variance associated with touch sensing. In Reference [33], semi-opaque grids of magenta and yellow makers, painted on the top and bottom surfaces of a transparent membrane are proposed, in which the mixture of the two colours is used to detect horizontal displacements of the elastomer. In Reference [34], green florescent particles are randomly distributed within the soft elastomer with black opaque coating so that a higher number of markers can be tracked and used to predict the interaction with the object, according to the authors. In Reference [35], a sensor with the same membrane construction method, 4 Raspberry PI cameras and fisheye lenses has been proposed for optical tactile skins. A summary of influential marker-based optical tactile sensors is shown in Table 10.1.

10.6.2 Image-based sensors

On the other side of the spectrum, the GelSight sensors, initially proposed in Reference [26], exploit the entire resolution of the tactile images captured by the sensor camera, instead of just tracking markers. Due to the soft opaque tactile membrane, the captured images are robust to external light variations and capture information about the touched

Table 10.1 A summary of influential marker-based optical tactile sensors

	Sensor structures	**Illumination and tactile membrane**
TacTip [25]	A domed (finger) shape, 40 × 40 × 85 mm^3, and tracks 127 pins; uses a Microsoft LifeCam HD webcam.	The membrane is black on the outside, with white pins and filled with transparent elastomer inside. Initially the membrane was cast from VytaFlex 60 silicone rubber, the pins painted by hand and the tip filled with the optically clear silicone gel (Techsil, RTV27905); however, currently the entire sensor can be 3D-printed using a multi-material printer (Stratasys Objet 260 Connex), with the rigid parts printed in Vero white material and the compliant skin in the rubber-like TangoBlack+.
TacTip-M2 [31]	A thumb-like or semi-cylindrical shape, 32 × 102 × 95 mm^3 and tracks 80 pins.	
TacTip-GR2 [30]	A cone shape with a flat sensing membrane, 40 × 40 × 44 mm^3, tracks 127 pins and uses an Adafruit SPY PI camera.	
TacCylinder [36]	A catadioptric mirror is used to track the 180 markers around the sensor cylindrical body.	
FingerVision [32]	It uses a ELP Co. USBFHD01M-L180 camera with a 180° fisheye lens. It has approximately 40 × 47 × 30 mm.	The membrane is transparent, made with Silicones Inc. XP-565, with 4 mm of thickness and markers spaced by 5 mm. No internal illumination is used, as its membrane is transparent.
Subtractive colour mixing [33]	N/A	Two layers of semi-opaque-coloured markers is proposed. Sorta-Clear 12 from Smooth-On, clear and with Ignite pigment, is used to make the inner and outer sides.
Green markers [34]	The sensor has a flat sensing surface, measures 50 × 50 × 37 mm and is equipped with a ELP USBFHD06H RGB camera with a fisheye lens.	It is composed of three layers: stiff elastomer, soft elastomer with randomly distributed green florescent particles in it and black opaque coating. The stiff layer is made of ELASTOSIL® RT 601 RTV-2 and is poured directly on top of the electronics, the soft layer is made of Ecoflex™ GEL (shore hardness 000-35) with the markers mixed in, and the final coat layer is made of ELASTOSIL® RT 601 RTV-2 (shore hardness 10A) black silicone. A custom board with an array of SMD white LEDs is mounted on the sensor base, around the camera.
Multi-camera skin [35]	It has a flat prismatic shape of 49 × 51 × 17.45 mm. Four Pi cameras are assembled in a 2 × 2 array and fish-eye lenses are used to enable its thin shape.	

surface's geometry structure, unlike most conventional tactile sensors that measure the touching force. Leveraging the high resolution of the captured tactile images, high-accuracy geometry reconstructions are produced in References [37–40]. In Reference [37], this sensor was used as the fingers of a robotic gripper to insert a USB cable into the correspondent port effectively. However, the sensor only measures a small flat area oriented towards the grasp closure.

Markers were also added to the membrane of the GelSight sensors, allowing the same set of methods that were explored in the TacTip sensors. There are some other sensor designs and adaptations for robotic fingers in References [41–43]. In Reference [41], matte aluminium powder was used for improved surface reconstruction, together with the LEDs being placed next to the elastomer, and the elastomer being slightly curved on the top/external side. In Reference [42], the GelSlim is proposed, a design, where a mirror is placed at a shallow and oblique angle for a slimmer design. The camera was placed on the side of the tactile membrane, such that it captures the tactile image reflected onto the mirror. A stretchy textured fabric was also placed on top of the tactile membrane to prevent damages to the elastomer and to improve tactile signal strength. Recently, an even more slim design has been proposed 2 mm [44], where a hexagonal prismatic shaping lens is used to ensure radially symmetrically illumination. In Reference [43], DIGIT is also proposed, an ease to manufacture and use sensor, with a USB 'plug-and-play' port and an easily replaceable elastomer secured with a single screw mount.

In these previous works on camera-based optical tactile sensors, multiple designs and two distinct working principles have been exploited. However, none of these sensors has the capability of sensing the entire surface of a robotic finger, i.e., both the sides and the tip of the finger. As a result, they are highly constrained in object manipulation tasks, due to the fact that the contacts can only be sensed when the manipulated object is within the grasp closure [37, 45]. To address this gap, we propose the fingertip-shaped sensor named GelTip that captures tactile images by a camera placed in the centre of a finger-shaped tactile membrane. It has a large sensing area of approximately 75 cm^2 (*vs.*) 4 cm^2 of the GelSight sensor) and a high resolution of 2.1 megapixels over both the sides and the tip of the finger, with a small diameter of 3 cm (vs. 4 cm of the TacTip sensor). More details of the main differences between the GelSight sensors, TacTip sensors and our GelTip sensor are given in Table 10.2.

With its compact design, the GelTip [48, 49] and other GelSight [37, 42–44] sensors are candidate sensors to be mounted on robotic grippers; however, custom grippers and sensors built using the GelSight working principle have also been proposed [50, 51]. Simulation models of such sensors have also been proposed [46, 47].

Two recent works [52, 53] also address the issue of the flat surface of previous GelSight sensors. However, their designs have large differences compared to ours. In Reference [52], the proposed design has a tactile membrane with a surface geometry close to a quarter of a sphere. Therefore, a great portion of contacts happening in the regions outside the grasp closure is undetectable. In Reference [53], this issue is mitigated using five endoscope micro cameras looking at different regions of the finger. However, this results in a significant increase in cost for the

Table 10.2 A summary of influential flat and finger-shaped GelSight sensors

	Sensor structures	**Illumination**	**Tactile membrane**
GelSight [37]	It has a cubic design with a flat square surface. A Logitech C310 (1280 × 720) camera is placed at its base pointing at the top membrane.	Four LEDs (RGB and white) are placed at the base. The emitted light is guided by the transparent hard surfaces on the sides, so that it enters the membrane tangentially.	A soft elastomer layer is placed on top of a rigid, flat and transparent acrylic sheet. It is painted using semi-specular aluminium flake powder.
GelSight [41]	It has a close-to hexagonal prism shape. The used webcam is also the Logitech C310.	Three sets of RGB LEDs are positioned (close to) tangent to the elastomer, with a 120° angle from each other.	A matte aluminium powder is proposed for improved surface reconstruction. Its elastomer has a flat bottom and a curved top.
GelSlim [42]	A mirror placed at a shallow oblique angle and a Raspberry Pi Spy (640 × 480) camera is used to capture the tactile image reflected by the mirror.	A single set of white LEDs is used. These are pointed at the mirror, so that the light is reflected directly onto the tactile membrane.	A stretchy and textured fabric on the tactile membrane prevents damages to the elastomer and results in improved tactile signal strength.
GelSlim v3 [44]	It is shaped similar to References [37, 41] however slimmer 20 mm of thickness, and a round sensing surface.	A custom hexagonal prism is constructed to ensure radially symmetric illumination.	An elastomer with Lambertian reflectance is used, as proposed in Reference [41].
DIGIT [43]	A prismatic design, with curved sides. An OmniVision OVM7692 (640 × 480) camera is embedded in the custom circuit board.	Three RGB LEDs are soldered directly into the circuit board, illuminating directly the tactile membrane.	The elastomer can be quickly replaced using a single screw mount.

(Continues)

Table 10.2 Continued

	Sensor structures	Illumination	Tactile membrane
Round Fingertip [46]	It has a round membrane, close to a quarter of sphere. A single 160° FoV Raspberry Pi (640 × 480) is installed on its base.	Two rings of LEDs are placed on the base of the sensor, with the light being guided through the elastomer.	Both rigid and soft parts of the membrane are cast, using SLA 3D-printed moulds.
OmniTact [47]	It has a domed shape. Five endoscope cameras (400 × 400) are installed on a core mount and placed orthogonally to each other: pointing at the tip and sides.	RGB LEDs are soldered both onto the top and sides of the sensor.	The elastomer gel is directly poured onto the core mount (and cameras) without any rigid surface or empty space in between.
GelTip [48, 49]	It has a domed (finger) shape, similar to a human finger. A Microsoft Lifecam Studio webcam (1920 × 1080) is used.	Three sets of LEDs, with a 120° angle from each other, are placed at the sensor base, and the light is guided through the elastomer	An acrylic test tube is used as the rigid part of the membrane. The deformable elastomer is cast using a three-part SLA/FFF 3D printed mould.

sensor, according to the authors, approximately US$3200 vs. only around US$100 for ours.

10.7 Conclusions

In this chapter, we introduced different aspects of the vision sensors for robotics applications, from their working principles to their applications to robotics, particularly on robot perception and their use in sensors of other modalities. As one of the most widely used sensors for robots, they have the advantages of low cost and high resolution. Vision sensors have also been used in other sensing modalities, and we have introduced the state-of-the-art research in optical tactile sensors using visual sensors. By having vision sensors, robots can sense and estimate the properties of the objects that they interact with, e.g., object shapes, appearances, textures and mechanical parameters. It can be forecast that vision sensors will be one of the most widely used sensors in the research of robotics, and new types of vision sensors, such as event sensors and more robust RGB-Dsensors, will emerge in the future research and development.

Acknowledgements

This work was supported by the EPSRC project "ViTac: Visual-Tactile Synergy for Handling Flexible Materials" (EP/T033517/1).

References

[1] Sheth B.R., Sharma J., Rao S.C., Sur M. 'Orientation maps of subjective contours in visual cortex'. *Science*. 1996, vol. 274(5295), pp. 2110–15.

[2] Szeliski R. 'Computer vision algorithms and applications'. London: Springer Science & Business Media. 2010.

[3] Zbontar J., LeCun Y. 'Stereo matching by training a convolutional neural network to compare image patches'. *Journal of Machine Learning Research*. 2016, vol. 17(1), pp. 2287–318.

[4] Zbontar J., LeCun Y. 'Computing the stereo matching cost with a convolutional neural network'. *IEEE Conference on Computer Vision and Pattern Recognition (CVPR)*; Boston, MA, IEEE, 2015. pp. 1592–99.

[5] Lichtsteiner P., Posch C., Delbruck T. 'A 128 × 128 120 dB 15 μs latency asynchronous temporal contrast vision sensor'. *IEEE Journal of Solid-State Circuits*. 2008, vol. 43(2), pp. 566–76.

[6] Posch C., Matolin D., Wohlgenannt R. 'A QVGA 143 dB dynamic range frame-free pwm image sensor with lossless pixel-level video compression and time-domain CDS'. *IEEE Journal of Solid-State Circuits*. 2010, vol. 46(1), pp. 259–75.

[7] Brandli C., Berner R., Yang M., Liu S.C., Delbruck T. 'A 240 × 180 130 dB 3 μs latency global shutter spatiotemporal vision sensor'. *IEEE Journal of Solid-State Circuits*. 2014, vol. 49(10), pp. 2333–41.

[8] Ramesh B., Zhang S., Lee Z.W., Gao Z., Orchard G., Xiang C. 'Long-term object tracking with a moving event camera' in BMVC; 2018. p. 241.

[9] Delbruck T., Lang M. 'Robotic goalie with 3 MS reaction time at 4 % CPU load using event-based dynamic vision sensor'. *Frontiers in Neuroscience*. 2013, vol. 7, 223.

[10] Liu M., Delbruck T. 'Adaptive time-slice block-matching optical flow algorithm for dynamic vision sensors'. *British Machine Vision Conference (BMVC), Newcastle upon Tyne, UK, Sep 2018. BMVC*. n.d.

[11] Zhu A.Z., Yuan L., Chaney K., Daniilidis K. 'EV-flownet: self-supervised optical flow estimation for event-based cameras' [online]. *Robotics*; 2018. Available from http://www.roboticsproceedings.org/rss14/index.html

[12] Rebecq H., Gallego G., Mueggler E., Scaramuzza D. 'EMVS: event-based multi-view stereo—3D reconstruction with an event camera in real-time'. *International Journal of Computer Vision*. 2018, vol. 126(12), pp. 1394–414.

[13] Kim H., Leutenegger S., Davison A.J. 'Real-time 3D reconstruction and 6-dof tracking with an event camera'. European Conference on Computer Vision; Berlin, Germany, Springer, 2016. pp. 349–64.

[14] Ghosh R., Mishra A., Orchard G., Thakor N.V. 'Real-time object recognition and orientation estimation using an event-based camera and CNN'. *2014 IEEE Biomedical Circuits and Systems Conference (BioCAS)*; Lausanne, Switzerland, IEEE, 2018. pp. 544–47.

[15] Amir A., Taba B., Berg D, *et al.* 'A low power, fully event-based gesture recognition system'. *IEEE Conference on Computer Vision and Pattern Recognition (CVPR) [online]*; Honolulu, HI, IEEE, 2018. pp. 7243–52. Available from https://ieeexplore.ieee.org/xpl/mostRecentIssue.jsp?punumber=8097368

[16] Wang Q., Zhang Y., Yuan J., Lu Y. 'Space-time event clouds for gesture recognition: from RGB cameras to event cameras'. *IEEE Winter Conference on Applications of Computer Vision (WACV)*; Waikoloa Village, HI, IEEE, 2002. pp. 1826–35.

[17] Gerstner W., Kistler W.M. Spiking Neuron Models [online]. New York: Cambridge University Press Aug 2002; Available from https://www.cambridge.org/core/product/identifier/9780511815706/type/book

[18] Taunyazov T., Sng W., Lim B., *et al.* 'Event-driven visual-tactile sensing and learning for robots' [online]. *Robotics*; 2002. Available from http://www.roboticsproceedings.org/rss16/index.html

[19] Massa R., Marchisio A., Martina M., Shafique M. 'An efficient spiking neural network for recognizing gestures with a DVS camera on the Loihi neuromorphic processor' [online]. *International Joint Conference on Neural Networks (IJCNN)*; Glasgow, UK, 2002. Available from https://ieeexplore.ieee.org/xpl/mostRecentIssue.jsp?punumber=9200848

[20] Gehrig M., Shrestha S.B., Mouritzen D., Scaramuzza D. 'Event-based angular velocity regression with spiking networks' [online]. *IEEE International Conference on Robotics and Automation (ICRA)*; Paris, France, IEEE, 2002. pp. 4195–202. Available from https://ieeexplore.ieee.org/xpl/mostRecentIssue.jsp?punumber=9187508

[21] Luo S., Mou W., Althoefer K., Liu H. 'Localizing the object contact through matching tactile features with visual MAP'. *IEEE International Conference on Robotics and Automation (ICRA)*; Seattle, WA, IEEE, 2015. pp. 3903–08.

[22] Luo S., Mou W., Althoefer K., Liu H. 'Novel tactile-SIFT descriptor for object shape recognition'. *IEEE Sensors Journal*. 2015, vol. 15(9), pp. 5001–09.

[23] Luo S., Mou W., Althoefer K., Liu H. 'Iterative closest labeled point for tactile object shape recognition'. *IEEE/RSJ International Conference on Intelligent Robots and Systems (IROS)*; Daejeon, South Korea, IEEE, 2016. pp. 3137–42.

[24] Luo S., Mou W., Althoefer K., Liu H. 'ICLAP: shape recognition by combining proprioception and touch sensing'. *Autonomous Robots*. 2019, vol. 43(4), pp. 993–1004.

[25] Chorley C., Melhuish C., Pipe T., Rossiter J. 'Development of a tactile sensor based on biologically inspired edge encoding'. *International Conference on Advanced Robotics (ICAR)*; IEEE, 2009.

[26] Johnson M.K., Adelson E.H. 'Retrographic sensing for the measurement of surface texture and shape retrographic sensing for the measurement of surface texture and shape'. *IEEE Conference on Computer Vision and Pattern Recognition*; IEEE, 2009.

[27] Dong S., Yuan W., Adelson E.H. 'Improved gelsight tactile sensor for measuring geometry and slip'. *IEEE/RSJ International Conference on Intelligent Robots and Systems (IROS)*; Vancouver, BC, IEEE, 2017.

[28] Vlack K., Kamiyama K., Mizota T., Kajimoto H., Kawakami N., Tachi S. 'GelForce: a traction field tactile sensor for rich human-computer interaction'. *IEEE Conference on Robotics and Automation. TExCRA Technical Exhibition Based*; Minato-ku, Tokyo, Japan, IEEE, 2004. pp. 11–12.

[29] Ward-Cherrier B., Pestell N., Cramphorn L, *et al.* 'The tactip family: soft optical tactile sensors with 3D-printed biomimetic morphologies'. *Soft Robotics*. 2018, vol. 5(2), pp. 216–27.

[30] Ward-Cherrier B., Rojas N., Lepora N.F. 'Model-free precise in-hand manipulation with a 3D-printed tactile gripper'. *IEEE Robotics and Automation Letters*. 2017, vol. 2(4), pp. 2056–63.

[31] Ward-Cherrier B., Cramphorn L., Lepora N.F. 'Tactile manipulation with a tacthub integrated on the open-hand M2 gripper'. *IEEE Robotics and Automation Letters*. 2016, vol. 1(1), pp. 169–75.

[32] Yamaguchi A., Atkeson C.G. 'Combining finger vision and optical tactile sensing: reducing and handling errors while cutting vegetables'. *IEEE-RAS 16th International Conference on Humanoid Robots (Humanoids)*; Cancun, Mexico, IEEE, 2016. pp. 1045–51.

[33] Lin X., Wiertlewski M. 'Sensing the frictional state of a robotic skin via subtractive color mixing'. *IEEE Robotics and Automation Letters*. 2019, vol. 4(3), pp. 2386–92.

[34] Sferrazza C., D'Andrea R. 'Design, motivation and evaluation of a full-resolution optical tactile sensor'. *Sensors (Basel, Switzerland)*. 2019, vol. 19(4), p. 4.

[35] Trueeb C., Sferrazza C., D'Andrea R. 'Towards vision-based robotic skins: a data-driven, multi-camera tactile sensor'. *3rd IEEE International Conference on Soft Robotics (RoboSoft)*; New Haven, CT, IEEE, 2020. pp. 333–38.

[36] Winstone B., Melhuish C., Pipe T., Callaway M., Dogramadzi S. 'Toward bio-inspired tactile sensing capsule endoscopy for detection of submucosal tumors'. *IEEE Sensors Journal*. 2017, vol. 17(3), pp. 848–57.

[37] Li R., Platt R., Yuan W, *et al.* 'Localization and manipulation of small parts using gelsight tactile sensing'. *IEEE International Conference on Intelligent Robots and Systems*; IEEE, 2014.

[38] Luo S., Yuan W., Adelson E., Cohn A.G., Fuentes R. 'ViTac: feature sharing between vision and tactile sensing for cloth texture recognition'. *IEEE International Conference on Robotics and Automation (ICRA)*; Brisbane, QLD, IEEE, 2018. pp. 2722–27.

[39] Lee J.-T., Bollegala D., Luo S. 'Touching to see and seeing to feel: robotic cross-modal sensory data generation for visual-tactile perception'. *International Conference on Robotics and Automation (ICRA)*; Montreal, QC, IEEE, 2019. pp. 4276–82.

[40] Cao G., Zhou Y., Bollegala D., Luo S. 'Spatio-temporal attention model for tactile texture recognition'. *IEEE/RSJ International Conference on Intelligent Robots and Systems (IROS)*; Las Vegas, NV, IEEE, 2020. Available from https://ieeexplore.ieee.org/xpl/mostRecentIssue.jsp?punumber=9340668

[41] Yuan W., Dong S., Adelson E.H. 'GelSight: high-resolution robot tactile sensors for estimating geometry and force'. *Sensors (Basel, Switzerland)*. 2017, vol. 17(12), p. 12.
[42] Donlon E., Dong S., Liu M., Li J., Adelson E., Rodriguez A. 'GelSlim: a high-resolution, compact, robust, and calibrated tactile-sensing finger'. *2018 IEEE/RSJ International Conference on Intelligent Robots and Systems (IROS)*; Madrid, IEEE, 2018. pp. 1927–34.
[43] Mike L., Po-Wei C., Stephen T. *et al.* 'Digit: a novel design for a low-cost compact high-resolution tactile sensor with application to in-hand manipulation 5'.n.d.
[44] Taylor I.H., Dong S., Rodriguez A. 'GelSlim 3.0: high-resolution measurement of shape, force and slip in a compact tactile-sensing finger'. *IEEE International Conference on Robotics and Automation (ICRA)*; Philadelphia, PA, IEEE, 2021.
[45] Dong S., Ma D., Donlon E., Rodriguez A. 'Maintaining grasps within slipping bounds by monitoring incipient slip'. *International Conference on Robotics and Automation (ICRA)*; Montreal, QC, IEEE, 2019. pp. 3818–24.
[46] Gomes D.F., Wilson A., Luo S. 'Gelsight simulation for sim 2 real learning'. in ICRA ViTac Workshop; 2019.
[47] Gomes D.F., Paoletti P., Luo S. 'Generation of gelsight tactile images for sim 2 real learning'. *IEEE Robotics and Automation Letters*. 2021, vol. 6(2), pp. 4177–84.
[48] Gomes D.F., Lin Z., Luo S. 'GelTip: a finger-shaped optical tactile sensor for robotic manipulation'. *IEEE/RSJ International Conference on Intelligent Robots and Systems (IROS)*; Las Vegas, NV, IEEE, 2020. Available from https://ieeexplore.ieee.org/xpl/mostRecentIssue.jsp?punumber=9340668
[49] Gomes D.F., Lin Z., Luo S. 'Blocks world of touch: exploiting the advantages of all-around finger sensing in robot grasping'. *Frontiers in Robotics and AI*. 2020, vol. 7, p. 541661.
[50] Cao G., Jiang J., Lu C., Gomes D.F., TouchRoller S.L. 'A rolling optical tactile sensor for rapid assessment of large surfaces'.2021. arXiv preprint arXiv:2103.00595.
[51] She Y., Wang S., Dong S., Sunil N., Rodriguez A., Adelson E. 'Cable manipulation with a tactile-reactive gripper'. *ArXiv Preprint*. 2019, 1910.02860.
[52] Romero B., Veiga F., Adelson E., Soft R. 'Soft, round, high resolution tactile fingertip sensors for dexterous robotic manipulation' [online]. *IEEE International Conference on Robotics and Automation (ICRA)*; Paris, France, 2020. Available from https://ieeexplore.ieee.org/xpl/mostRecentIssue.jsp?punumber=9187508
[53] Padmanabha A., Ebert F., Tian S., Calandra R., Finn C., Levine S. 'OmniTact: a multi-directional high-resolution touch sensor' [online]. *IEEE International Conference on Robotics and Automation (ICRA)*; Paris, France, 2020. Available from https://ieeexplore.ieee.org/xpl/mostRecentIssue.jsp?punumber=9187508

Chapter 11
Audio sensors

Kazuhiro Nakadai[1], Hirofumi Nakajima[2], and Hiroshi G. Okuno[3,4]

Audio information is crucial to interface with robots and systems. There are mainly three reasons: (1) Speech is used in human–human verbal communication, and thus it is natural that a robot or a system has auditory functions to interface with humans. (2) It is said that humans use multimodal information for perception. Mehrabian [1] stated that three main factors in communication are verbal, vocal, and facial information, and their contributions are 7%, 38%, and 55%, respectively, in terms of understanding attitudes and characters. This reveals that auditory functions are essential to a robot or a system in addition to visual and other sensory functions. (3) Visual sensing is generally more accurate than audio sensing, but sound can be detected from transparent/occluded/extravisual objects which are difficult to be detected only with vision. Audiovisual sensing will disambiguate missing information from each other [2]. This will be helpful for a robot or a system to implement situation awareness such as anomaly and danger detection functions. This chapter, therefore, widely describes using audio sensors, i.e. "microphones" from devices to auditory processing. Note that this chapter focuses mainly on microphones as audio sensors, although loud speakers and transducers are audio devices which are often discussed together with microphones.

11.1 Audio sensors

Audio sensors are categorized in various ways. From the viewpoint of robotics, it is intuitive to categorize them by media such as air, underwater, underground/structures, and biological bodies, shown in Table 11.1.

[1]Honda Research Institute Japan Co., Ltd., Tokyo Institute of Technology Tokyo, Japan
[2]Kogakuin University, Kogakuin, Japan
[3]Kyoto University, Kyoto, Japan
[4]Waseda University, Waseda, Japan

Table 11.1 Types of audio sensors by media

Media	Sensor type	Frequency range	Robotics applications
Air	Ultrasonic microphones	>20 kHz	Object detection, parametric speakers [3]
	Microphones (audible range)	20 Hz–20 kHz	Speech recognition beamforming [4]
	Infrasound microphones	<20 Hz	Vibration pollution measurement, footstep attributes detection [5, 6]
Underwater	Hydrophones	1–100 kHz	Sound navigation and ranging, fish finders
		10–100 Hz	Ocean acoustic tomography [7]
Underground/ structures	Ground-penetrating radar	1–3,000 MHz	Nondestructive inspection
		300–600 Hz	Underground imaging
Biological body	Ultrasonic microphones	1–40 MHz	Medical inspection, ultrasonography
	Bone conduction microphones	100 Hz–5 kHz	Hearing aid

11.1.1 Airborne microphones

There are three types of airborne microphones according to the target frequency ranges.

For ultrasonic sounds, which are defined as sounds above audible frequency, i.e. over 20 kHz, an ultrasonic microphone is used. In robotics, this type of microphone is often used with ultrasonic transducers to measure distance to obstacles. It is a reasonable and cheap solution compared with a laser scanner or a laser range finder, while its measurement accuracy is lower. In addition, the ultrasonic transducer repeatedly produces impulsive pulses, which make jarring sounds. As for ultrasonic transducers, there is another robotic application for them, i.e. a parametric speaker array, which produces directional sound beams by using intermodulation of ultrasonic sound beams and nonlinearity in the air [8]. It has been applied to a humanoid robot to achieve "hearing during speaking" by utilizing characteristics in which a speaking directional sound beam does not interfere with a robot's listening function as a noise source [3].

Sensors to capture audible sounds ranging from 20 Hz to 20 kHz are simply called "microphones." They are used for automatic speech recognition (ASR) to achieve verbal communication with humans and microphone arrays to realize noise reduction of target sounds and auditory scene analysis (ASA). The details will be explained in the later sections.

Another type of airborne microphone for low frequencies is called an infrasound microphone. In particular, sounds with frequency below 20 Hz are inaudible, and thus such sounds are considered to be vibration. Since humans feel that high-power vibrations are annoying although it is inaudible, infrasound microphones are adopted to measure vibration pollution. As another application, they are used for footstep attributes detection, which provides not only footstep detection but also walking direction, pattern, speed, and pedestrian weight estimation [5, 6].

11.1.2 Microphones for underwater

Electromagnetic waves are hardly transmitted in water, and thus for underwater remote sensing, acoustic waves are effective. Microphones for underwater are called hydrophones. They are often used in marine robotics as sensors for autonomous underwater vehicles and unmanned underwater vehicles. The target frequency ranges vary according to applications. For sound navigation and ranging, higher frequencies such as 1–100 kHz are commonly used, while for ocean acoustic tomography [7], lower frequency sounds such as 10–100 Hz to maintain a high signal-to-noise ratio (SNR) are used. Ocean acoustic tomography is a technique to investigate the internal conditions of the ocean by measuring the propagation time of acoustic waves in the ocean. It provides ocean conditions in the range of 100 km with high temporal resolution. Since sound travels much faster in water, say about 1,500 m/s, the intermicrophone distance should be much wider in water than on the ground, when a microphone array is used.

11.1.3 Microphones for underground and structures

Because remote sensing for underground and structures generally uses electromagnetic waves of 1–3,000 MHz, such sensors can be called a radar rather than a microphone. In disaster robotics, they are used for nondestructive inspections of structures of buildings, tunnels, and bridges. On the other hand, for the detection of objects at shallow depths such as relics and ruins, it is reported that underground imaging with shear waves ranging from 300 to 600 Hz is effective [9], which suggests that audio signals are still applicable to underground remote sensing.

11.1.4 Microphones for biological bodies

Because an ultrasonic microphone is helpful for noninvasive inspections, it is often applied to medical use such as ultrasonography. As another application to a biological body, to support hearing difficulties and elderly people, bone conduction microphones and speakers are available. Basically, they are the same as ordinary ones except that the sound of kilohertz order is amplified to compensate for the attenuation by the bones and the body.

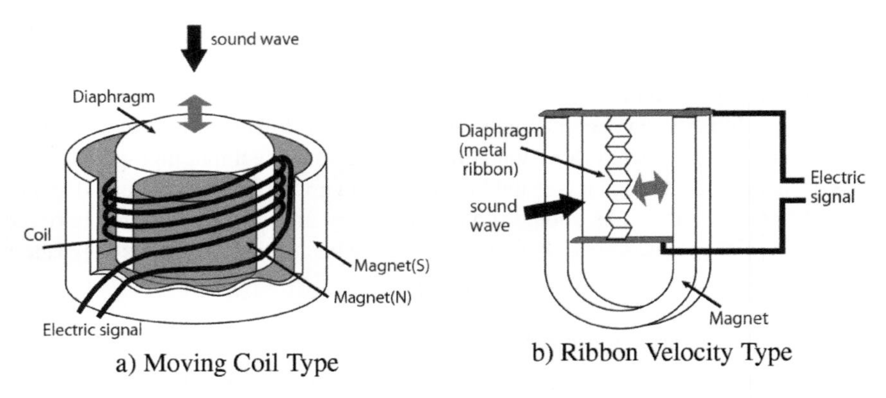

Figure 11.1 Dynamic microphones

11.2 Microphones for audible sounds

This section explains normal airborne microphones for audible sounds. They mechanically convert air pressure changes into electrical signals. There are five types of microphones as follows:

- Dynamic microphones

A dynamic microphone has a simple mechanism without using a battery or power supply, which makes it heavy duty. Compared with other microphones, it generates small distortions for high-power sound input, and thus it is often used in public address and music stages that require robustness. On the other hand, it captures mechanical noises such as friction noise generated when holding the microphone, and it should be carefully introduced to robots.

In dynamic microphones, acoustic signals are converted into electrical signals using electromagnetic induction generated by the vibration of the microphone diaphragm that is generally made of thin plastic film. There are two types of dynamic microphones: a moving-coil microphone shown in Figure 11.1a and a ribbon velocity microphone shown in Figure 11.1b. Both use a permanent magnet to generate the magnetic field. For electromagnetic induction, coil and ribbon-shaped metal are used in moving-coils and ribbon velocity microphones, respectively.

- Condenser microphones

Condenser microphones have quick response and flat frequency response caused by their light weight and super thin diaphragm, and thus they are widely used for many applications such as mobile phones. Most robots with listening functions are also equipped with this type of microphone.

In a condenser microphone, the vibration of the microphone diaphragm produces the capacitance changes, and the changes are converted into electrical signals shown in Figure 11.2. This means that the condenser microphone needs a power supply for the capacitor. A direct current-biased condenser microphone, which is used for professional audio devices, 48 V phantom power is commonly used. Electret condenser microphones reduce the voltage to 3–5 V using permanently charged materials for a

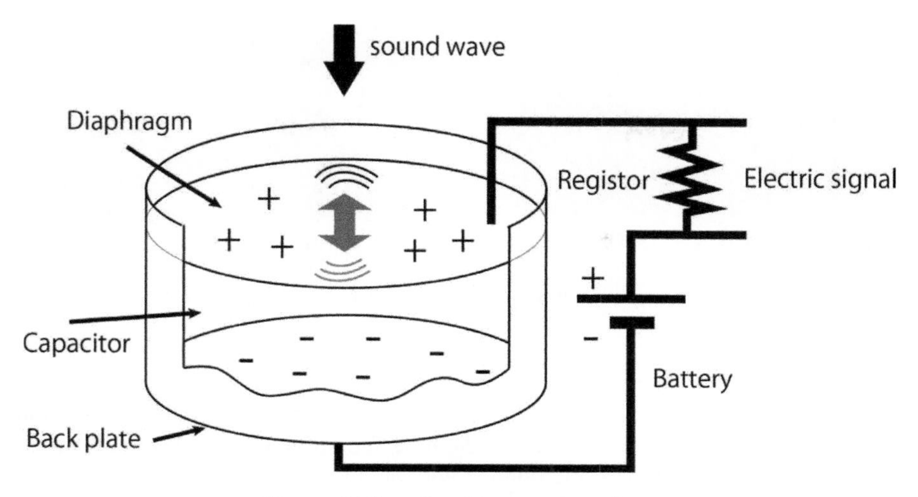

Figure 11.2 Condenser microphone

diaphragm or back plate of the microphone capsule. To provide this small power, a plug-in power supply is commonly adopted from the microphone jack on consumer devices such as personal computers (PCs). Due to different voltages, an appropriate adapter is necessary to interconnect consumer and professional audio equipment.

As another type of condenser microphone, microelectromechanical systems (MEMS) microphones shown in Figure 11.3 are getting more popular. Basically, it is an electret microphone on a silicon wafer, and thus, it is known as a silicon microphone. Because MEMS technology is used, the size of each microphone is about 5 × 5 mm or even smaller. Many MEMS microphones which have uniform characteristics are created from a single silicon wafer. These uniform characteristics make them free from individual calibrations. This is advantageous when a microphone array is designed. The microphone package has a hole called a port at the top or bottom. Figure 11.4 shows sample photos for both types of microphones. Since

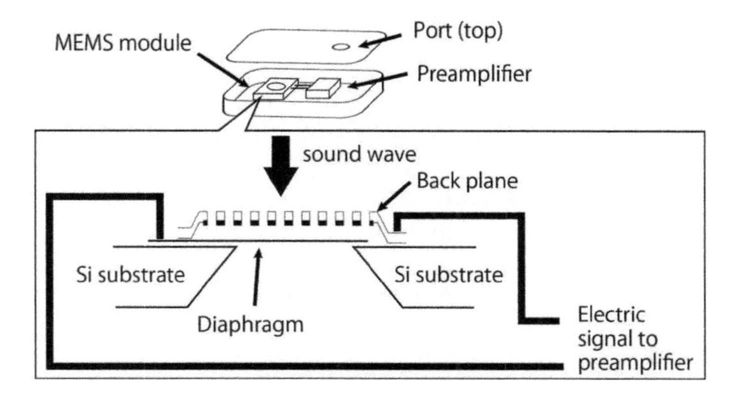

Figure 11.3 MEMS microphone

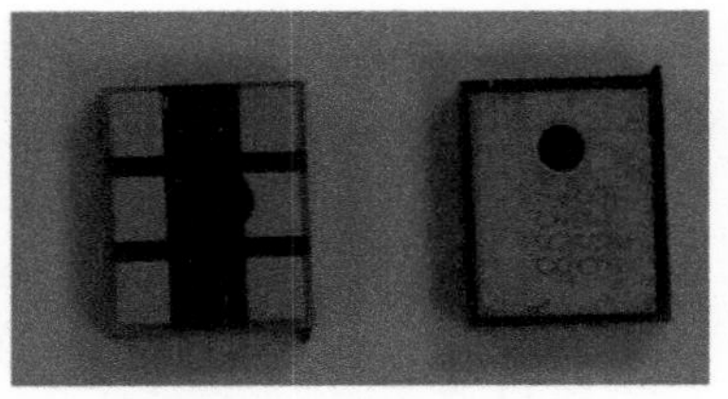

a) HOSHIDEN KRM5303, bottom port

b) Knowles SPM0406, top port

Figure 11.4 Samples of MEMS microphone: (a) the left side shows the bottom and (b) the right side shows the top

air pressure is propagated through the hole, it has a directivity in the direction of the hole. When it is embedded to a robot, the hole should be designed to be exposed outside the robot. A MEMS microphone with an audio-digital converter (ADC) is also available, which is called a digital MEMS microphone. The size of the digital MEMS microphone is slightly larger than a normal analog MEMS microphone. However, it has an advantage in that an external ADC is not necessary, which makes the size of the entire system smaller. In addition, the characteristics of the ADCs are almost identical, especially when made on the same wafer. This reduces the cost of calibrating the entire microphone system.

- Piezoelectric microphones

These types of microphones shown in Figure 11.5 utilize the piezoelectric effect that is a phenomenon in which a voltage is generated when stress is applied to a dielectric material, and the material deforms when a voltage is applied. It is also called a crystal or ceramic microphone, which is derived from the piezoelectric materials used in the microphone. Piezoelectric microphones have frequency

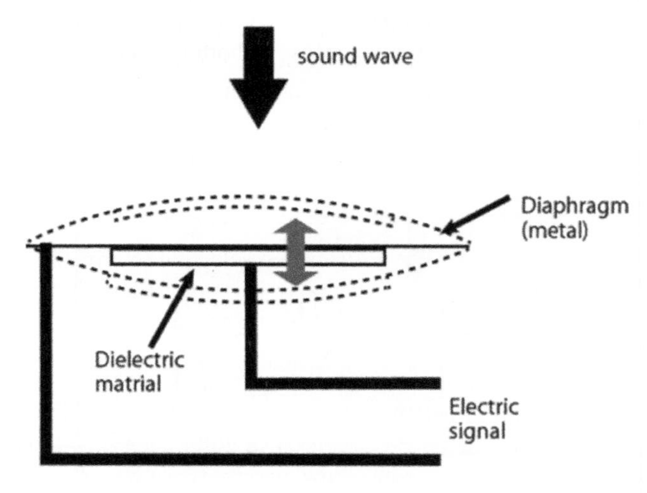

Figure 11.5 Piezoelectric microphone

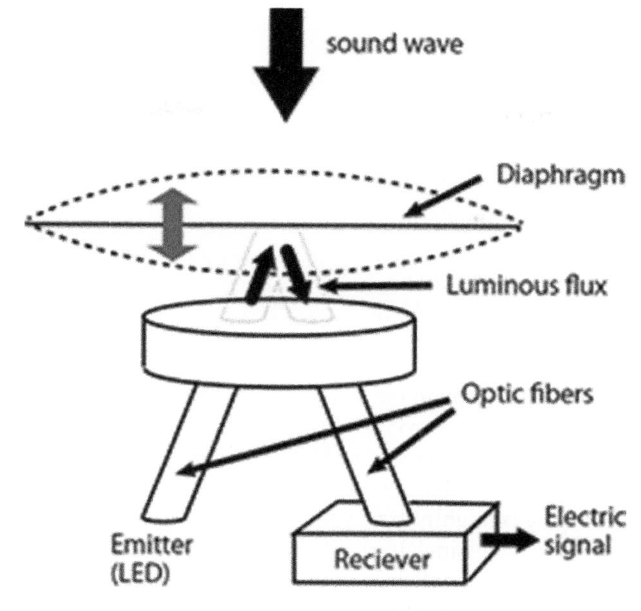

Figure 11.6 Optical microphone

response with a peak of 3–5 kHz. Due to this narrow band characteristic, it is used for wireless communication.

- Optical microphones

These microphones shown in Figure 11.6 detect the vibration of the diaphragm as the changes of luminous flux happen, which is emitted by a laser or light emitting diode. Structurally, the diaphragm can be lightened, and it has a high fidelity property for input sound. Since light is propagated via an optic fiber, it is free from contamination by electromagnetic noises. In addition, it is heat-resistant, and it can be designed with high directivity. On the other hand, other noises such as friction noise to the optic fiber affect the detection of the luminous flux changes as noise. These properties should be carefully considered to introduce this microphone to a robot.

11.2.1 Indicators for microphone characteristics

This section explains important properties when selecting microphones. The specification of a microphone describes the characteristics of a microphone mainly using indicators like directivity, sensitivity, frequency response, dynamic range, and SNR.

- Directivity

A directivity pattern is often represented by a circular diagram with the angle indicating the input sound direction and the radius indicating the microphone gain illustrated in Figure 11.7. The characteristics also differ according to the frequency of the input sound, and it is common to show directivity patterns for multiple typical frequencies.

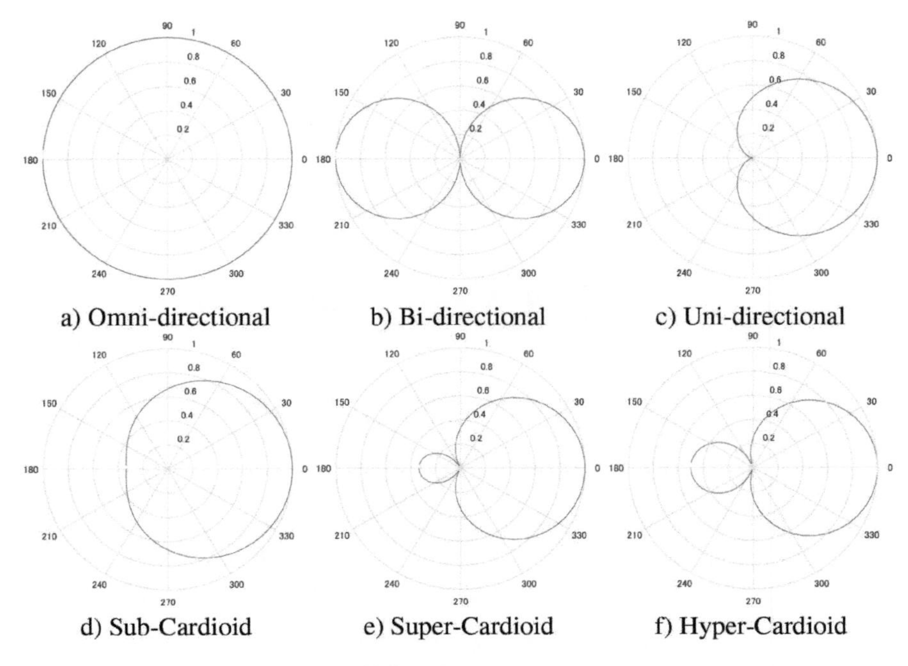

Figure 11.7 Directivity patterns

Figure 11.7a shows the directivity pattern for an omnidirectional microphone, which has equal gain for all directions. Basically, a microphone's directivity is formed by its shape and mechanism. Because the diaphragm of a microphone is a flat membrane, it originally has high directivity in the direction perpendicular to the plane of the diaphragm, while it has low directivity in the direction parallel to the plane. In addition, the material has a specific frequency characteristic. In this sense, there is no perfect omnidirectional microphone for all three-dimensional (3D) directions and frequencies. By devising the structure of the microphone, on a specific plane or region, an omnidirectional property can be achieved. For instance, the orientation of diaphragm changes depending on microphones. In a side-address microphone, the normal direction of the diaphragm plane directs out of the side of the microphone. This is typical of large-diaphragm microphones. In a top-address microphone, the normal direction of the diaphragm plane is parallel to the outer direction of the top of the microphone. This is typical of small-diaphragm microphones. When such an omnidirectional microphone is installed in a robot, it should be considered that the robot's auditory system with the microphone is not omnidirectional any more due to the influence of a robot's body and head. The actual directivity of the system can be revealed by measuring transfer functions to the microphone. For a microphone array discussed in section 11.3, omnidirectional microphones are commonly used, because in microphone array processing, directivity can be controlled by a software algorithm, and thus it is convenient that each microphone of the microphone array does not have a specific directivity in terms of hardware.

Figure 11.7b shows the directivity pattern for a bidirectional microphone. Its gain becomes high for sounds originating from specific and opposite directions. The phase information between the sound coming from these two directions is reversed (180° difference). This microphone is sometimes used by being placed between two people facing each other to record their voices at once.

Figure 11.7c shows a unidirectional microphone also known as cardioid derived from its heart shape. Since this microphone suppresses sounds originating from non-target directions, it is robust for howling and noisy conditions.

An intermediate pattern between omnidirectional patterns and cardioid is called a subcardioid shown in Figure 11.7d. It increases the gain from the side directions and mainly suppresses sounds from back directions. When this microphone is attached to a robot, it may be helpful to suppress the robot's ego-noise to some extent, since such noise comes only from the back directions.

Supercardioid (Figure 11.7e) and hypercardioid (Figure 11.7f) are recognized as patterns between cardioid and bidirectional. These patterns have higher directivity to the target sound direction than a cardioid, and instead they produce gains for sounds coming from the opposite direction, which is a similar side effect to sidelobes in beamforming (section 11.4). Due to their high directivity, these types of microphones should be placed in consideration of the geometrical relationship to the target sound source. In other words, when these types of microphones are used for a robot, the robot needs to be well controlled to track the target sound source.

- Sensitivity

The sensitivity of a microphone, G, is defined as 0 dB or 0 V/Pa when the microphone receives a sound with a sound pressure of 1 Pa (94 dB SPL) and its output voltage is 1 V.

$$G = 20 \log_{10} \frac{V}{P} \tag{11.1}$$

where V is the output voltage of the microphone in volts, and P is the sound pressure of the input sound in pascal. As this indicator shows how easy a microphone can detect a sound, in general, the larger the value is, the better the microphone is. However, in practice, it is not always the case when the microphone is connected to another device such as a preamplifier, because their differences in signal levels and impedance should be considered. The output impedance of a microphone has been traditionally designed to be 600 Ω, but currently much smaller impedance such as 50 Ω is becoming familiar. The input impedance of a preamplifier should be several times higher than the output impedance of the microphone.

- Frequency response

This is represented as a frequency sensitivity graph when the sensitivity at 1 kHz is 0 dB. The frequency axis is often expressed on a log scale. This graph is helpful, since various characteristics of a microphone can be seen at a glance.

- Dynamic range

This indicates the range of input sound pressure that can be recorded with lower than a certain distortion level. Total harmonic distortion is used as a measure of distortion, and 0.5% or 1.0% is typically set as the threshold of the distortion level. Dynamic range D is denoted in decibel as

$$D = 20\log_{10}\frac{P_{max}}{P_{min}} \tag{11.2}$$

where P_{max} and P_{min} are the maximum and minimum input sound pressure, respectively. When the target sound has a wide dynamic range, e.g. in an outdoor environment, microphones with wide dynamic range should be selected. In most cases, the captured sounds with the microphones are digitized, and thus an ADC with a large number of quantization bits should be selected for sounds recorded by a microphone with a wide dynamic range. Generally, 16-bit ADCs are normally used in indoor situations, and 24-bit and 32-bit ADCs will be appropriate outdoors. Note that the size of recorded data increases in proportion to the number of quantization bits of ADC.

- SNR
  This is defined as the voltage ratio of target and noise signals in recorded signals.

$$SNR = 20\log_{10}\frac{S}{N} \tag{11.3}$$

Generally, microphones should have an SNR of 60 dB or better, and normal microphones fulfill this requirement. The SNR of MEMS microphones was slightly low at 50–60 dB in the beginning, but currently they have been improved to an SNR of 60 dB or more. When considering the introduction of a microphone into a robot, the SNR of the entire auditory system is more important than the SNR of the microphone itself. This should be carefully considered when designing a robot.

11.3 Microphone array

A microphone array consists of multiple microphones. It can provide the functions of sound source localization that estimates sound source directions, and sound source separation that extracts each sound source from a mixture of sound sources, by utilizing the phenomenon that the time and amplitude of the sound originating from a sound source are different. Because a robot should be operated in a noisy environment and the robot itself generally makes high-power ego-noise, a microphone array is useful. Therefore, in the field of robot audition [10], technology using microphone arrays has been actively developed as discussed in sections 11.4 and 11.5.

A microphone array using analog microphones can be illustrated in Figure 11.8. Although commercial microphone arrays are available, a microphone array can be designed to fit a robot. In this case, the following three factors should be carefully considered.

1. Microphone characteristics
2. ADC
3. Microphone array layout

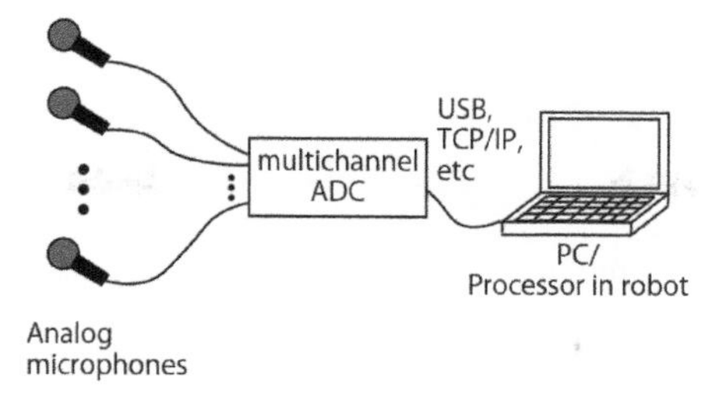

Figure 11.8 Conventional microphone array with analog microphones

The first factor is essential for maintaining the performance of microphone array processing. Since microphone array processing utilizes the time and/or amplitude differences between microphones, the characteristics of all microphones should be uniform. In this sense, the same model number microphone should be selected for all microphones making up a microphone array. Even when the model number is fixed to be the same, there are still individual differences between microphones. To mitigate this issue, it has traditionally been practiced to manually select a microphone with the same characteristics or to calibrate the microphone based on acoustic measurements. Recently, the simple use of MEMS microphones provides a better solution, because these microphones originally have the same characteristics as mentioned above.

The second factor is indispensable to achieve microphone array processing. The time difference between microphones is typically less than 1 μs, which is on the order of subsamples after analog–digital conversion. This means that a sound must be recorded synchronously on all microphones. Here, note that there are two types of ADCs for multichannel recording shown in Figure 11.9. Figure 11.9a provides perfect synchronous recording because all channels are sampled synchronously. This type of ADC is suitable for microphone array processing. On the other hand, Figure 11.9b provides a multichannel recording with time-division multiplexing. The output from this type of ADC includes a time difference between channels. Although it is small, it is not negligible for microphone array processing. The

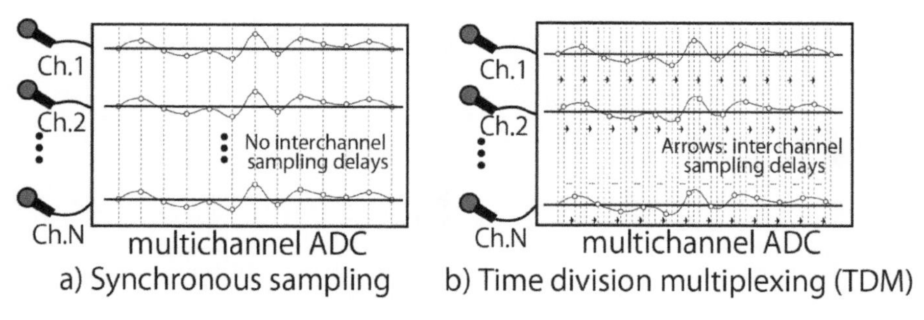

Figure 11.9 Two kinds of sampling for multichannel ADC. Dotted bars indicate sampling timing

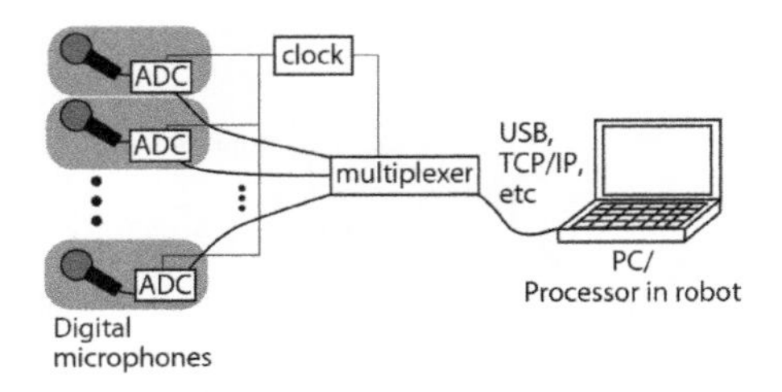

Figure 11.10 Microphone array with digital microphones

former type of ADC is generally expensive, which adds to the cost of a microphone array system. This conventional microphone array has a centralized ADC shown in Figure 11.8. This structure causes the cost problem. Digital MEMS microphones can provide a simpler and more reasonable solution. Figure 11.10 shows the system structure of a microphone array with digital MEMS microphones. Each digital MEMS microphone has an ADC, and the characteristics of the ADC are uniform. Although ADCs are distributed in this structure, synchronous recording is realized by feeding the same clock signals.

The last factor should be decided by taking target applications into account. For example, a linear microphone array provides one-dimensional sound source localization as shown in Figure 11.11a. However, it is difficult to decide if the sound comes from the front or the back direction, because both have the same time and amplitude differences. This is called the front–back problem. Circular microphone arrays can solve this problem as shown in Figure 11.11b. However, it is difficult to estimate elevation of target sound sources, because all microphones are aligned on the same plane. When the target application needs localization in both azimuth and elevation, the microphones are three dimensionally placed as shown in Figure 11.11c. Even in this case, distance information is still difficult to be estimated. The size of a microphone array and the distance between microphones are also tunable parameters. Generally, the large size of a microphone array leads to high angular resolution of sound source localization and separation (narrow mainlobe in beamforming), and

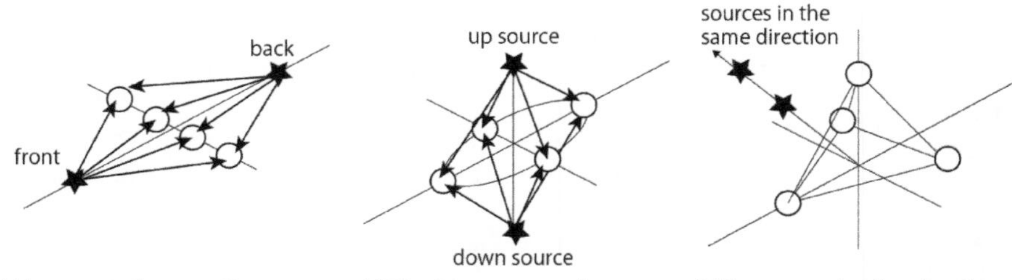

Figure 11.11 Microphone layouts in microphone array. Circles and stars indicate microphones and sound sources, respectively

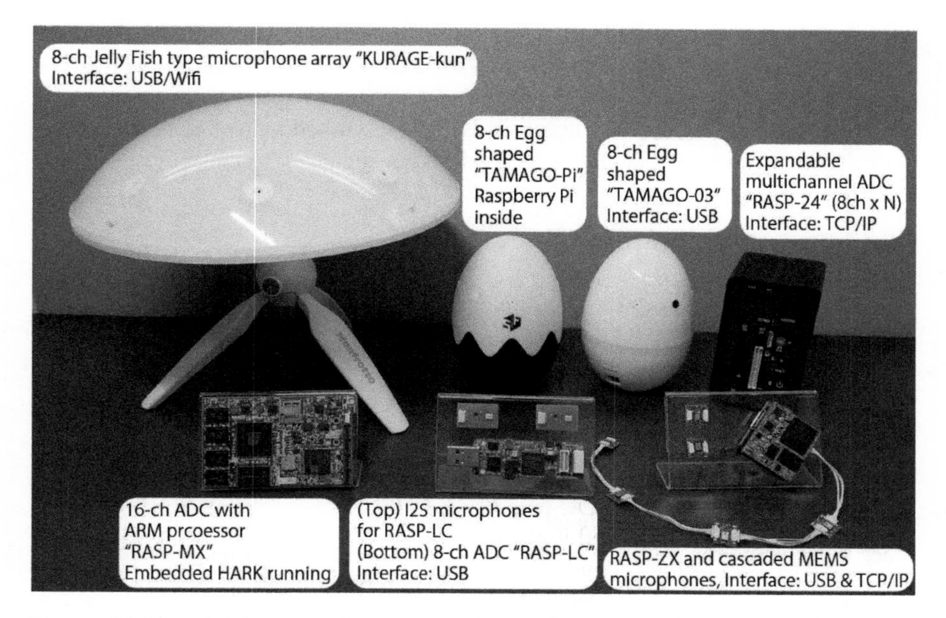

Figure 11.12 RASP series for microphone arrays, the photo was provided by System in Frontier, Inc. with permission to be used

the smaller distance between microphones can contribute to produce less artifacts (less sidelobes in beamforming). This means that the performance of a microphone array improves as the number of microphones increases. The best size and distance depend on the target application and microphone array algorithms.

There are many commercial microphone arrays. Kinect and PlayStation Eye are off-the-shelf microphone arrays. However, these microphone arrays need to be hacked to install them into a system. Furthermore, once a new version is released, the old version is discontinued, and new hacks will be required. The real-time array signal processor (RASP) series are long-term support microphone arrays shown in Figure 11.12. They generally have a universal serial bus (USB) and/or a network interface to be connected with Linux and Windows PCs. USB-interface-type devices support Advanced Linux Sound Architecture (ALSA) and Windows Audio Session Application Programming Interface (API), and network-interface-type devices provide APIs for Linux and Windows. In any cases, no hacks are necessary to use them. All the three factors discussed above are well considered. For the first and second factors, they offer solutions with analog–digital MEMS microphones. For the last factor, in addition to various types of layouts which are originally prepared, they offer a solution which allows us to design an arbitrary layout of a microphone array. One of the most advantageous characteristics of this solution is to construct a microphone array by a cascade connection (RASP-ZX). Conventional microphone arrays need star wiring with a centralized ADC or a clock element. This limits the microphone layout and can make the wiring itself difficult. The cascade connections drastically reduce this problem. In addition to the RASP series, professional audio devices that support multichannel recording such as RME's products [11] can be used for microphone arrays. Recently, embedded microphone array boards are also available. Note that some of these boards include embedded software for microphone array processing that cannot be modified by the user.

11.4 Robot audition

In addition to audio sensor devices such as microphones and microphone arrays, audio sensing is essential, which has been studied in the research area of "robot audition." It has been proposed in 2000 to realize listening functions for robots in a noisy environment using microphones embedded on the robot [10]. Before that, although auditory functions are essential to a robot to achieve natural human–robot interactions, studies reporting auditory processing for robots had limitations, a microphone was attached to the position of a human mouth like a headset microphone [12, 13], and it was assumed that the level of a target speech was high enough to ignore other noise sources [14–16]. A robot, thus, had to follow a strategy called "stop-perceive-act" to avoid motion noise generation on listening [10]. In robot audition, three primary issues have been discussed under the context of online and real-time processing.

1. Understanding a general sound—computational ASA (CASA)
2. Active audition
3. Multimodal integration

The first issue is that a robot should understand a general sound. Since a target sound source and robot's embedded microphones are distant from each other (usually 1–2 m), a robot captures "general sounds," defined as a mixture of sounds from various sources including a target source. Therefore, the ability to understand a general sound is indispensable for the robot operating in the real environment. Such a robot is surrounded by various types of sounds including directional sound sources such as human utterances, diffuse sound sources such as background noise, reverberations reflected by the wall and floor, and self-noise generated by a robot's actuators and voice. These sound sources have different characteristics, and methods of suppressing them also depend on the type of sound source. Moreover, target sound sources are situation-dependent. For example, when a robot listens to a user's utterances, nonspeech sources should be considered noise. However, when the target source is music, speech sources should be considered noise. Thus, the concept of CASA [17] should be considered in the robot.

CASA was proposed as a constructivist approach to elucidate human auditory functions in the 1990s inspired by "ASA" written by Bregman in 1990 [18]. ASA aims to elucidate human auditory functions psychophysically based on the idea that humans perceive each sound as a sound stream. In a general environment where there are multiple sound sources, ASA claims that a mixture of sound sources are perceived as multiple streams by stream segregation based on various cues. This concept of human perception when multiple sound sources exist simultaneously has drawn attention because mainly a single sound source has been considered before that.

Therefore, robot audition considers understanding a general sound as a primary issue, and technology for its realization, such as sound source localization and sound source separation, has been studied together with ASR extensively and deeply [2, 19–25].

The second issue is unique to robots. Since a robot can move, active movements can improve its auditory functions. This is similar to the concept of active vision and active perception [26]. For active audition, two subissues should be considered. One is that motions for better perception result in more noise, in particular, ego-noise. The other is that motion planning for better perception is not obvious. The former is regarded as ego-motion noise cancellation. Techniques based on spectral subtraction (SS) [27] have been often used [28–30]. Sound source separation techniques, such as independent component analysis (ICA) [31] and non-negative matrix factorization (NMF) [32], have been proposed; these methods result in less distortion and thus are more appropriate for ASR. The latter is mainly regarded as sound source localization. In binaural robots, active motion has been used to solve the front–back problem [10, 33]. Mobile robots with a microphone array have been reported to localize sound sources in three dimensions by triangulation [34]. Motion planning strategies were studied to better listen to a target source by taking noise source positions into account [35, 36].

The last issue is related to maintaining robustness of a robot's perception. Sound localization and separation using audio-visual integration has been reported. Actually, even in humans, sound source localization has an error of several degrees, with auditory perception reported affected by visual cues [37, 38]. Thus, robots can use audio-visual integration for sound source localization. Real-time multiple human tracking systems using a pair of microphones and a pair of cameras have been reported for humanoid robots [2, 19]. Sound source localization and tracking have also involved a microphone array and cameras embedded in a robot's head [40–42].

11.5 Acoustic signal processing

As discussed in the previous section, sound source localization and separation are key technologies in robot audition. These technologies have been developed with microphone array processing, which is a hot research topic in acoustic signal processing. Since many algorithms for sound source localization and separation have been developed, this section overviews microphone array processing from the perspective of robot audition and introduces commonly used methods for localization and separation together with *Open Source Software* (*OSS*) for robot audition called Honda Research Institute Japan Audition for Robots with Kyoto university (HARK).

Deep learning techniques have first come to the forefront in the field of computer vision such as the ImageNet Large Scale Visual Recognition Challenge [43] that accelerated the third Artificial Intelligence (AI) boom. It has been actively

Table 11.2 Models for observation signal, separation, and suppression signals—from multiply-accumulate model to function model

Multiply-accumulate model

a) Model to deal with additive noise

$X = S + N \rightarrow \|\hat{S}\|^p = \|X\|^p - \|\hat{N}\|^p$,

where X, S, N, $\hat{S}$, and $\hat{N}$ are observation, source, noise, noise-suppressed, and estimated noise signals in the frequency domain, respectively. p values indicate amplitude and power for 1 and 2. More generally, p could be a noninteger value, and p-norm operation is performed according to p value

b) Model based on spatial information

$\mathbf{X} = \mathbf{HS} \rightarrow \hat{\mathbf{S}} = \mathbf{WX}$

$\mathbf{X}$ is an M-dimensional observation signal vector, where M and N are the number of microphones and sound sources, respectively. $\mathbf{S}$ is an N-dimensional sound source signal vector. $\mathbf{H}$ is an $M \times N$ transfer function matrix. $\mathbf{W}$ is an $N \times M$ separation matrix. $\hat{\mathbf{S}}$ is a separated signal vector. All are represented in the frequency domain

c) Model based on sparseness

$\mathbf{X} = \mathbf{HU}$

$\mathbf{X}$ shows a matrix for the amplitude spectrogram of the observation signal. $\mathbf{H}$ is a basis matrix consisting of spectral basis vectors, where each spectral basis vector corresponds to each sound source and/or a part of a sound source. $\mathbf{U}$ is an activation matrix for $\mathbf{H}$

Function model

d) Model based on deep learning

$\mathbf{X} = h(\mathbf{S}) \rightarrow \hat{\mathbf{S}} = f(\mathbf{X})$

$\mathbf{S}$ is a signal vector (in most cases, amplitude spectral vector). $\mathbf{X}$ is an observation signal vector for $\mathbf{S}$. h and f are functions representing a signal model and noise suppression model, respectively. $\hat{\mathbf{S}}$ is a noise suppressed signal vector.

introduced to ASR [44] and the performance of ASR drastically improved. Recently, for sound source localization and separation, deep learning techniques have been actively explored. When considering the changes in sound source localization and separation by deep learning, it could be summarized as an extension of multiply-accumulate models in traditional acoustic signal processing to function models mainly with deep learning shown in Table 11.2.

The simplest observation signal model to deal with noise is a multiply-accumulate model only with simple additions that considers additive noise shown in Table 11.2a. In this model, noise suppression can be achieved by subtracting estimated additive noise from the observation signal. One of the typical methods is called SS [27], and many methods are derived from this method. For example, in CASA [17], the mainstream methods have been used to estimate the sound source using various heuristics obtained from human auditory studies such as harmonic structures and perform SS-based separation [45]. Common methods for speech enhancement such as minima controlled recursive averaging (MCRA) [46] and minimum mean square error (MMSE) [47] are derived from SS by introducing advanced noise estimation. In addition, many dereverberation methods are considered as SS, because they estimate the late reverberation component and subtract it from the input. This type of multiply-accumulate model is, thus, effective to suppress additive noise and reverberation and has been studied extensively for half a century.

Another type of multiply-accumulate model including multiplications in the frequency domain also has been extensively studied. There are mainly two approaches using spatial information (Table 11.2b) and sparseness (Table 11.2c). In the approach using spatial information, multichannel signals recorded with a microphone array are used as input to be able to use the position information of sound sources. As the product of this multichannel input vector and a separation matrix, the separated signals originating from each sound source are obtained. The idea behind this is that because the observed signal can be modeled as the product of a target sound source and a transfer function in the frequency domain, the sound source can be extracted by multiplying the inverse function of the transfer function to the observed signal. There are mainly two sound source separation categories based on this idea: "beamforming" uses the transfer function or its corresponding information as known information to estimate a separation matrix, and "blind separation" estimates the separation matrix in a blind way, i.e. without explicitly using the transfer function. ICA [48] is a method of estimating the separation matrix by assuming statistical independence between sound sources instead of using transfer functions. Since the separation matrix estimated by ICA is shown to be a spatial separation filter in the same way as beamforming, ICA is also considered an approach to use spatial information [49]. Many extensions of ICA have been proposed such as independent vector analysis [50, 51] and independent low-rank matrix analysis [52].

NMF [53] is a typical method based on another approach to use the sparseness of sound sources. In NMF, the observed signal is expressed as a power spectrogram, i.e. a matrix of frequency × time frame, and sound source separation is performed by decomposing this matrix into a product of two matrices in which each element is non-negative. By using non-negativeness as a constraint, matrix factorization is performed such that the element vectors of each decomposed matrix are less correlated, and the decomposed matrices are sparse. Thus, the decomposed matrices $\mathbf{H}$ and $\mathbf{U}$ are called a basis matrix and activation matrix, respectively. Compared with the approach to use spatial information, this approach has an advantage that it can perform sound source separation from a single channel input without using a microphone array. There are various extensions such as multichannel extensions [54] and complex number extensions [55]. Many sound source separation methods are actually based on a multiply-accumulate model.

There are also attempts to incorporate with statistical signal processing and AI technologies such as Bayesian estimation and nonparametric Bayesian models. Although the algorithms become complicated in these cases, their models for observation signal, separation, and suppression do not deviate from the multiply-accumulate model.

On the other hand, deep learning extended the multiply-accumulate model to be more general, i.e. a function model shown in Table 11.2d. Since a general function can be applied to a separation process, nonlinear separation can be dealt with. Traditional shallow networks, such as three-layer neural networks, are also based on a function model in principal, although the model is less flexible due to the small number of layers. Because deep learning uses more layers, the model becomes flexible (or redundant), and further improvements in the performance can be expected.

For example, "denoising" applies deep learning as a regression problem to suppress noise, estimates the target signal as the output from an observation signal and the input by using a black box separation function of a denoising neural network. Various methods are applied, ranging from relatively simple ones with general deep neural networks and deep autoencoder [56] to those with convolutional neural networks (CNN) [57] and recurrent neural networks [58]. These studies reported higher performance than the conventional sound source separation methods based on the multiply-accumulate model. However, from the perspective of robot audition, there are three issues to move on deep learning:

- Adaptation to dynamic environmental changes
- Data collection and annotation for various configurations of recording devices
- Processing speed and model size

The first issue is crucial when considering applications to robots. Generally, trained neural networks are static and retraining of the networks is time-consuming. An idea is to train the network to include all the expected situations from the beginning. However, robots are real-world applications, and the risk of coping with unexpected situations is unavoidable.

The second issue is also crucial when considering practical use. Usually deep learning requires a huge amount of annotated data. Data collection and annotation are generally expensive and time-consuming. In addition, whenever the recording device configuration changes, e.g., the change of the number of microphones, data recollection and annotation will be necessary. Although unsupervised learning approaches have been studied, data collection is still necessary.

The last issue can be solved by new hardware, since hardware is evolving day by day. Although it takes a lot of time for training networks, processing like decoding requires less computational power. Nevertheless, a dynamic range of acoustic signal processing is wider than that of image processing, and thus floating point operations are necessary. In a robot, besides audio processing, many modules should be embedded in an integrated way. It is recommended to reduce the size of a network to make the processing speed faster. Model compression techniques such as knowledge distillation [59–61], node pruning [62, 63], and model factorization [64–68] will be helpful to relax this problem.

In practice, it may require more time to entirely move sound source localization and separation to deep learning–based methods. Refer to the literature [69–71] for the detailed sound source localization and separation algorithms, as shown in Table 11.2. The following section will explain OSS for robot audition called HARK that provides practical robot audition functions based on traditional signal processing.

11.6 OSS for robot audition

Since robotic research needs a wide spectrum of technologies, it is difficult to cover with all of such research fields by a researcher or a research group. As a research achievement, it is effective to directly share information by providing software/hardware modules as open sources. In the field of robot audition, there are several OSS packages [89] to provide auditory functions described above. BeamformIT developed at UC Berkeley consists of filter-and-sum beamformers that are implemented using C++ and can be linked to the *ASR* engine Kaldi. BTK developed at Karlsruhe University, Saarland University, and Carnegie Mellon University, is a tool kit for distant speech recognition that is implemented using C++ with a Python interface, including several beamforming methods such as delay-and-sum beamforming. ManyEars developed at the University of Sherbrooke includes sound source localization by high-speed two-dimensional beamforming and sound source separation by geometric source separation in which a hybrid algorithm between blind separation and beamforming is implemented. Model-based Expectation-Maxmization Source Separation and Localization developed at Columbia University is a sound source localization and separation package based on a binaural model. Flexible Audio Source Separation Toolbox developed at Institut National de Recherche en Informatique et en Automatique supports single-channel sound source separation such as NMF, which is implemented using C++ with a Python interface.

These packages have two problems: they do not provide whole robot audition but only a part of the robot audition functions. Many were open sourced as achievements of projects and Ph.D.'s work, and they have not been maintained for a long time. *HARK*[*] can solve these problems [90].

On the first problem, it provides all the processing running on Linux and Windows from microphones to ASR such as 12 types of sound source separation algorithms and noise robust sound source localization algorithms shown in Table 11.3. All algorithms generally work online, and they are provided as modules encapsulated for different functions; even researchers who lack sufficient knowledge of signal processing or speech processing can easily combine and embed modules in their own systems. HARK can directly connect with multichannel ADCs that support ALSA, direct X, and audio stream input output, in addition to several

[*] Hark is a medieval word that means "listeners."

Table 11.3 Major algorithms in HARK

Sound source localization
• Multiple signal classification (MUSIC) [72] • Generalized eigenvalue decomposition based MUSIC [73] • Generalized singular value decomposition based MUSIC [74]
Sound source separation
Fixed beamformers • Delay-and-sum beamformer (DS-BF) • Null beamformer • Weighted DS-BF • Indefinite term and least square estimator-based beamformer [75] Explicit use of noise information • Maximum likelihood beamformer [76, 77] • Maximum SNR beamformer [78] Minimum variance • Minimum variance distortion-less response [79] • Linear constrained minimum variance beamformer [80] • Griffith-Jim beamformer [81] Linearly constrained blind separation • Geometric source separation [82] • Geometric ICA [83] • Geometric high-order decorrelation-based source separation [84]
Other noise estimation and speech enhancement
• MCRA [46] • MMSE [47] • Histogram-based recursive level estimation [85] • Semi-Blind ICA (SB-ICA) [86] • Template-based ego-noise estimation [87] • Online robust principal component analysis [88]

commercially available microphone arrays such as Microsoft Kinect for Windows (four microphones), Sony PlayStation® Eye (four microphones), and System in Frontier TAMAGO-03 (eight microphones). Furthermore, HARK can be integrated into a user's system using a function to seamlessly connect to *Robot Operating System (ROS)* [92]. HARK also provides a function to create a user's own module by using Python, which can be programmed with *Graphical User Interface* (*GUI*) programming environment called HARK Designer.

On the second problem, since HARK has been released to the public in 2008 to share robot audition research results obtained over a period of more than 10 years [4], it has continued to be developed and promoted. The updated versions of HARK are released almost every year with e-mail support, detailed documents in Japanese and English, free tutorials, and hackathons [91]. In particular, over 300-page rich documentations such as the HARK Document and the HARK Cook Book including some tutorials will be helpful, when users need to set certain parameters of HARK modules (e.g. in order to use a user's original microphone array, the parameters must be adjusted).

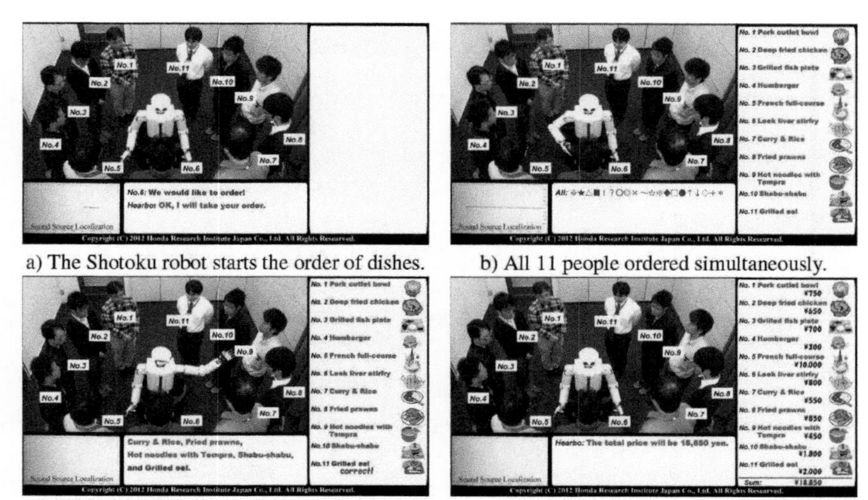

a) The Shotoku robot starts the order of dishes. b) All 11 people ordered simultaneously.

c) Each speech is extracted and recognized. d) The total price was displayed.

Figure 11.13 *Prince Shotoku Robot understands simultaneous orders from 11 people. Reproduced with permission [91]. Copyright 2017, Fuji Technologies.*

11.7 Applications of robot audition

Robot audition was proposed for human–robot verbal communication and is currently expanding to many research fields such as human search in a disaster-stricken area, noise detection, hands-free application in automobiles and mobile devices, support for persons having hearing difficulty, and sound monitoring in a natural environment for the behavior analysis of animals. This section introduces three case studies: a Prince Shotoku robot†, a drone audition system, and a virtual reality (VR) system based on bird song analysis.

11.7.1 Prince Shotoku robot

Prince Shotoku robot takes meal orders from 11 people, and they answer different dishes simultaneously. The robot understands all orders, confirms each order, and the total price was announced based on the recognized orders. The flow of the task is shown in Figure 11.13. This task is achieved using a microphone array mounted on the head of a robot, which consists of 16 MEMS microphones connected in a cascade manner. With the microphone array, simultaneous speech by 11 people is captured, which is sent to HARK online. In a HARK network shown in Figure 11.14, frequency analysis is performed, and the directions of 11 people are estimated with sound source localization. A sound originating from each estimated direction is extracted with sound source separation, and the extracted sound was converted into

†Prince Shotoku is a legendary person in Japan who could listen to ten petitions simultaneously.

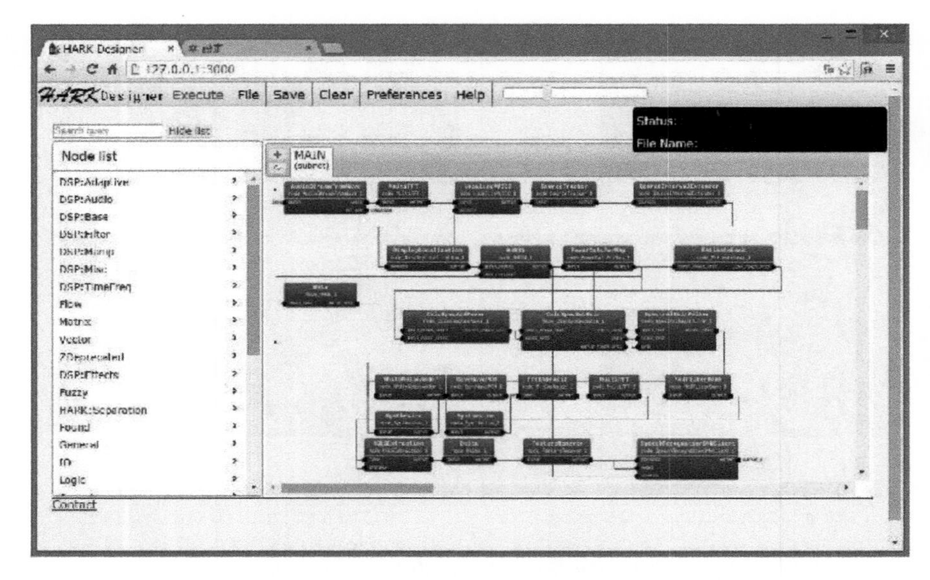

Figure 11.14 *HARK network for Prince Shotoku robot. Reproduced with permission [91]. Copyright 2017, Fuji Technologies.*

acoustic features for ASR. All acoustic features are recognized, and a dialog system connected to HARK using ROS deals with the recognized texts and controls the task of the Prince Shotoku robot.

11.7.2 Drone audition system

Drone audition has been actively studied as an extension of robot audition to search for survivors in disaster situations using a drone embedded microphone array [95, 96]. In 2014, the Tough Robotics Challenge (TRC) [97] which is a 5-year project of the Impulsing Paradigm Challenge through Disruptive Technologies Program was launched by the Japanese Cabinet and Japan Science and Technology Agency. Drone audition was strongly supported by TRC, and sound source localization and extraction with a microphone array mounted on a drone were achieved in real time under highly noisy outdoor situations where drone noise, i.e. ego-noise, and other environmental sounds exist [97–99]. Figure 11.15 shows a drone with a microphone array, and Figure 11.16 provides a set of snapshots of a drone audition demo using the drone [100]. In Figure 11.16a, two operators control the drone flying. A person wearing a red jacket (bottom-left) and another in a pipe (right-bottom) are target sound sources. In Figure 11.16b, the operator's computer shows sound localization results as blue circles corresponding to two targets on the point cloud map measured in advance. In Figure 11.16c, the subject in the pipe is successfully detected even when the lid is closed. Figure 11.16d shows a view for rescue operators. The left panel shows the radar view in the drone's coordinates. The angle and radius of the circle represent azimuth and elevation, respectively, and the white fan shape shows the area to be ignored because

Figure 11.15 *Drone with a microphone array: ACSL PF1 connected to a spherical microphone array. The other two spheres are dummies to balance the weight. Adapted with permission. Copyright 2020 Nakadai Lab., Tokyo Tech.*

Figure 11.16 *Drone audition demo: search and rescue task from the sky to search for people outdoors. Adapted with permission. Copyright 2020 Nakadai Lab., Tokyo Tech.*

Figure 11.17 A 16-ch microphone array for bird song analysis, supporting long-time and timer recording. The photo was provided by System in Frontier, Inc. with permission to be used.

extremely high-power noise generated by drone's propellers exists in this area. When sound sources are detected, a bright area will appear in the black part. The right panel illustrates a top view including frame-based sound source candidates as red dots, integrated estimation of sound source positions as blue dots (corresponding to blue circles in Figure 11.16b), and a drone position and its trajectory as a black dot and line. A 3D point cloud map was generated in advance, but real-time point cloud map generation will be realized in the near future [101].

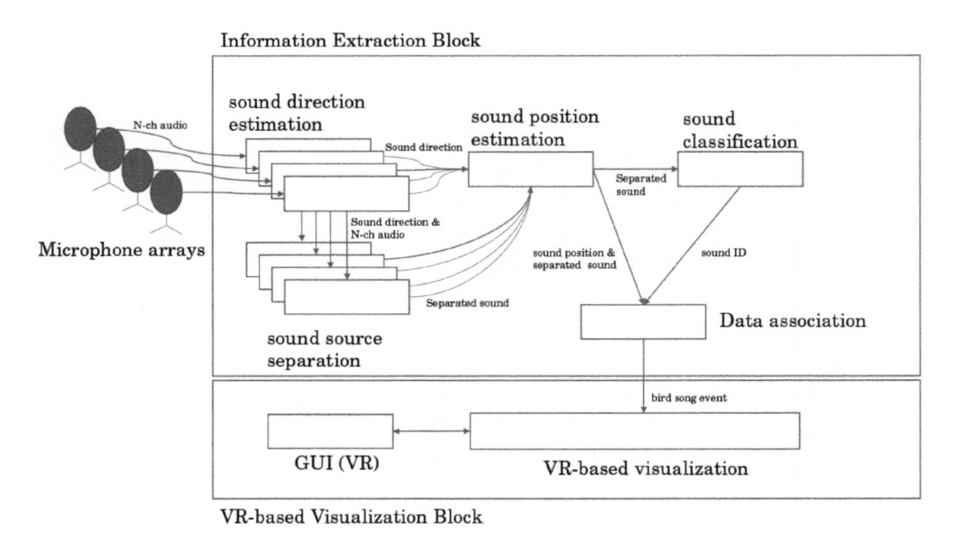

Figure 11.18 System architecture of bird song scene analysis system. Reproduced with permission [94]. Copyright 2019Taylor & Francis.

The developed technology can detect a sound source up to the distance of 12–15 m even when an SNR is around −15 dB [95, 96]. Over 80% of time frames are successfully localized in a low SNR of −20dB [98], and it significantly improves with temporal integration using Kalman filtering, the localization errors are at most 3 m [102] even when two speakers utter simultaneously. The system uses a 16-ch spherical microphone array with an ARM processor, which is placed at the end of a drone's arm. A sound mixture of drone noise and target speech are captured with the microphone array, and it is sent to the ARM processor. The embedded version of HARK is running on the processor, and sound source localization is performed. The localized directions are sent to a PC in a base station via a wireless connection. Temporal integration is performed in the PC, and the results are visualized in the PC display.

11.7.3 VR system based on bird song scene analysis

Bird song scene analysis is an essential research topic in ethology. Researchers in this field manually record information on bird songs such as "when," "where," and "what kind of birds" by listening. It is a tough task, and the recording results can have some flaws such as poor reproducibility and inevitable human errors. Automation of bird song scene analysis as an application of robot audition technologies is helpful. This section introduces VR-based visualization of bird song scene analysis so that people can understand the extracted information on bird songs.

Figure 11.17 shows a microphone array for bird song scene analysis, which has 16 microphones distributed three dimensionally. The bird song recordings were performed with four 16-ch microphone array outdoors in an asynchronous manner. The analysis was conducted in an offline manner using the system shown in Figure 11.18. From the recorded bird song sounds with the four microphone arrays, the information extraction block extracts bird song information on duration, position, and species from the recorded sound and integrates them into a bird song event. In this block, sound source localization and separation implemented in HARK were adopted. For the localized sound directions, triangulation-based sound position estimation was applied [94]. A bird song classification network based on a conventional CNN was trained and connected to the system via a Transmission Control Protocol/Internet Protocol (TCP/IP) socket interface. The generated bird song events were sent to the VR-based visualization block, which was implemented using Unity [103]. HTC Vive [104] was selected for the VR device.

For the classification with CNN, we achieved 81.52% accuracy for 12-class classification using a CNN trained with 2-hour bird song data as a result of five-fold cross-validation. The snapshots of views of the VR system are shown in Figure 11.19a–c. Users can see a singing bird (Figure 11.19a) and a flying bird (Figure 11.19b) from his/her viewing angle. In addition, they can intuitively control scene playback with a GUI displayed in their views as shown in Figure 11.19c.

a) Visualization of bush warbler's song

b) Visualization of flying crow's call

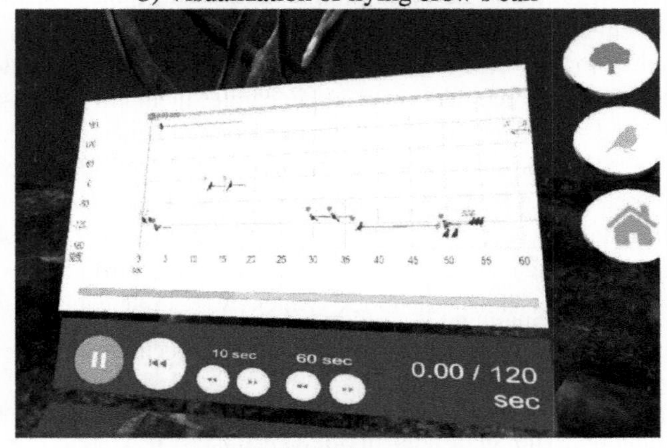

c)GUI of VR system

Figure 11.19 VR-based visualization of bird songs. Reproduced with permission [94]. Copyright 2019, Taylor & Francis.

11.8 Summary

This chapter described topics related to audio sensors for robot systems such as audio devices (microphones and microphone arrays), the concept of robot audition, acoustic signal processing for robots, OSS packages for robot audition, and applications of robot audition. Since audio devices for robot systems should be selected by taking target applications and entire systems into consideration, this chapter covering with a wide spectrum of audio sensing will be helpful for readers.

References

[1] Mehrabian A. *Silent messeges*. Belmont CA: Wadsworth Publishing Company, Inc; 1971.

[2] Nakadai K., Matsuura D., Okuno H.G., Tsujino H. "Improvement of recognition of simultaneous speech signals using AV integration and scattering theory for humanoid robots". *Speech Communication*. 2004, vol. 44(1–4), pp. 97–112.

[3] Nakadai K., Tsujino H. "Towards new human-humanoid communication: listening during speaking by using ultrasonic directional speaker". *2005 IEEE International Conference on Robotics and Automation*; Barcelona, Spain, 2005. pp. 1495–500.

[4] Nakadai K., Takahashi T., Okuno H.G., Nakajima H., Hasegawa Y., Tsujino H. 'Design and implementation of robot audition system "HARK"'. *Advanced Robotics*. 2010, vol. 24, pp. 739–61.

[5] Asano F., Fukushima M. 'Classification of footstep attributes using a vibration sensor'. *'Asia-Pacific Signal and Information Processing Association Annual Summit and Conference (APSIPA)'*; Jeju, South Korea, 2016. pp. 1–5.

[6] Tsubaki S., Asano F., Nakadai K. 'Estimation of walking speed using a vibration sensor and multiple regression submission number'. *Proceedings of SICE Annual Conference 2020*; Chiang Mai, Thailand, 2020.

[7] Munk W., Worcester P., Wunsch C. *Ocean Acoustic Tomograpthy*. Cambridge: Cambridge University Press; 1995. Available from http://ebooks.cambridge.org/ref/id/CBO9780511666926

[8] Westervelt P.J. 'Parametric acoustic array'. *Journal of the Acoustical Society of America*. 1963, vol. 35(4), pp. 535–37.

[9] Kawasaki H., Sugimoto T. 'Underground imaging using shear waves' in Akiyama I. (ed.). *Acoustical Imaging*. Dordrecht: Springer; 2009. pp. 339–45.

[10] Nakadai K., Lourens T., Okuno H.G., Kitano H. 'Active audition for humanoid'. *Proceedings of 17th National Conference on Artificial Intelligence (AAAI)*; Austin, Texas, 2000. pp. 832–39.

[11] *RME [homepage on the internet]*. 1996. Available from https://www.rme-audio.de/home.html

[12] Matsusaka Y., Tojo T., Kubota S, *et al.* 'Multi-person conversation via multi-modal interface-a robot who communicate with multi-user'. *6th European Conference on Speech Communication and Technology (Eurospeech 1999)*; Budapest, Hungary, ISCA, 1999. pp. 1723–26. Available from https://www.isca-speech.org/archive/eurospeech_1999
[13] Breazeal C., Scassellati B. ''A context-dependent attention system for A social robot''. *Proceedings of International Joint Conferences on Artificial Intelligence (IJCAI-99)*; 1999. pp. 1146–51.
[14] Kurata T., Chang D., Hashimoto S. ''Multimedia sensing system for robot''. *Proceedings of 4th IEEE International Workshop on Robot and Human Communication*; Tokyo, Japan, 1995. pp. 83–88.
[15] Huang J., Ohnishi N., Sugie N. ''Building ears for robots: sound localization and separation''. *Artificial Life and Robotics*. 1997, vol. 1(4), pp. 157–63.
[16] Asoh H., Hayamizu S., Hara I., Motomura Y., Akaho S., Matsui T. ''Socially embedded learning of the office-conversant mobile robot Jijo-2''. *Proceedings of 15th International Joint Conference on Artificial Intelligence (IJCAI-97)*; 1997. pp. 880–85.
[17] Rosenthal D., Okuno H.G. (eds.) *Computational Auditory Scene Analysis*. Mahawah, NJ: Lawrence Erlbaum Associates; 1998.
[18] Bregman A.S. *Auditory scene analysis* [online]. Cambridge: The MIT Press; 1990. Available from https://direct.mit.edu/books/book/3887/auditory-scene-analysisthe-perceptual-organization
[19] Kim H.-D., Komatani K., Ogata T., Okuno H.G. 'Human tracking system integrating sound and face localization using an expectation-maximization algorithm in real environments'. *Advanced Robotics*. 2009, vol. 23(6), pp. 629–53.
[20] Rodemann T., Heckmann M., Joublin F., Goerick C., Scholling B. 'Real-time sound localization with a binaural head-system using a biologically-inspired cue-triple mapping'. *IEEE/RSJ International Conference on Intelligent Robots and Systems*; Beijing, China, IEEE Press, 2006. pp. 860–65.
[21] Portello A., Danès P., Argentieri S., Pledel S. 'HRTF-based source azimuth estimation and activity detection from a binaural sensor'. *IEEE/RSJ International Conference on Intelligent Robots and Systems (IROS 2013) international conference on intelligent robots and systems (IROS 2013)*; Tokyo, Japan, IEEE, 2013. pp. 2908–13.
[22] Youssef K., Argentieri S., Zarader J.L. 'A learning-based approach to robust binaural sound localization'. *IEEE/RSJ International Conference on Intelligent Robots and Systems (IROS 2013)international conference on intelligent robots and systems (IROS 2013)*; Tokyo, Japan, IEEE, 2013. pp. 2927–32.
[23] Valin J.M., Michaud F., Rouat J. 'Robust localization and tracking of simultaneous moving sound sources using beamforming and particle filtering'. *Robotics and Autonomous Systems*. 2007, vol. 55(3), pp. 216–28.

[24] Sasaki Y., Kaneyoshi M., Kagami S., Mizoguchi H., Enomoto T. 'IEEE/RSJ International Conference on Intelligent Robots and Systems (Iros 2009); St. Louis, MO, 2009. pp. 2724–29'.

[25] IshiC.T., EvenJ., HagitaN. 'Using multiple microphone arrays and reflectionsfor 3D localization of sound sources'. *IEEE/RSJ International Conference on Intelligent Robots and Systems (IROS 2013)international conference on intelligent robots and systems (IROS 2013)*; Tokyo, Japan, 2013. pp. 3937–42.

[26] Aloimonos J., Weiss I., Bandyopadhyay A. 'Active vision'. *International Journal of Computer Vision*. 1988, vol. 1(4), pp. 333–56.

[27] Boll S.F. 'A spectral subtraction algorithm for suppression of acoustic noise in speech'. *IEEE International Conference on Acoustics, Speech, and Signal Processing*; Washington, DC, IEEE, 1979. pp. 200–03.

[28] Ito A., Kanayama T., Suzuki M., Makino S. 'Internal noise suppression for speech recognition by small robots'. *Proceedings of European Conferenceon Speech Communication and Technology (Eurospeech-2005)*; Lisbon, Portugal, 2005. pp. 2685–88. Available from https://www.isca-speech.org/archive/interspeech_2005

[29] Nishimura Y., Ishizuka M., Nakadai K., Nakano M., Tsujino H. '6th IEEE-RAS International Conference on Humanoid Robots; University of Genova, Genova, Italy, IEEE, 2006. pp. 26–33'.

[30] Ince G., Nakadai K., Rodemann T., Imura J., Nakamura K., Nakajima H. 'Incremental learning for ego noise estimation of a robot'. *IEEE/RSJ International Conference on Intelligent Robots and Systems (Iros 2011)*; San Francisco, CA, IEEE, 2011. pp. 131–36.

[31] Even J., Sawada H., Saruwatari H., Shikano K., Takatani T. 'Semi-blind suppression of internal noise for hands-free robot spoken dialog system'. *IEEE/RSJ International Conference on Intelligent Robots and Systems (Iros 2009)*; St. Louis, MO, IEEE, 2009. pp. 658–63.

[32] Tezuka T., Yoshida T., Nakadai K. 'Ego-motion noise suppression for robots based on semi-blind infinite non-negative matrix factorization'. *IEEE International Conference on Robotics and Automation (ICRA)*; Hong Kong, China, 2009. pp. 6293–98.

[33] Portello A., Dans P., Argentieri S. 'Active binaural localization of intermittent moving sources in the presence of false measurements'. *IEEE/RSJ International Conference on Intelligent Robots and Systems (Iros 2012)*; Vilamoura-Algarve, Portugal, IEEE, 2012. pp. 3294–99.

[34] Sasaki Y., Thompson S., Kaneyoshi M., Kagami S. 'Map-generation and identification of multiple sound sources from robot in motion'. *IEEE/RSJ International Conference on Intelligent Robots and Systems (Iros 2010)*; Taipei, Taiwan, IEEE, 2010. pp. 437–43.

[35] Martinson E., Schultz A.C. 'Discovery of sound sources by an autonomous mobile robot'. *Autonomous Robots*. 2009, vol. 27(3), pp. 221–37.

[36] Yoshida T., Nakadai K. 'Active audio-visual integration for voice activity detection based on a causal bayesian network'. *12th IEEE-RAS International*

Conference on Humanoid Robots (Humanoids 2012); Osaka, Japan, IEEE, 2012. pp. 370–75.

[37] Blauert J. *Spatial hearing*. Cambridge: The MIT Press; 1997.

[38] Cavaco S., Hallam J. 'A biologically plausible acoustic azimuth estimation system'. *Proceedings of IJCAI-99 Workshop on Computational Auditory Scene Analysis (CASA'99)*; 1999. pp. 78–87.

[40] Ando S. 'An autonomous three–dimensional vision sensor with ears'. *IEICE Transactions on Information and Systems*. 1995, vol. E78–D(12), pp. 1621–29.

[41] Hara I., Asano F., Asoh H. 'Robust speech interface based on audio and video information fusion for humanoid HRP-2'. *Proceedings of the IEEE/RSJ International Conference on Intelligent Robots and Systems (IROS 2004)*; 2004. pp. 2404–10.

[42] Nakamura K., Nakadai K., Asano F., Ince G. 'Intelligent sound source localization and its application to multimodal human tracking'. *IEEE/RSJ International Conference on Intelligent Robots and Systems (IROS 2011) international conference on intelligent robots and systems (IROS 2011)*; San Francisco, CA, IEEE, 2011. pp. 143–48.

[43] *ImageNet large scale visual recognition competition [homepage on the internet]*. 2010. Available from http://image-net.org/

[44] Seide F., Li G., Yu D. 'Conversational speech transcription using context-dependent deep neural networks'. *INTERSPEECH 2011*; Florence, Italy, ISCA, 2011. pp. 437–40. Available from https://www.isca-speech.org/archive/interspeech_2011

[45] Nakatani T., Okuno H.G., Kawabata T. 'Residue-driven architecture for computational auditory scene analysis'. *Proceedings of 14th International Joint Conference on Artificial Intelligence (IJCAI-95)*; ACM, 1995. pp. 165–72.

[46] Cohen I., Berdugo B. 'Microphone array post-filtering for non-stationary noise suppression'. *ICASSP-2002*; Orlando, FL, IEEE, 2002. pp. 901–04.

[47] Ephraim Y., Malah D. 'Speech enhancement using a minimum mean-square error log-spectral amplitude estimator'. *IEEE Transactions on Acoustics, Speech, and Signal Processing*. 1985, vol. 33(2), pp. 443–45.

[48] Jutten C., Herault J. 'Blind separation of sources, part I: an adaptive algorithm based on neuromimetic architecture'. *Signal Processing*. 1991, vol. 24(1), pp. 1–10.

[49] Araki S., Makino S., Mukai R., Saruwatari H. 'Equivalence between frequency domain blind source separation and frequency domain adaptive null beamformers'. *7th European Conference on Speech Communication and Technology (Eurospeech 2001)*; Aalborg, Denmark, 2001. pp. 2595–98. Available from https://www.isca-speech.org/archive/eurospeech_2001

[50] Hiroe A. 'Solution of permutation problem in frequency domain ICA, using multivariate probability density functions' in Rosca J., Erdogmus D., Príncipe J.C., Haykin S. (eds.). *Independent Component Analysis and Blind Signal Separation*. Berlin, Heidelberg: Springer; 2006. pp. 601–08.

[51] Lee I., Kim T., Lee T.W. 'Fast fixed-point independent vector analysis algorithms for convolutive blind source separation'. *Signal Processing*. 2007, vol. 87(8), pp. 1859–71.
[52] Kitamura D., Ono N., Sawada H., Kameoka H., Saruwatari H. 'Determined blind source separation unifying independent vector analysis and nonnegative matrix factorization'. *IEEE/ACM Transactions on Audio, Speech, and Language Processing*. , vol. 24(9), pp. 1626–41.n.d
[53] Virtanen T. 'Monaural sound source separation by nonnegative matrix factorization with temporal continuity and sparseness criteria'. *IEEE Transactions on Audio, Speech and Language Processing*. 2006, vol. 15(3), pp. 1066–74.
[54] Sawada H., Kameoka H., Araki S., Ueda N. 'Multichannel extensions of nonnegative matrix factorization with complex-valued data'. *IEEE Transactions on Audio, Speech, and Language Processing*. 2013, vol. 21(5), pp. 971–82.
[55] Kameoka H., Ono N., Kushino K., Sagayama S. 'Complex NMF: a new sparse representation for acoustic signals'. *IEEE International Conference on Acoustics, Speech and Signal Processing*; Taipei, Taiwan, IEEE, 2009. pp. 3437–40.
[56] Noda K., Hashimoto N., Nakadai K., Ogata T. 'Sound source separation for robot audition using deep learning'. *IEEE-RAS 15th International Conference on Humanoid Robots (Humanoids)*; Seoul, South Korea, IEEE, 2015. pp. 389–94.
[57] Chandna P., Miron M., Janer J. 'Monoaural audio source separation using deep convolutional neural networks' in Tichavský P., Babaie-Zadeh M., Michel O.J.J., et al (eds.). *Latent Variable Analysis and Signal Separation*. Cham: Springer International Publishing; 2017. pp. 258–66.
[58] Huang P.S., Kim M., Hasegawa-Johnson M., Smaragdis P. 'Singing-voice separation from monaural recordings using deep recurrent neural networks'. *15th International Society for Music Information Retrieval Conference (ISMIR)*; 2014. pp. 477–82.
[59] Ba J., Caruana R. 'Do deep nets really need to be deep'. *Advances in Neural Information Processing Systems 27: Annual Conference on Neural Information Processing Systems*; 2014. pp. 2654–62. Available from http://papers.nips.cc/paper/5484-do-deep-nets-really-need-to-be-deep
[60] Hinton G., Vinyals O., Dean J. 'Distilling the knowledge in a neural network'. *NIPS Deep Learning and Representation Learning Workshop*; 2015. pp. 1–9. Available from http://arxiv.org/abs/1503.02531
[61] Kim Y., Rush A.M. 'Sequence-level knowledge distillation'. *Proceedings of the 2016 Conference on Empirical Methods in Natural Language Processing*; Stroudsburg, PA, 2017. pp. 1317–27. Available from http://aclweb.org/anthology/D16-1
[62] Han S., Mao H., Dally W.J., LeCun Y., Bengio Y. *Proceedings of 4th International Conference on Learning Representations*; 2016. pp. 1–14.
[63] Takeda R., Nakadai K., Komatani K. 'Acoustic model training based on node-wise weight boundary model for fast and small-footprint deep neural networks'. *Computer Speech & Language*. 2017, vol. 46(461), pp. 461–80.

[64] Xue J., Li J., Gong Y. 'Restructuring of deep neural network acoustic models with singular value decomposition'. *Interspeech 2013*; ISCA, IEEE, 2017. pp. 2365–69. Available from https://www.isca-speech.org/archive/interspeech_2013

[65] Sainath T.N., Kingsbury B., Sindhwani V., Arisoy E., Ramabhadran B. '*ICASSP 2013 -IEEE international conference on acoustics, speech and signal processing (ICASSP)*; vancouver, BC, 2017. pp. 6655–59'.

[66] Gong Y., Liu L., Yang M, *et al.* 'Compressing deep convolutional networks using vector quantization'. *CoRR*. 2014.

[67] Sindhwani V., Sainath T., Kumar S. 'Structured transforms for small-footprint deep learning' in Cortes C., Lawrence N.D., Lee D.D., Garnett R. (eds.). *Advances in neural information processing systems 28advances in neural information processing systems 28*. New york, United States: Curran Associates, Inc; 2015. pp. 3088–96.

[68] Pollot M., Zhang R., Kaup A. 'An efficient alternative to network pruning through ensemble learning'. *ICASSP 2020 - IEEE International Conference on Acoustics, Speech and Signal Processing (ICASSP)*; Barcelona, Spain, 2020. pp. 4022–26. Available from https://ieeexplore.ieee.org/xpl/mostRecentIssue.jsp?punumber=9040208

[69] Nakadai K., Nakamura K. *Wiley Encyclopedia of Electrical and Electronics Engineering*. John Wiley & Sons, Inc; 2015. pp. 1–18. Available from https://onlinelibrary.wiley.com/doi/abs/10.1002/047134608X.W8266

[70] Rascon C., Meza I. 'Localization of sound sources in robotics: A review'. *Robotics and Autonomous Systems*. 2017, vol. 96, pp. 184–210.

[71] Argentieri S., Danès P., Souères P. 'A survey on sound source localization in robotics: from binaural to array processing methods'. *Computer Speech & Language*. 2015, vol. 34(1), pp. 87–112.

[72] Schmidt R.O. 'Multiple emitter location and signal parameter estimation'. *IEEE Trans Antennas Propag*. 1986, vol. 34(3), pp. 276–80.

[73] Nakamura K., Nakadai K., Asano F., Hasegawa Y., Tsujino H. 'Intelligent sound source localization for dynamic environments'. *IEEE/RSJ International Conference on Intelligent Robots and Systems (Iros 2009)*; St. Louis, MO, IEEE, 2009. pp. 664–69.

[74] Nakamura K., Nakadai K., Okuno H.G. 'A real-time super-resolution robot audition system that improves the robustness of simultaneous speech recognition'. *Advanced Robotics*. 2013, vol. 27(12), pp. 933–45.

[75] Nakajima H., Tanaka N., Tsuru H. 'Minimum sidelobe beamforming based on mini-max criterion'. *Acoustical Science and Technology*. 2004, vol. 25(6), pp. 486–88.

[76] Barroso V.A.N., Moura J.M.F. 'Maximum likelihood beamforming in the presence of outliers'. *Proceedings of 1991 International Conference on Acoustics, Speech, and Signal Processing (ICASSP-91)*; Toronto, ON, 1991. pp. 1409–12.

[77] Seltzer M.L., Raj B., Stern R.M. 'A bayesian classifier for spectrographic mask estimation for missing feature speech recognition'. *Speech Communication*. 2004, vol. 43(4), pp. 379–93.

[78] Monzingo R.A., Miller T.W. *Introduction to Adaptive Arrays*. New York: SciTech Publishing; 1980.
[79] Capon J. 'High-resolution frequency-wavenumber spectrum analysis'. *Proceedings of the IEEE*. 1969, vol. 57(8), pp. 1408–18.
[80] Frost O.L. 'An algorithm for linearly constrained adaptive array processing'. *Proceedings of the IEEE*. 1972, vol. 60(8), pp. 926–35.
[81] Griffiths L.J., Jim C.W. 'An alternative approach to linearly constrained adaptive beamforming'. *IEEE Trans Antennas Propag*. 1982, vol. 30(1), pp. 27–34.
[82] Parra L.C., Alvino C.V. 'Geometric source separation: merging convolutive source separation with geometric beamforming'. *IEEE Transactions on Speech and Audio Processing*. 2002, vol. 10(6), pp. 352–62.
[83] Knaak M., Araki S., Makino S. 'Geometrically constrained independent component analysis'. *IEEE Transactions on Audio, Speech and Language Processing*. 2007, vol. 15(2), pp. 715–26.
[84] Nakajima H., Nakadai K., Hasegawa Y., Tsujino H. 'Blind source separation with parameter-free adaptive step-size method for robot audition'. *IEEE Transactions on Audio, Speech, and Language Processing*. 2010, vol. 18(6), pp. 1476–85.
[85] NakajimaH., InceG., NakadaiK., HasegawaY. 'An easily-configurable robot audition system using histogram-based recursive level estimation'. *IEEE/RSJ International Conference on Intelligent Robots and Systems (Iros 2010)*; Taipei, Taiwan, IEEE, 2010. pp. 958–63.
[86] Takeda R., Nakadai K., Takahashi T., Komatani K., Ogata T., Okuno H.G. 'Efficient blind dereverberation and echo cancellation based on independent component analysis for actual acoustic signals'. *Neural Computation*. 2012, vol. 24(1), pp. 234–72.
[87] Ince G., Nakadai K., Rodemann T., Tsujino H., Imura J.-I. 'Whole body motion noise cancellation of a robot for improved automatic speech recognition'. *Advanced Robotics*. 2011, vol. 25(11–12), pp. 1405–26.
[88] Bando Y., Itoyama K., Konyo M. 'Human-voice enhancement based on online RPCA for a hose-shaped rescue robot with a microphone array'. *IEEE International Symposium on Safety, Security, and Rescue Robotics (SSRR)*; West Lafayette, IN, 2015. pp. 1–6.
[89] *Wiki of ISCA's special interest group (SIG) on robust speech processing [wiki on the internet]*. 2016. Available from https://wiki.inria.fr/rosp/Software#Speech_enhancement_and_separation
[90] *Homepage of HARK [homepage on the internet]*. 2008. Available from https://www.hark.jp/
[91] Nakadai K., Okuno H.G., Mizumoto T. 'Development, deployment and applications of robot audition open source software HARK'. *Journal of Robotics and Mechatronics*. 2019, vol. 29(1), pp. 16–25.
[92] *Homepage of ROS [homepage on the internet]*. 2007. Available from https://www.ros.org/
[93] *Prince shotoku robot [youtube movie]*. 2012. Available from https://youtu.be/NoiwXa6D3Uc

[94] Gabriel D., Kojima R., Hoshiba K., Itoyama K., Nishida K., Nakadai K. '2D sound source position estimation using microphone arrays and its application to a VR-based bird song analysis system'. *Advanced Robotics*. 2019, vol. 33(7–8), pp. 403–14.

[95] Okutani K., Yoshida T., Nakamura K., Nakadai K. 'Outdoor auditory scene analysis using a moving microphone array embedded in a quadrocopter'. *IEEE/RSJ International Conference on Intelligent Robots and Systems (IROS 2012)international conference on intelligent robots and systems (IROS 2012)*; Vilamoura-Algarve, Portugal, IEEE, 2012. pp. 3288–93.

[96] Ohata T., Nakamura K., Nagamine A, *et al*. 'Outdoor sound source detection using a quadcopter with microphone array'. *Journal of Robotics and Mechatronics*. 2017, vol. 29(1), pp. 177–87.

[97] Nonami K., Hoshiba K., Nakadai K. 'Recent R&D technologies and future prospective of flying robot in tough robotics challenge' in Tadokoro S. (ed.). *Disaster Robotics - Results from the Impact Toughrobotics Challenge*. Vol. 128. Springer Tracts in Advanced Robotics. Springer; 2019. pp. 77–142.

[98] Hoshiba K., Washizaki K., Wakabayashi M, *et al*. 'Design of UAV-embedded microphone array system for sound source localization in outdoor environments'. *Sensors (Basel, Switzerland)*. 2017, vol. 17(11), pp. 1–16.

[99] Nakadai K., Kumon M., Okuno H.G. 'Development of microphone-array-embedded UAV for search and rescue task'. *IEEE/RSJ International Conference on Intelligent Robots and Systems (IROS)*; Vancouver, BC, IEEE, 2017. pp. 5985–90.

[100] *Drone audition demo movie [youtube movie]*. 2018. Available from https://youtu.be/71jEYKAEgnc

[101] Suzuki T., Inoue D., Amano Y. 'Robust UAV position and attitude estimation using multiple GNSS receivers for laser-based 3D mapping'. *IEEE/RSJ International Conference on Intelligent Robots and Systems (IROS)*; Macau, China, 2019. pp. 4402–08. Available from https://ieeexplore.ieee.org/xpl/mostRecentIssue.jsp?punumber=8957008

[102] Wakabayashi M., Okuno H.G., Kumon M. 'Multiple sound source position estimation by drone audition based on data association between sound source localization and identification'. *IEEE Robotics and Automation Letters*. 2020, vol. 5(2), pp. 782–89.

[103] *Unity technologies*. 2004. Available from https://unity3d.com

[104] *HTC vive*. 2011. Available from https://www.vive.com

Chapter 12
Audio and gas sensors

Caleb Rascon[1]

12.1 Audio sensors

Audio is propagated through a medium (such as air or water) as continuous differences of air pressure caused by vibrational mechanisms that originated at the source [1]. These can be the human trachea which houses vibrating vocal cords that emit speech, or an electronic speaker which houses a vibrating metallic plate that emits previously recorded sound.

Acoustic sensory can be considered as the counterpart of this process, which aims to capture these vibrations from the air and interface them into electric signals, which can then be further analyzed to obtain information of the auditory scene.

Audio sensors in robotic applications [2] have been applied in a wide variety of case scenarios, from human–robot interaction to search and rescue efforts. Although their use is not as widespread as other sensors explored in this work, they are worth describing so as to offer a complete survey of sensory systems for robotic applications.

The structure of this section is divided into two parts. The first explores the hardware side of auditory sensing in robotic applications, where microphones and audio interfacing are concerned. The second part explores the software side, where current implementations are described to offer the reader alternatives to best choose what is more appropriate in their case scenario.

12.1.1 Hardware

The hardware presented in this section has been narrowed down to only those that are relevant for robotic applications and, thus, should not be considered as an extensive review of the overall state of audio hardware available. However, it is worth observing that the hardware that is relevant in robotic applications is usually also relevant in other types of scenarios [3] such as mobile applications and domestic security [4].

[1]Instituto deInvestigaciones en MatematicaAplicadas y en Sistemas (IIMAS), Universidad NacionaAutonoma de Mexico (UNAM), Mexico City, Mexico

Most case scenarios in which auditory sensors are used make use of complex algorithmic solutions which require computational equipment. To this effect, auditory sensory for robotic applications is divided into two main categories: the acoustic sensors (microphones) and the digital interfaces that convert the signal captured from the sensors and transfers it to the computational equipment.

12.1.1.1 Microphones

As mentioned before, audio sensors aim to capture the differences in air pressure (or vibrations) in the environment and convert them into electric signals. Currently, this effort is mostly carried out by microphones [5] which, for the most part, employ a material that responds to such vibration, and a built-in subsystem that aims to convert such responses into an electric signal. There is a wide variety of microphones which vary in terms of the type of vibrational material as well as in its electric subsystem. These variations impact how sensitive the microphone is across the acoustic frequency range, how much space it occupies, and how much power it requires to work.

The types of microphones that are relevant to robotic applications are those that have a relatively high sensitivity in the acoustic frequency range, occupy a small amount of space, and do not require as much power. To this effect, there are two types of microphones that are discussed in this section: miniature condenser microphones and MEMS microphones.

12.1.1.1.1 Miniature condenser microphones

Typical condenser microphones [6] employ two plates set-up in a capacitive module; see Figure 12.1 for a summarized diagram of its inner circuitry. The two plates are generally of two different materials, or of the same material but with different mass, as one needs to be considerably lighter than the other. The lighter plate (generally referred to as the *diaphragm*) is poised to be the material that will vibrate when differences of air pressure hit it. When the distance of the diaphragm to the other plate varies, the capacitance also varies. Since the capacitance signal carries too little current to be used robustly outside of the capacitive module, it is passed through an impedance converter to offer it more current. The impedance converter is basically an electronic resistor and a voltage source connected in parallel to the capacitive module, and the output electric signal is the voltage measured from the impedance.

As it can be seen, this type of microphone requires a power source to function which can be considered impractical in some case scenarios. However, the amount of current that condenser microphones require is quite small. Additionally, engineering advances have solved this issue by incorporating a permanently charged ferroelectric material (known as an *electrect*) [7] into the inner circuitry of the microphone. It is important to mention, though, that the voltage range of the signal provided by the condenser microphone is usually quite small and requires amplification [8]. This is carried out by integrating a preamplification inside the circuitry of the microphone and powering it through the audio interface that the microphone is connected to (commonly known as *Phantom Power*) [9]. The amount of amplification varies depending on the size of the metal plates inside the capacitive module

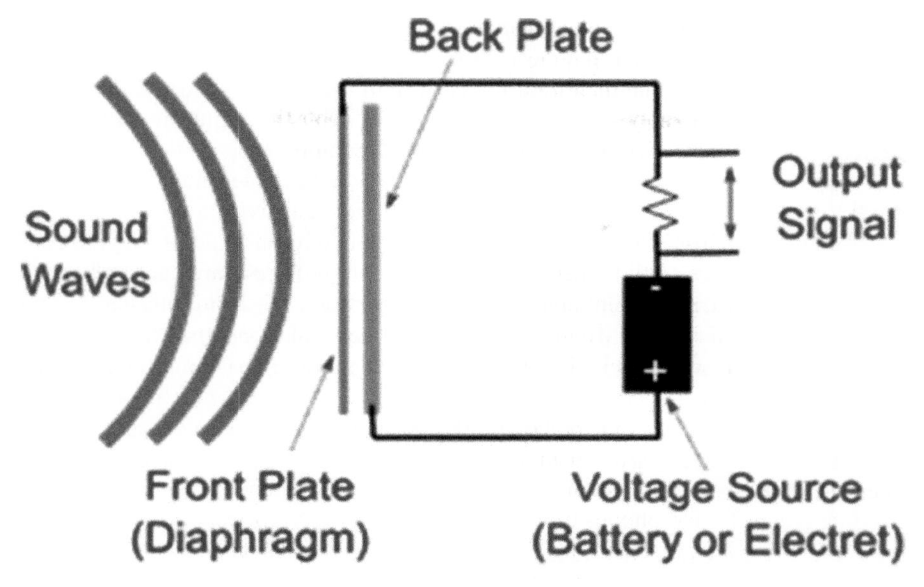

Figure 12.1 *Diagram of a condenser microphone*

(since it needs to compensate for the resulting capacitance) and can range from 5 to 48 V [10].

However, one important issue that is relevant to robotic applications is that of their size, as they usually occupy an impractically large amount of space considering the size of a typical robotic platform. To this effect, efforts have been made to miniaturize the capacitive capsules with measurable success [11]. It has been found, however, that the size of the diaphragm is directly correlated to the sensitivity of the microphone in the acoustic frequency range [12]. Fortunately, the trade-off has been acceptable for human–robot interaction and flying-drone-based security applications.

There are several examples currently available in the market, such as the CMA-4544PF-W capsule [13], which is employed in the 8SoundsUSB [14] effort later discussed.

12.1.1.1.2 MEMS microphones

Microphones based on microelectromechanical systems (MEMS) [15], for the most part, can be considered as a variation of the condenser microphone, as they rely on the use of a type of diaphragm (also known as the *membrane*) whose vibrations change the capacitance between it and another plate, which is then converted to an electric signal by an integrated amplifier. The main difference between a MEMS microphone and an actual condenser microphone is its size. The membrane of a typical MEMS microphone has a diameter of 500 μm [16], orders of magnitude smaller than that of a typical condenser microphone with diameters ranging from 12.7 ("miniature") to 25.4 ("large") mm [17]. And it is because of their size that

MEMS microphones have garnered so much interest in mobile applications and are quite relevant to robotic applications.

It is important to note that initially, because of their small size, MEMS microphones were considerably less sensitive across the acoustic frequency range compared to their condenser counterparts [16]. However, more recently, their sensitivity has improved drastically by building the membrane with lighter materials [18] and modifying its structure [19] to account for this lack of sensitivity.

Additionally, a MEMS microphone is accompanied by an analog-to-digital converter which simplifies further digital processing but requires hardware-embedded solutions to provide a usable digital signal [20]. However, it is not unusual to find in the market all-in-one MEMS-microphone-based solutions that take care of these issues [21], which can trivially replace current condenser-microphone-based solutions.

There are several examples currently available in the market, such as the MP34DB02 [22] audio sensor, which is employed in the Matrix Voice [23] effort later discussed.

12.1.1.2 Portable audio interfaces

Having acquired an electric audio signal, it is then the job of an audio interface to amplify it (optionally) and convert it to a digital representation [24] with which it can be analyzed by computational equipment. The most popular protocol with which these digital signals are fed to the computer is via the USB protocol [25].

Most current audio interfaces require external power to feed phantom power to the microphones and power their preamplifiers. This makes them impractical to use in robotic platforms, which may require a considerable number of microphones (as explained later) which, in turn, requires a considerable amount of power to function that may be required by other sensory modules. Fortunately, the USB protocol can provide a nonnegligible amount of power [25], and some audio interfaces have been able to use it efficiently enough to power a considerable number of microphones. Two examples that are worth mentioning:

- 8SoundsUSB [14]: Built by IntRoLab of the University of Sherbrooke, Canada. It employs eight miniature condenser microphones with RJ11 connectors, wired in crossover for ground floor noise removal. There is another variation of this audio interface called 16SoundsUSB which employs 16 microphones and can be used with both electret condenser microphones or MEMS microphones.
- Matrix Voice [23]: Built by Matrix Labs. It is a development board that houses eight MEMS microphones and is compatible with the expansion interface of a Raspberry PI. This makes it not only portable but also quite flexible.

12.1.1.3 Recommendations

Given the recent surge of their sensitivity, MEMS microphones should be the first consideration when deciding upon an audio sensor for a robotic platform. Their power consumption, portability, practicality, size, and robustness against electrical noise are superior to their miniature condenser counterparts. However, MEMS offer less mounting flexibility, with reflow soldering being the most widely used option [26].

In terms of cost, although miniature condenser microphones are less costly than MEMS microphones, it is important to consider that the former requires an analog-to-digital converter to work while the latter can directly provide the digital signal to be analyzed. To this effect, as of this writing, the overall cost of using any of these two sensors is roughly around the same [27], with MEMS microphones being slightly more cost-effective.

12.1.1.4 Other acoustic sensors

It is important to state that the audio sensors reviewed here are the most relevant because of their current use in robotic applications. However, there are other types of acoustic sensors that, although not currently used in robotic applications, could be in the foreseeable future. It is important to mention that the inclusion of these sensors in this chapter is for completeness sake; there are not any variations in existence that are practical for robotic applications.

Laser-based vibrometers [28] aim to measure the acoustic signals of a target by measuring the changes of a beam of light when reflected from such a target. The Doppler shift (for laser Doppler vibrometers) or changes in sensitivity of pulsed laser beams are two of such measured changes. These vibrometers have been used for acoustic analysis of instruments and electronic speakers as well as long-range microphones. Because of their high directivity, it is quite possible to use this type of acoustic sensor, in tandem with a direction estimator of the target (by visual or acoustic means), to produce an audio signal with a high signal-to-noise ratio.

A fiber optic sensor [29] measures the impact of a mechanical signal (such as an acoustic one) onto a fiber optic wire. The most frequently used paradigm uses a reference fiber optic signal that is isolated from the environment, which is compared to the fiber optic signal that is impacted by the acoustic signal. From these changes, the acoustic signal is estimated. Since the whole fiber optic wire acts as the sensor, they provide impressive sensitivity compared to current technologies. Additionally, they can be used in different types of mediums (air, water, gas, oil, etc.) without requiring any modifications and only trivial re-calibrations.

As stated before, these types of acoustic sensors have not been used in robotic applications. High cost and technology immaturity are the main reasons for this. However, there are some benefits to them that are worth for further development and may well replace miniature condenser or MEMS microphones in future robotic applications.

12.1.2 Software

Audio sensory in robotic applications has recently been linked to the area of robot audition [30], which aims to emulate human hearing in nonrobotic entities. Because of this broad objective, auditory functionalities cannot solely be prescribed to the sensory/hardware side of robotic development. Thus, although the main focus of

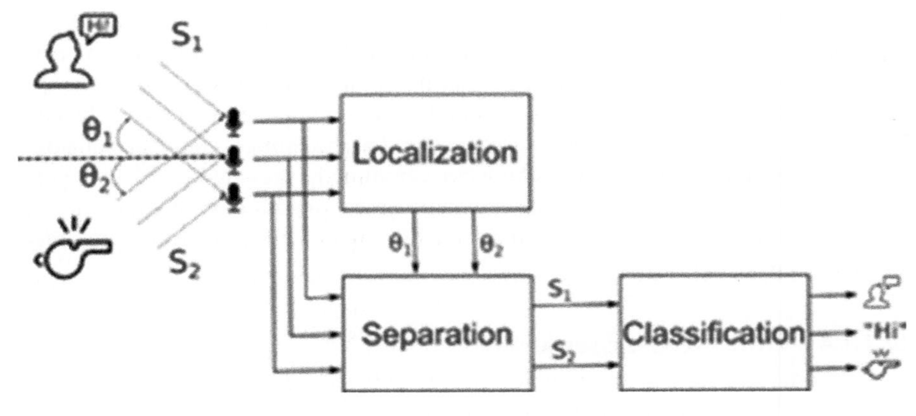

Figure 12.2 Flow of data used for robot audition

this work is this side, it is relevant to include a brief summary of the software side of audio sensory as to contextualize the reader of how the signals captured by audio sensors are to be used.

Although several data processing paradigms have been proposed to carry out robot audition, the most popular is the one presented in Figure 12.2.

The signals are fed to the localization module that aims to estimate the position of all the sound sources in the environment. This localization is usually presented as the direction of arrival (DOA) [1] of each sound source, either in only the horizontal plane or in both horizontal and vertical planes. Estimating distance is challenging via sound alone, and as there are other sensors that can estimate such information in much more trivial manner, this part of the localization is rarely carried out.

The estimated localizations, along with the captured audio signals, are fed to the separation module [31] that aims to isolate the sound from each of the sound sources in the environment from the mixture captured by the microphones into separate audio channels. This is popularly known as the cocktail party problem [32].

Finally, the isolated channels are each fed to the classification module that aims to extract relevant information such as the type of sound source (human, fan hum, dog barking, etc.) [33], the identity of the user (if the source is a human) [34], the words that the user is saying [35], their mood [36], etc.

It is important to mention that even though the aim of robot audition is to emulate human hearing, the paradigm shown in Figure 12.2 may not be the most close-to-human representation of human audition (the paradigm of which is currently unknown in its entirety). However, this paradigm covers most of the ground that is required for service robotics and search and rescue in a practical manner.

In the rest of this section, some popular methods used for the localization and separation modules are briefly reviewed. Because of the broad nature of the classification module, methods for this module are outside the scope of this writing.

12.1.2.1 Localization

Localization methodologies can be classified in a wide variety of manners [1] such as how many dimensions are estimated from the sound source location (horizontal, vertical, and/or distance), the number of microphones employed, etc. For the sake of simplicity, the methods reviewed here are divided between those that aim to locate one source and those that aim to locate multiple simultaneous sources.

In the case of locating one source, the generalized cross-correlation with phase transform (GCC-PHAT) [37] is the most popular. It estimates the time difference of arrival (TDOA) of the sound source signal between pairs of microphones. Then, using an assumed propagation model (the free-field, far-field one being the most popular), it maps the estimated TDOA to the source's DOA.

In the case of locating multiple sources, two types of methods have been mostly used: subspace methods and beamforming. Multiple signal classification (MUSIC) [38] is the most used subspace method, and it aims to represent all of the sound sources in a signal domain calculated from their eigen decomposition. In this space, at least one sound source is considered as noise, with which a measure of orthogonality is used to test if a source is located at each possible candidate DOA. Beamforming methods [39] are similar in the sense that a set of DOA candidates are proposed, but instead of measuring orthogonality from a signal subspace, the energy being received from each candidate DOA is estimated. Beamforming is also utilized as a form of sound source separation; thus, more detail is provided in the following subsection.

It is important to state that both types of multiple-source location methods require a considerable quantity of microphones to provide robust results. In the case of MUSIC, it is imperative that the number of microphones exceeds by at least one the number of sources in the environment (so that one source in the signal domain can be considered noise). Since the number of sources is rarely known *a-priori*, the rule of thumb is to use many microphones to make certain this requirement is satisfied. As for beamforming methods, as discussed in the next subsection, the more microphones are used, the better the separation quality, and, thus, the better resolution the results will have. This is important as in a low-resolution scenario, sound sources that are located near each other may be estimated as just one sound source.

12.1.2.2 Separation

There is a considerable amount of source separation methodologies in the literature. In this review, these are divided into two big groups: blind source separation and spatial filtering (or beamforming).

Blind source separation methodologies assume certain characteristics from the sound sources and aim to estimate a de-mixing matrix that will separate them from the captured signals. Low correlation (principal component analysis) [40], statistical independence (independent component analysis) [41], or nonnegative components (nonnegative matrix factorization) [42] are the most popular characteristics currently employed. However, because of the nature of the estimated de-mixing matrix, these methods require that the number of sources in the environment are equal or less than the number of microphones employed. Thus, to make certain of this, a frequent recommendation is to use a high number of microphones.

Beamformers [39] aim to isolate the signal that arrives at a certain direction of interest (DOI). If a localization process is carried out in parallel, one beamformer is run for each of the estimated DOAs. There are several ways to carry out this isolation, the most popular of which is to align the capture signals to compensate for the TDOA suffered by the signal arriving at the established DOI. Then, the aligned captured signals are mixed into one signal. Because of the interferometry nature of this method, the parts of the captured signals that are in phase (which, because of the alignment, are the ones arriving from the DOI, meaning, the source of interest) will be amplified, while those that are not (which can be considered as interferences) will be unchanged. Usually, this result is also divided by the number of microphones, so that the energy from the source of interest is unchanged, while the energy from the interferences is reduced by the number of microphones. To this effect, it is frequently recommended to use a high number of microphones so that the presence of interferences in the resulting signal is diminished as much as possible.

12.1.2.3 Currently implemented examples

Currently, there are fully integrated implemented solutions that carry out localization and separation of sound sources aimed for robotic applications.

Honda Research Institute Japan Audition for Robots with Kyoto University (HARK) [43] is a joint effort from both institutions to provide a complete implementation of a robot audition solution. It carries out localization via a variation of MUSIC and can perform separation using a wide variety of methods, spanning both types of separation: beamforming and blind source separation.

ManyEars [14], and its successor ODAS [44], is an effort from the IntRoLab Group of the University of Sherbrooke, Canada, to provide a lightweight alternative for Robot Audition. It carries out localization via beamforming using the steered response power with phase transform technique (SRP-PHAT) [45] and separation via a hybrid method called geometric source separation [46] that aims to combine the benefits of blind source separation and beamforming.

Finally, asteroid [47] is a deep-learning-based sound source separation toolkit (based on PyTorch) developed in a joint effort of eight different institutions located in France, Italy, USA, Germany, Spain, and Israel, most of whom belong to the joint research team known as MULTISPEECH.

References

[1] Fahy F., Thompson D. *Fundamentals of Sound and Vibration*[online]. Boca Raton, FL:CRC Press; 2015. Available from https://www.taylorfrancis.com/books/9781482266634

[2] Rascon C., Meza I. 'Localization of sound sources in robotics: a review'. *Robotics and Autonomous Systems*. 2017, vol. 96, pp. 184–210.

[3] Shah M.A., Shah I.A., Lee D.G., Hur S. 'Design approaches of MEMS microphones for enhanced performance'. *Journal of Sensors*. 2019, vol. 2019, pp. 1–26.

[4] Bogue R. 'Recent developments in MEMS sensors: a review of applications, markets and technologies'. *Sensor Review*. 2013, vol. 33(4), pp. 300–04.

[5] Réveillac J. *Recording and voice processing 1 [online]*. Hoboken, New Jersey, United States: Wiley Telecom; 2021 Dec 28. pp. 59–110. Available from https://onlinelibrary.wiley.com/doi/book/10.1002/9781119885061

[6] Chen J.Y., Hsu Y.C., Lee S.S., Mukherjee T., Fedder G.K. 'Modeling and simulation of a condenser microphone'. *Sensors and Actuators A: Physical*. 2008, vol. 145, pp. 224–30.

[7] Kressmann R., Klaiber M., Hess G. 'Silicon condenser microphones with corrugated silicon oxide/nitride electret membranes'. *Sensors and Actuators A*. 2002, vol. 100(2–3), pp. 301–09.

[8] Cittadini R., Poulin F. 'TS971 based electret condenser microphone amplifier'. *Application Note*. 2002, vol. AN1534.

[9] Edwards J. 'Choosing the right microphone'. *Choral Journal*. 1980, vol. 21(3), p. 5.

[10] Zaim M., Kikutani T., Green J. 'Phantom powering the modern condenser microphone: A practical look at conditions for optimized performance'. *Audio Engineering Society*. 2008, vol. 125.

[11] Zou Q., Li Z., Liu L. 'Theoretical and experimental studies of single-chip-processed miniature silicon condenser microphone with corrugated diaphragm'. *Sensors and Actuators A*. 1997, vol. 63(3), pp. 209–15.

[12] Harrison H.C., Flanders P.B. 'An efficient miniature condenser microphone system* [online]'. *Bell System Technical Journal*. 1932, vol. 11(3), pp. 451–61. Available from http://doi.wiley.com/10.1002/bltj.1932.11.issue-3

[13] CUI Devices. *cma-4544PF-W electret condenser microphone technical sheet*. Available from https://www.cuidevices.com/product/resource/cma-4544pf-w.pdf

[14] Grondin F., Létourneau D., Ferland F., Rousseau V., Michaud F. 'The many-ears open framework'. *Autonomous Robots*. 2013, vol. 34(3), pp. 217–32.

[15] Weigold J.W., Brosnihan T.J., Bergeron J., Zhang X. 'A MEMS condenser microphone for consumer applications'. *19th IEEE International Conference on Micro Electro Mechanical Systems*; IEEE, 2006.

[16] Loeppert P.V., Lee S.B. 'SISONIC – the first commercialized MEMS microphone [online]'. *Solid-State, Actuators, and Microsystems Workshop*; USA, San Diego, CA, 2006. Available from https://transducer-research-foundation.org/technical_digests/HiltonHead_2006_TechnicalDigest.pdf

[17] Bergqvist J., Rudolf F. 'A new condenser microphone in silicon'. *Sensors and Actuators A*. 1990, vol. 21(1–3), pp. 123–25.

[18] Dehe A., Wurzer M., Fuldner M., Krumbein U. 'Design of a poly silicon MEMS microphone for high signal-to-noise ratio'. *ESSDERC 2013 – 43rd European Solid State Device Research Conference*; IEEE, 2013.

[19] Segovia-Fernandez J., Sonmezoglu S., Block S.T, *et al.* 'Monolithic piezoelectric aluminum nitride MEMS-CMOS microphone'. *19th International Conference on Solid-State Sensors, Actuators and Microsystems (TRANSDUCERS)*; Kaohsiung, IEEE, 2017.

[20] Lewis J. 'Analog and digital MEMS microphone design considerations'. *Technical Article MS-2472. Analog Devices*. 2013.

[21] *Adafruit I2S MEMS microphone breakout - SPH0645LM4H* [online]. 2022 Apr 25. Available from https://www.adafruit.com/product/3421

[22] *st.MP34DB02 MEMS audio sensor omnidirectional digital microphone [online]*. Available from https://www.st.com/resource/en/datasheet/mp34db02.pdf

[23] *MATRIX voice*. 2022 Apr 25 [online]. Available from https://www.matrix.one/products/voice

[24] Oppenheim A.V. 'Applications of digital signal processing'. Englewood Cliffs. 1978.

[25] Anderson D. *USB system architecture*. Boston, US: Addison-Wesley Professional; 1997.

[26] Tilmans H.A.C., van de Peer D.J., Beyne E. 'The indent reflow sealing (IRS) technique a method for the fabrication of sealed cavities for MEMS devices'. *Journal of Microelectromechanical Systems*. 2000, vol. 9(2), pp. 206–17.

[27] *MEMS versus ECM: comparing microphone technologies* [online]. 2022 Apr 25. Available from https://www.digikey.com/en/articles/mems-vs-ecm-comparing-microphone-technologies

[28] Lutzmann P., Kamerman G.W., Steinvall O, *et al.* 'Laser vibration sensing: overview and applications'. *SPIE Security + Defence*; Prague, Czech Republic, 2011. pp. 11–26.

[29] Teixeira J.G.V., Leite I.T., Silva S., Frazão O. 'Advanced fiber-optic acoustic sensors'. *Photonic Sensors*. 2014, vol. 4(3), pp. 198–208.

[30] Okuno H.G., Nakadai K. 'Robot audition: its rise and perspectives'. *IEEE International Conference on Acoustics, Speech and Signal Processing (ICASSP)*; IEEE, 2015.

[31] Vincent E., Bertin N., Gribonval R., Bimbot F. 'From blind to guided audio source separation: how models and side information can improve the separation of sound'. *IEEE Signal Processing Magazine*. 2014, vol. 31(3), pp. 107–15.

[32] Haykin S., Chen Z. 'The cocktail party problem'. *Neural Computation*. 2005, vol. 17(9), pp. 1875–902.

[33] Karol J. 'ESC: dataset for environmental sound classification'. *Proceedings of the 23rd ACM International Conference on Multimedia*; 2015.

[34] Vélez I., Rascon C., Fuentes-Pineda G. 'Lightweight speaker verification for online identification of new speakers with short segments'. *Applied Soft Computing*. 2020, vol. 95, 106704.

[35] Nassif A.B., Shahin I., Attili I., Azzeh M., Shaalan K. 'Speech recognition using deep neural networks: a systematic review'. *IEEE Access: Practical Innovations, Open Solutions*. 2019, vol. 7, pp. 19143–65.

[36] Hossain N., Mahmuda N. 'Emovoice: finding my mood from my voice signal'. *Proceedings of the 2018 ACM International Joint Conference and 2018 International Symposium on Pervasive and Ubiquitous Computing and Wearable Computers*; 2018.

[37] Charles K., Carter G. 'The generalized correlation method for estimation of time delay'. *IEEE Transactions on Acoustics, Speech, and Signal Processing*. 1976, vol. 24(4), pp. 320–27.

[38] Ralph S. 'Multiple emitter location and signal parameter estimation'. *Transactions on Antennas and Propagation*. 1986, vol. 34(3), pp. 276–80.

[39] Rascon C. 'A corpus-based evaluation of beamforming techniques and phase-based frequency masking'. *Sensors (Basel, Switzerland)*. 2021, vol. 21(15), 5005.

[40] Hervé A., Williams L. 'Principal component analysis'. *Wiley Interdisciplinary Reviews*. 2010, vol. 2(4), pp. 433–59.

[41] Comon P. 'Independent component analysis, a new concept?'. *Signal Processing*. 1994, vol. 36(3), pp. 287–314.

[42] Daniel L., Seung H.S. 'Algorithms for non-negative matrix factorization'. *Advances in Neural Information Processing Systems*. 2000, vol. 13.

[43] Nakadai K., Okuno H.G., Mizumoto T., Honda Research Institute Japan Co., Ltd, Graduate Program for Embodiment Informatics, Waseda University, Graduate School of Information Science and Engineering, Tokyo Institute of Technology 'Development, deployment and applications of robot audition open source software HARK'. *Journal of Robotics and Mechatronics*. 2017, vol. 29(1), pp. 16–25.

[44] Grondin F., Michaud F. 'Lightweight and optimized sound source localization and tracking methods for open and closed microphone array configurations'. *Robotics and Autonomous Systems*. 2019, vol. 113, pp. 63–80.

[45] Hoang D., Silverman H.F. 'SRP-PHAT methods of locating simultaneous multiple talkers using a frame of microphone array data'. *IEEE International Conference on Acoustics, Speech and Signal Processing*; IEEE, 2010.

[46] Parra L.C., Alvino C.V. 'Geometric source separation: merging convolutive source separation with geometric beamforming'. *IEEE Transactions on Speech and Audio Processing*. 2002, vol. 10(6), pp. 352–62.

[47] Pariente M., Cornell S., Cosentino J, *et al*. 'Asteroid: the pytorch-based audio source separation toolkit for researchers [online]'. *Interspeech 2020*; ISCA, 2002. Available from https://www.isca-speech.org/archive/interspeech_2020

Index